W9-CZS-530

encyclopedia of
marine invertebrates

Edited by Jerry G. Walls

Front Cover: *Stenopus hispidus*. Photo by Dr. D. Terver. Nancy Aquarium, France.

Spine Photo: *Anthopleura midori*. Photo by Takemura and Susuki.

Title Page: *Stenopus hispidus*. Photo by G. Marcuse.

Line drawings, except for sections on Sea Slugs, Cephalopods, and Crustaceans (provided by authors), by Bea Gagliano.

ISBN 0-87666-495-8

© 1982 T.F.H. Publications Inc., Ltd.

Distributed in the U.S. by T.F.H. Publications, Inc., 211 West Sylvania Avenue, PO Box 427, Neptune, NJ 07753; in England by T.F.H. (Gt. Britain) Ltd., 13 Nutley Lane, Reigate, Surrey; in Canada to the pet trade by Rolf C. Hagen Ltd., 3225 Sartelon Street, Montreal 382, Quebec; in Canada to the book trade by H & L Pet Supplies, Inc., 27 Kingston Crescent, Kitchener, Ontario N28 2T6; in Southeast Asia by Y.W. Ong, 9 Lorong 36 Geylang, Singapore 14; in Australia and the South Pacific by Pet Imports Pty. Ltd., P.O. Box 149, Brookvale 2100, N.S.W. Australia; in South Africa by Valid Agencies, P.O. Box 51901, Randburg 2125 South Africa. Published by T.F.H. Publications, Inc., Ltd., the British Crown Colony of Hong Kong.

Contents

Introduction

Any aquarist is familiar with the great diversity of form and color in marine fishes and the great amount of literature available on them. He probably also knows of only two or three shrimp or anemones that are sold for marine aquaria at his local dealer. Yet there are only some 20,000 species of fishes (including both marine and freshwater species) but probably over 500,000 species of marine invertebrates, many of relatively large size and often bizarre form. Although marine invertebrates are abundant in almost every habitat and often form the most conspicuous portion of the fauna, it is difficult for the average aquarist to find readable information that will give him a working knowledge of the animals being observed. There are many college textbooks covering the invertebrates, often at length, but these tend to concentrate on morphology and systematics or on biochemistry, with little mention of the ecology and habits of even the most common types of crabs or anemones. The typical layman reader finds himself wading through a morass of technical terms and anatomical diagrams without absorbing any practical information.

The *Encyclopedia of Marine Invertebrates* was designed to serve as a bridge between such technical literature and the small 'seashore life' books more suitable for children. Each non-parasitic phylum or major group of marine invertebrates has been covered using terms that should be familiar to the average intelligent reader with a dictionary; more obscure terms are defined in the text. Through the liberal use of color photographs, the reader can readily become familiar with the appearance of the animals in life and thus have a better understanding of their structure than is possible from just dry words and diagrams. After all, the aquarist is only interested in the living animal, not a dissection preparation. Of course, some technical language is unavoidable in a book of this type, but we have tried to keep it to a minimum.

Before the discussions of most phyla will be found a short summary (*in italic type*) of the main characters of the group and an overview of the higher classification. These are purely the work of the editor and may not necessarily reflect the opinions of the authors of the phylum discussions. Commonly I have included in these summaries a few notes on the suitability of the group to marine aquaria, with mention of any special problems or merits that go along with the group. The phylum and class discussions themselves were written by a panel of experts on the invertebrates (identified in the table of contents) and draw heavily on the literature as well as, in most cases, the personal experiences of the authors. Although of course not exhaustive accounts, each major group is given coverage somewhat proportionate to its importance to the aquarium hobby. Thus the discussions of the crustaceans and molluscs are long and rather detailed, while the nematodes and brachiopods get only token coverage. A very few groups of obscure invertebrates are not covered at all, such as the mesozoans (parasites in octopus and squid kidneys) and gnathostomulids (microscopic worms found between sand grains), while the strictly parasitic phylum Acanthocephala is also ignored. I doubt if any of these few omissions will even be noticed by the hobbyist.

After the reader manages to absorb the information presented here, he should be ready to go on to more technical literature dealing with individual groups or more detailed surveys. A selection of such works is given at the end of the volume, though it is barely an introduction. The literature on invertebrates is tremendous, but it requires that the reader be versed in the fundamentals if he is to gain anything from it. A check of the invertebrate section at your local college library will quickly show you just how much information is available to the informed reader. The *Encyclopedia of Marine Invertebrates* should provide this background and at the same time foster an appreciation of the marine invertebrates as more than just poor cousins of damsels, tangs, and butterflyfishes.

Jerry G. Walls

Phylum Protozoa

About 30,000 species comprise this group of saltwater, freshwater, and terrestrial, as well as parasitic, animals. Their principal recognition character is that all members, regardless of size or appearance, are comprised of a single cell; a few are colonial, but the colonies are comprised of individual cells not closely connected by cytoplasm or reacting as a unit.

Class Flagellata (also Mastigophora): Locomotion by means of one or more flagella. Euglena, *dinoflagellates, many parasites.*

Class Sarcodina (also Rhizopoda): Locomotion by means of pseudopodia. Amoeba, *foraminiferans, radiolarians.*

Class Sporozoa: Parasitic within cells and undergoing a complicated life cycle. Malarial parasites, myxosporidians, coccidians.

Class Ciliata: Locomotion by means of cilia. Paramecium, Vorticella, *tintinnids, many others.*

Identification within this large phylum requires careful microscopic preparations and considerable skill plus a large library and is not normally feasible at less than the professional level. The freshwater protozoans are fairly well known, and illustrated accounts such as How to Know the Protozoa *(Jahn, 1949:* Wm. C. Brown Co., Dubuque, Iowa) *and the protozoan sections of* Ward and Whipple's Fresh-Water Biology *(Edmundson, 1959: J. Wiley & Sons, N.Y.) will be useful in placing marine representatives in their proper groups. There is a very large amount of literature on the foraminiferans because this group has proved useful in geological testing; most college libraries will contain well-illustrated technical and non-technical literature on this group.*

MARINE PROTOZOA

The protozoans are a vast assemblage of organisms with one thing in common: they consist of a single cell. Yet they are very varied in type and there are enormous numbers of different protozoans. This is because of the variation possible within a single cell. One should, in fact, think of protozoans as whole organisms rather than as single cells, and many distinquished zoologists refer to them as acellular animals. Thus they have structures known as *organelles* which are the equivalent of organs in multicellular organisms. Locomotory organelles include the usually tiny and hair-like *cilia* and the larger whip-like *flagella*. Some protozoans have modified the cilia into sense organs for picking up chemical stimuli. Certain protozoans have structures comparable to the nerves, and eyespots are quite common. Some large protozoans found in the intestines of cows and other ruminants possess complicated organ-like systems and are hard to recognize as being composed of only a single cell.

Protozoans can be classified according to their method of locomotion. There are four main classes: Sarcodina, Flagellata, Ciliata, and Sporozoa. The Sarcodina are those protozoans which move around by means of *pseudopodia* ('false feet'), which are flowing extensions of the body of the protozoan. The familiar amoeba fits into this group.

Perhaps the most important marine representatives of the Sarcodina are the **foraminiferans**. These secrete a shell which is made up almost entirely of calcium carbonate with small amounts of other inorganic compounds. Some species live in radially symmetrical single-chambered shells, but most have many-chambered shells. At first each young foraminiferan has a single-chambered shell, but as it grows it spills out of the first chamber and a new one is secreted. This process may be repeated several times as the animal grows, and the resulting shell may be of quite intricate design. Some form long chains like a string of beads, whereas others coil so that the shell is almost snail-like. However it is formed, the shell is perforated with numerous tiny holes through which the pseudopodia can be extended.

Foraminiferans are very common. Most are benthic, but one genus (*Globigerina*) is so common in the plankton that the empty shells of dead individuals collect in such numbers on the ocean bottom as to form an ooze. Deep ocean bottoms are covered with this globigerina ooze, and the term has come to mean a sterile, lifeless substrate.

The **radiolarians** are another group of the Sarcodina. As the name suggests, they are radially symmetrical forms. They are large as protozoans go, some species being several millimeters in diameter and colonial forms reaching several centimeters. Nearly all are planktonic. Their bodies are usually spherical and are composed of two layers. The inner nucleated region is bounded by a membranous capsule which is perforated with regularly spaced openings. These allow the cytoplasm to flow into the outer region and to project as pseudopodia. The gelatinous outer layer or *calymma* contains many vacuoles (fluid-filled cavities) giving it a frothy appearance. In some species there are symbiotic dinoflagellates inside the calymma. Oil droplets may color the cell blue, red, or yellow.

Radiolarians nearly always have a skeleton. This is usually siliceous (made of silicon dioxide, the chemical composing sand) in the radiolarians but consists largely of strontium sulphate in the specialized acantharians. Strontium is present in only minute traces in seawater, and it is remarkable that acantharians can accumulate sufficient quantities to build a skeleton.

There are two skeletal types. One is composed of long needles radiating from the center of the body and extending beyond the calymma. Where they leave the body they are surrounded by contractile fibrils which allow the needles to be moved. The other skeletal type is in the form of a lattice sphere. Several spheres may be present, arranged concentrically around and within the body. The lattice is frequently sculptured so that such skeletons are very beautiful, being ornamented with spines and barbs of varying sizes and numbers.

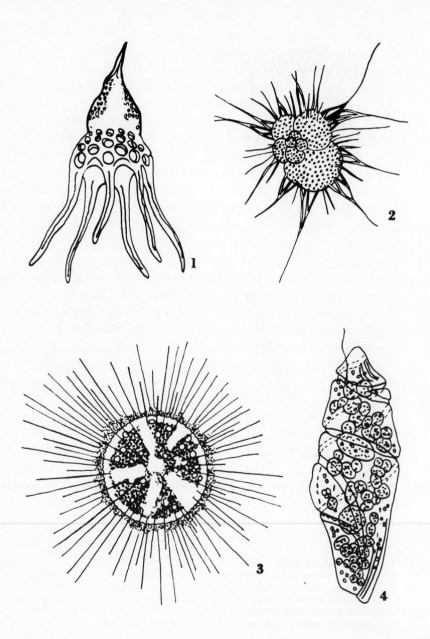

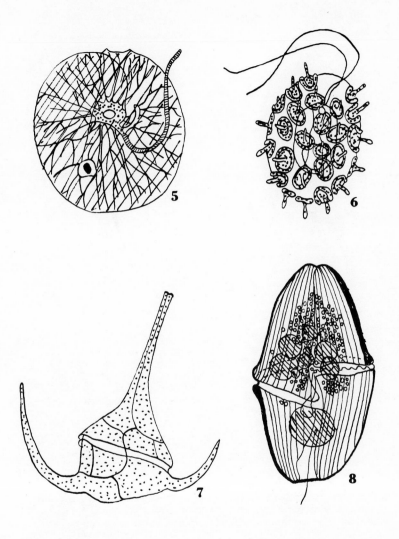

Above and facing page: Common marine protozoans. 1) Test of a radiolarian. 2) *Globigerina*, a foraminiferan. 3) *Acanthocystis*, a heliozoan. 4) *Cochlodinium*, an unarmed dinoflagellate. 5) *Noctiluca*, a luminescent dinoflagellate. 6) *Coccochrysis*, a flagellate known as a coccolithophorid. 7) *Ceratium*, a typical armored dinoflagellate. 8) *Gymnodinium*, an unarmed dinoflagellate; several species of this genus cause red tides.

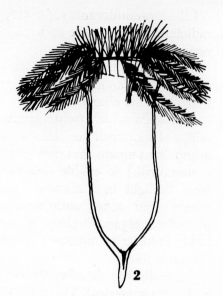

Some unusual marine ciliates.
1) *Folliculina* in its case. 2)
Favella, a rather typical case-
building tintinnid. 3) *Foetten-
geria*, a parasite of sea
anemones.

The phylum Mesozoa consists of microscopic animals closely resembling colonies of ciliated protozoans or highly degenerate larvae of flatworms. All known species are internal parasites, mostly in cephalopods.

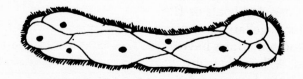

Like foraminiferans, radiolarians are very numerous and radiolarian oozes made up of their skeletal components are very common.

The second class of the Protozoa is the Flagellata, which contains an important marine group, the **dinoflagellates**. Flagellates are characterized by the presence of one or more long *flagella* which twist through the water rather like a whip. Dinoflagellates belong to the 'phytoflagellate' group. This means that they possess the necessary pigments (chlorophylls) to enable them to produce energy directly from sunlight in a manner similar to that of the green plants, and some authorities class phytoflagellates as unicellular algae. *Euglena*, a freshwater flagellate, is probably familiar to most readers as the major component of 'green water.'

Typically the dinoflagellate body is oval in shape and rather asymmetrical. There are two flagella; one is attached near the middle of the body and is directed posteriorly, while the other lies transversely in a groove which rings the body. This groove is known as the *girdle* in simple forms and as the *annulus* in those species where it is spiralled.

There are armored and naked dinoflagellates. The latter have no protective housing other than a simple cellulose cuticle. The common *Gymnodinium* belongs to this group. Armored species have a skeleton of cellulose plates or two cellulose valves. The basic armor is often sculptured, giving interesting shapes. It is thought that the projections may serve as flotation devices, increasing the surface area of the cell. *Ceratium*, for example, has three large horns giving the body the appearance of an anchor.

Many dinoflagellates are phosphorescent, giving off light at night. *Noctiluca*, a large and rather aberrant dinoflagellate, is the commonest of these. It is often abundant in coastal waters and may produce shining waves or washes about moving boats or fish.

Dinoflagellates may occur in great numbers in the plankton and sometimes have a detrimental effect on other organisms. When they are too common they may color the water; the famous red tides off California and Florida are caused by this. Where they occur in very large numbers oxygen may be depleted and cause the death of larger organisms, including fish. Some dinoflagellates, such as *Gonyaulax*, produce toxic metabolites or waste products, and a red tide of this organism results in the death of many marine invertebrates and fish. There have also been cases of people being poisoned by this protozoan as a result of eating affected shellfish such as oysters, which filter and accumulate toxic microorganisms under certain conditions.

The third class of protozoans is the **ciliates**, which, as the name suggests, move around by means of cilia. They are highly developed organisms showing many advanced features, especially a form of sexual reproduction whereby individuals with different types of nuclei mate together and undergo transverse fission.

There are many marine ciliates, some of them living in the plankton where they feed on phytoplankton. Some of the most common are the tintinnids, which live inside a vase-shaped case known as a lorica which is made up of tiny particles cemented together by the protozoan and carried around as a protective house. More typically ciliates live on the bottom and on algae and resemble the familiar freshwater *Paramecium*.

The fourth class of protozoans, the Sporozoa, need not concern us here since it is made up of only parasitic forms commonly found within the cells of fish and crustaceans.

The aquarist will find the protozoans of little more than academic interest. Although a few attached foraminiferans are large enough to be seen with the naked eye and may form small colonies in marine aquariums, most marine protozoans do not survive well in the non-flowing water of the aquarium. However, several types of flagellates and ciliates are valuable as food for larger invertebrates and may be purchased as pure cultures containing only one species. These can then be raised to produce a dependable source of microscopic food for small crustaceans, coelenterates, and sponges.

Phylum Porifera

This relatively small phylum (about 5,000 species) is typically marine, although a few species are found in freshwater and some in brackish environments. None present a typically animal-like appearance and all are sessile. Characteristics of the group are the lack of distinct organs, the penetration of the body walls by water canals, and the presence of at least one osculum through which water exits from the body cavity. Water currents are produced by special collar cells. Typically there are spicules of calcium carbonate or silica developed between the two layers of the body wall.

Class Calcarea: Spicules of calcium carbonate; small, relatively simple sponges. Sycon, Grantia.

Class Sclerospongia: Spicules of calcium carbonate, strongly consolidated. Deep reef-cave sponges.

Class Hexactinellida: Spicules of silica, the individual spicules with six points or sides; deep water sponges with relatively sturdy and symmetrical skeletons. Venus' flower basket.

Class Demospongia: Spicules of silica, spicules mixed with spongin, or spicules absent; spicules with other than six points; usually rather formless and asymmetrical. Bath sponges, freshwater sponges, boring sponges.

Taxonomically this is probably one of the most confused of the phyla. Because of the variability of individual sponges, it is difficult to determine what characters are genetic and which environmentally induced. Those with a casual interest in the phylum will probably not be able to accurately identify their finds, and even experts are often *unable to identify all specimens collected. Color and form are seldom usable, the identification depending largely on spicule form and details of canal systems. There are no guides available to the aquarist which will help identify most of the world's sponges even to family group, but local marine guides often allow the collector to come fairly close when identifying common species within a limited geographic area.*

PORIFERA

The Porifera, or sponges, form a group of primitive multicellular organisms which are isolated from the main evolutionary lines of other invertebrates. They are so unlike other animals, in fact, that for centuries they were believed to be a strange plant form, and it was not until the mid-eighteenth century that their true affinities were realized. The confusion is not surprising if one examines the life style of sponges. Most animals can move around and those that are stationary, such as the sea anemones, have moving parts, whereas plants are characterized by a basic immobility. Yet sponges show no signs of movement at all. The vast majority are attached in some way to the substratum and there are no tentacles or feelers to wave around. Furthermore, the internal organization of sponges is so very different from that shown in other animal groups that characteristic invertebrate features are not evident.

The discovery that sponges are really animals was made when it was observed that they produce water currents which pass through the body and that these currents are used for feeding purposes. Detailed anatomical studies followed, and we now have a fair understanding of the morphology and physiology of the group although taxonomic and ecological investigations are still in their infancy.

Basically, sponges consist of a thin bag-shaped structure which is perforated by many pores known as *ostia*. These are the openings through which the incurrent water is drawn. In addition each individual has a single exhalant opening or *osculum* which is larger than the ostia.

All other multicellular animals, or metazoans, show an organization of parts into tissues, which are made up of like cells, and in most cases into definite organs. In the Porifera, however, the organization is on a cellular level, there being no true tissues or organs. Indeed, in some respects they are little more advanced than the colonial protozoans which themselves may show some division of labor. Nevertheless, sponges may be considered as metazoans because in at least one function, the production of specific skeletal types, they show a high level of cellular cooperation and a marked species individuality.

In view of their unusual organization, it is not surprising that some of their cell types are unique. Most obvious in this respect are the *choanocytes* or collar cells, which consist of a flagellum encircled by a collar; these cells bear a remarkable resemblance to a group of colonial protozoans, the choano-

Diagrammatic cross-sections through the wall of an asconoid sponge (left) and a simple syconoid sponge (right). The osculum is at the top. Compare the number of ostia and the surface area available for absorption of nutrients.

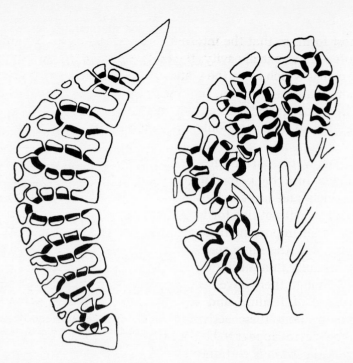

Diagrammatic cross-sections through the wall of an advanced syconoid sponge (left) and a leuconoid sponge (right). The osculum is at the top.

flagellates. These cells are responsible for producing the current of water through the sponge. There is no nervous system of any description and consequently no coordination of activity. This means that the flagella beat randomly with respect to each other, which tends to reduce the efficiency. However, the general internal morphology channels the water such that a powerful current can be achieved.

The porocytes are another group of special cells which occur in the sponge wall. The ostia are intracellular channels formed within these cells.

The area between the internal layer of collar cells and the outer wall is filled with a gelatinous *mesenchyme* containing several types of amoebocytes (free amoeboid cells) as well as skeletal elements where present.

All sponges follow this basic pattern, known as *asconoid*, but the original design has become more complicated in most forms so that the internal surface is much folded and, instead of a central cavity, the body is permeated by a series of canals which carry the water. In such forms there is often more than one osculum. In these cases it is difficult to decide where one animal ends and the next begins, but as the basic unit for survival needs only one osculum, those sponges with several openings are probably best thought of as colonies.

The water current forms the major life force for the sponge, bringing in food and oxygen and carrying out waste products and larvae. It is so important in fact that most sponges will not live for long if the current is stopped, and this is an important factor making laboratory studies difficult. For the same reason very few sponges can be kept successfully in aquariums.

Although there have been a few attempts to ascertain the manner of feeding and digestion, very little is actually known about these activities, although the mesenchymal amoebocytes appear to be the prime agents involved. In any event digestion is certainly intracellular.

The complete lack of any nervous system has intrigued scientists for many years. Recently suggestions have been made that some of the cells in the mesenchyme are anatomically like nerve cells, but there is no physiological evidence to corroborate this. It is perhaps difficult for us to imagine how an animal can survive with no sense organs and no method of rapidly transmitting information, but at such low grades of construction the irritability of the cells themselves is apparently sufficient.

One consequence of this is that nervous conductivity is very slow, thus any response to stimulation is very slow as well. It does occur, however, and most sponges are not completely rigid structures as once imagined. Contraction of the whole body is achieved by altering the shape of the cells and will occur if the sponge is handled or disturbed. More commonly movement is restricted to the closure of the oscula; this takes place at low tide or when the water becomes contaminated in some way. Light has no effect and may not be perceived, but touch produces a localized reaction. In all cases the movement is very slow indeed, the oscula taking several minutes to close.

Many invertebrates are able to regenerate missing parts, but the sponges are supreme in this art as one might expect from their loose construction. Any piece is capable of growing into a new individual, although, depending on the size, this may take several years. Early in this century some classic experiments into the regenerative powers of sponges were carried out whereby a sponge was pushed through a fine muslin or silk sieve so that it was broken up into very small cell aggregates or even individual cells. Yet in spite of this disruption the cells were able to reorganize themselves so that eventually a typical sponge construction was developed. Furthermore, the experiments showed that sponges are able to distinguish between 'self' and 'non-self.' Despite the loose association of the cells in the sponge body, when the above experiment was repeated with two sponges the cells from each individual soon separated and formed two new sponges.

The ability to regenerate is made use of as a guard against adverse conditions. Under such circumstances the body of the sponge collapses, leaving a 'reduction body' which contains the complete array of amoebocytes present in the sponge together with some modified collar cells. When conditions become more favorable these bodies can develop into complete sponges again.

A similar body occurs in those sponges that incorporate an asexual phase into their life cycle. The spores, or *gemmules*, are produced by an aggregation of amoebocytes enclosed within an epidermal layer. This then develops flagella for locomotory purposes and leaves the adult as a larva which in some instances may be virtually indistinguishable from those produced sexually.

Sexual reproduction occurs in all sponges, but precise details are not known. In any event the fertilization is believed to be internal, with the sperm entering via the water current and the resulting larvae expelled through the oscula. They remain as free-swimming larvae for a few hours to a few days and then attach and develop into small sponges.

Like many other sessile organisms, sponges are frequently used by other animals as a point of attachment. Their porous nature makes them even more susceptible to this, and in addition many animals may seek the protection of a sponge since the latter is so seldom attacked by fish or other predators. This is undoubtedly due in part to the prickly skeleton, but many sponges are known to produce a very unpleasant odor which is probably defensive.

Sessile animals such as sea anemones, bryozoans, and barnacles are often found growing on the surface of sponges, but the association may be accidental. There is at least one case of a more definite relationship, however, this being between the glass sponge *Hyalonema* and the anthozoan *Epizoanthus*. This sponge is frequently dredged from deep waters and nearly always has a growth of these little coelenterates on its root spicules. As described below under the glass sponges, many crustaceans live within sponges, where they gain protection from predators. A more mutually beneficial relationship exists between some hermit crabs and the suberitid sponges such as *Suberites*. Species of these sponges live on the outside of mollusc shells occupied by hermit crabs. As the sponge grows it completely envelops the shell so that the crustacean is apparently living inside a sponge house. The crab gains protection from predators by means of this disguise and the sponge benefits from the water currents produced as the crab moves around.

A less particular association exists with some other species of crab which break off pieces of sponge (and other organisms) and attach them to their backs as camouflage.

Because of the extreme intraspecific variability in sponges, the taxonomy of the group is very difficult. Only a few sponges exhibit a standard regular morphology, and most are greatly modified by the environment. This is to be expected in an organism which is so dependent on a current of water for its very existence. The shape of the sponge is thus affected by the velocity of the current, by wave action, and to a lesser extent by the nature of the substrate.

Some sponges show very bright coloration, orange, red, and yellow being particularly common, but many are characteristically brownish so that color is not a good feature for taxonomic purposes in most cases.

As can be seen from a quick examination of any taxonomic treatment of sponges, the principal taxonomic feature is the skeleton. This is secreted by mesenchymal cells and takes a variety of forms. Most sponges have a skeleton consisting of *spicules*, which are spiny bodies with a crystalline appearance. They consist of an internal organic deposit surrounded by either calcium carbonate or silica. The spicules have a very precise form and a large terminology has been built up around them based mainly on the number of rays present. In some sponges the spicules are interlocked to form a lattice which remains long after decay of the fleshy part of the animal.

Many sponges have a *spongin* skeleton instead of, or in addition to, the spicules. Spongin is a proteinaceous secretion which has a fibrous appearance; it may be arranged as a network throughout the mesenchyme or as a branching system. Most of the external variation between sponges is due to differences in the skeleton, and it is not surprising that the classification is based, in some instances rather arbitrarily, on this feature of the anatomy.

The class **Calcarea** is sharply defined, being marked off from the other sponges by its calcareous spicules. They are relatively primitive and are exclusively marine and almost entirely confined to shallow waters, with a few species found intertidally. They occur around all the major land masses of the world.

Perhaps reflecting a feature of their skeletal structure, the Calcarea are relatively small as sponges go, few attaining more than ten centimeters in height. The bright colors of some sponges seem to have overlooked this class and most are dull and inconspicuous. A further unattractive aspect is the prickly feel of the surface in some species, which is due to the spicules projecting through the body.

All the main types of sponge structure are present in this

class, including the asconoid, syconoid, and leuconoid types. Asconoids have a simple canal system connecting the ostia to the central cavity. In syconoids the wall is folded so there are ridges increasing the area. Leuconoid sponges have chambers branching off from the canals. All intermediate stages occur.

Most of the asconoid sponges are in the genus *Leucosolenia* and are typically colonial, each colony being made up of simple vase-shaped individuals reminiscent of the basic sponge plan. In some species the individuals are less easily recognized, the colony taking on a branching, twisted appearance.

The intermediate, or syconoid, sponges are also calcareous, *Sycon* being the typical example used in college zoology classes. Again the basic structure is vase-shaped, and a range of complexities of form may be observed imperceptibly grading into the most advanced or leuconoid sponges as displayed by the many species of *Leuconia*.

The second class of sponges is the Hexactinellida, so named because of their six-pointed spicules. The common name **glass sponges** also relates to the skeleton, which takes on the appearance of spun glass in dried specimens. Glass sponges are exclusively marine and in contrast to the Calcarea are found only in deep waters, often at abyssal depths. Most live at 500-1000 meters, but a few have been dredged from over 5000 meters. They are most commonly encountered in tropical waters, being particularly abundant around the West Indies and throughout the warmer waters of the western Pacific. Because of their habitat almost all our knowledge of the group is based on preserved material.

Glass sponges are typically radially symmetrical, a feature which may be associated with their environment. Conditions are virtually constant at great depths so that there are fewer variables to act as modifying influences on the sponge construction. In most cases the individual sponges can be easily recognized, the gradual submergence of the individual within the colony being absent from this class. They are usually vase- or urn-shaped and are attached to the substratum by a tuft of root spicules protruding from the base. The root spicules are curved at the end to ensure a firm anchorage. At great depths the ocean floor is covered with an ooze resulting from the continuous rain of material (especially foraminiferans) downward through the water, and the roots serve to keep the sponge clear of this sediment so that the water current is clean and the pores do not get blocked with particles.

Glass sponges are usually pale, bright colors being atypical in deep waters, and are moderate in size, ranging from about ten to thirty centimeters. Some are much larger, however, and in *Hyalonema*, for example, there is a very long tuft of root spicules giving the sponge a stalked appearance.

The skeleton is very beautiful, consisting as it does of silica spicules which have a very precise arrangement in the body. Some are free and others are loosely bound together or fused to give a lattice effect. It is this part of the skeleton which remains intact after drying. Such a skeleton makes

Representative spicules from sponges. 1) Amphidisc of a hexactinellid. 2) Style. 3) Oxyasters. 4) Trirachete. 5) Anatriaene. 6) Sanidaster. 7) Sigma.

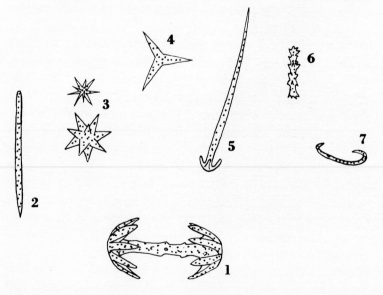

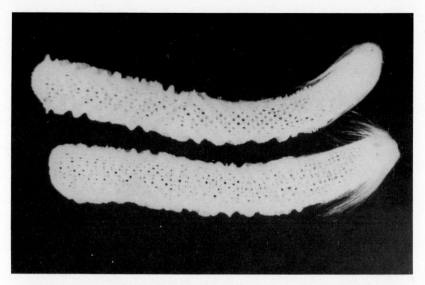

Sponges vary greatly in form, from the exquisite meshwork of Venus' flower basket (*Euplectella*), above, to the simple amorphous appearance of various encrusting sponges, below (K. Lucas photo at Steinhart Aquarium above).

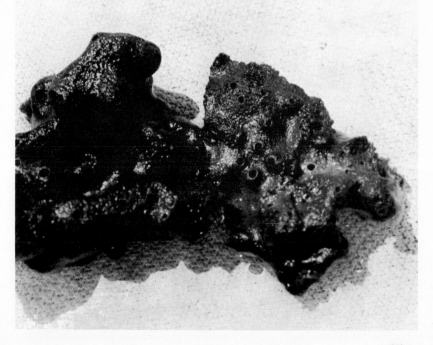

the sponge very rigid, and it is thought that this may also be associated with conditions in deep water. The open construction of the body certainly seems to be correlated with the very slow currents at great depths.

One of the most beautiful of the glass sponges is *Euplectella*, the Venus' flower basket, a very common sponge often seen in its dried form in museums. The body is a long, slightly curved cylinder and the skeleton is very light and open. The top consists of a spicular sieve plate which covers the osculum in life. In some species there are ridges down the side of the skeleton.

Members of this genus are hosts for commensal shrimp of the genus *Spongicola* which live in pairs inside the central cavity of the sponge. They enter through the sieve plate when young and become trapped inside the sponge as they grow. Because of this, dried *Euplectella* specimens were once commonly seen at Japanese weddings, the imprisoned crustaceans symbolizing the idea of marriage until death.

The third class of sponges, the Demospongiae, is much more heterogeneous in nature, consisting of those sponges with silica spicules which are not hexactinal, sponges with spicules and spongin fibers, and sponges without any spicules at all. Those with spicules are commonly known as *siliceous sponges* and those without a spicular skeleton are called *horny sponges*.

The siliceous sponges are divided for convenience into those with four-rayed spicules, the tetractinellids, and those with spicules with a single axis, the monaxonids. Tetractinellids are exclusively marine forms preferring shallow coastal waters where they may be found attached directly to rocks or lying free on the bottom. Most are rounded in shape, branched forms being atypical, but a few are encrusting. They are rather unattractive looking, being small and, with a few exceptions, rather drab. The body surface may be bristly, hard, or even leathery, depending on the nature of the underlying skeleton.

The monaxonids are the most common sponges, being found in shallow waters throughout the world where they are attached to the substrate by a spongin secretion. One group is found in fresh water. There is an enormous variety of shape and form, with rounded masses and branching bushes or slender cylinders and fans. This group includes the world's largest sponges, the loggerhead sponges (*Speciospongia*) and Neptune's goblet (*Poterion*), both of which extend for one or two meters. Most of the red and orange sponges fall into this group, and the body surface is characteristically bumpy.

Members of the family Clionidae (such as *Clione*) are unusual monaxonids which deserve a mention. These sponges bore into mollusc shells and other calcareous structures such as coral. It is not known how the sponge penetrates the shell, but chemical action has been suggested. Whatever the mechanism, boring sponges can play an important part in coral ecosystems and are responsible for large-scale degradation in the Caribbean.

The horny sponges, or Keratosa, are a very atypical subclass of Demospongiae in which the skeleton is formed entirely of spongin fibers. The familiar bath sponges belong to this group.

Horny sponges are often quite large and are usually rather amorphous, a rounded mass being the most common form, although *Phyllospongia* is leaf-like. The colors are generally black or dark brown. These sponges are lovers of warm waters, occurring tropically and subtropically, although small, rather stunted individuals may be found in colder seas. They live only in shallow waters and are attached by a spongin secretion to some hard object, usually a rock.

The bath sponges (*Spongia* and *Hippospongia*) are typical of the group. The best quality ones are harvested from the Adriatic and eastern Mediterranean, but they may also be found in the Gulf of Mexico and in the Caribbean. The sponges are traditionally collected by hooks, but more recently divers have been used. The living sponge is left lying in shallow water until the flesh rots away, leaving only the spongin skeleton which is then bleached and possibly dyed. In Florida a more scientific method of farming was developed in an attempt to compensate for over-fishing. Sponge cuttings were taken and set out on cement blocks, where they were then left to grow. This is a slow process, however, as large bath sponges are believed to be about fifty years old.

Bath sponges are the only members of this phylum to be well known to the general public, but the development of synthetic replacements means that even this will soon be a thing of the past. Sponges are likely to remain obscure plant-like animals for most people.

Recently a fourth class of sponges, Sclerospongia, has been recognized by some authors for a few bizarre sponges typically found in deep reef-caves in the Caribbean and other warm waters. In these sponges the calcium carbonate spicules are consolidated into a nearly solid mass covered with a thin layer of living tissue. There is a strong tendency to incorporate iron and other heavy metals into the structure, resulting in a sponge that can only be cut with a rock saw. Most species are rather small but are strongly attached in caves and crevices where light does not penetrate. Shapes vary from spherical to ear-like, from very coarse and heavy to thin and relatively fragile. The group is very poorly known.

Given the usual closed-system aquarium, the hobbyist should not attempt to keep living sponges in a community tank. As the entire group feeds by filtering very fine particles (debris, bacteria, algae, protozoans) from the water, they are almost impossible to keep alive. Many species also have the annoying ability to secrete irritating or toxic substances into the water, and many simply smell terrible when taken from the water. Few species have colors brilliant enough to even make their inclusion in the aquarium desirable, and they serve as food only for a few nudibranchs and even fewer fish. If the aquarist has a spare tank he might consider keeping one of the common encrusting sponges alive as an experiment, but even this requires very fine food and continually new water. As a general rule: no sponges in the tank.

Phylum Cnidaria

The coelenterates are the simplest of the multicelled animals that have a stomach or gastrovascular cavity. They were probably the first group of animals to make use of muscles for locomotion and to evolve elaborate sense organs to gain information about the environment around them. With the evolution of a stomach, large food particles could be consumed. A muscular system and sense organs allowed them to leave the sea bottom in search of food elsewhere. Some coelenterates (the groups is best called Cnidaria) are familiar to all of us. The corals, jellyfishes, and sea anemones are a few of the more common animals that belong to the phylum Cnidaria, formerly called Coelenterata.

Because of their plant-like appearance, the Cnidaria were called the Zoophyta, meaning animal-plant, for many years. They were first considered as an intermediate between plants and animals by the Greek philosopher Aristotle, and their animal nature was only established in the 1700's by the biologist Peysonnel, who observed the behavior of corals. Leuckart recognized that the cnidarian digestive tube, or coelenteron, was the only body cavity, and he coined the term Coelenterata for this animal group in the early 1800's. Finally, in 1888 the zoologist Hatschek separated the Cnidaria from the other groups of invertebrate animals. Through the course of history the Cnidaria have been classified with many other groups of animals.

The Cnidaria are a successful group with over 9,000 living species. There have probably been many fossil species, but because of their soft bodies, few fossil cnidarians (other than corals) have been found by paleontologists.

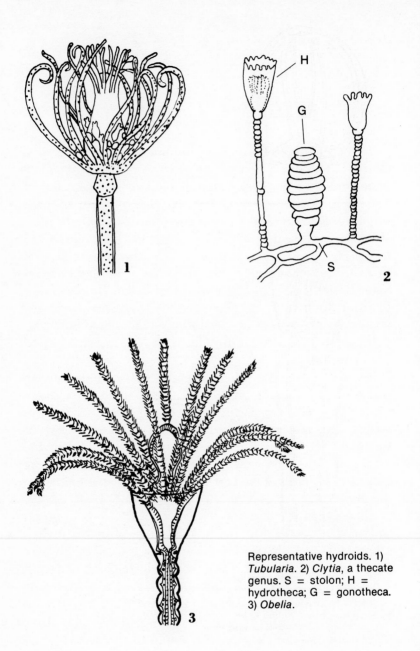

Representative hydroids. 1)
Tubularia. 2) *Clytia*, a thecate
genus. S = stolon; H =
hydrotheca; G = gonotheca.
3) *Obelia*.

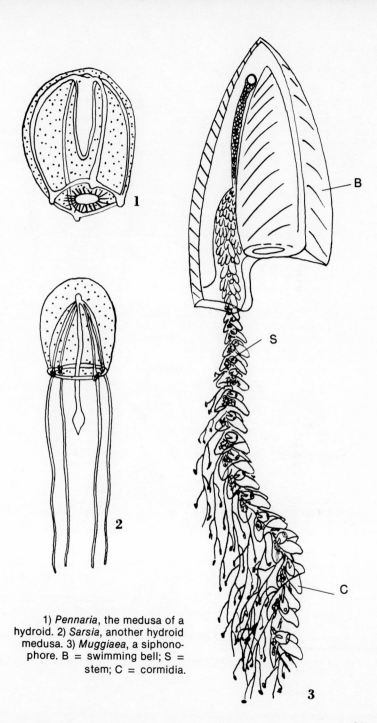

1) *Pennaria*, the medusa of a hydroid. 2) *Sarsia*, another hydroid medusa. 3) *Muggiaea*, a siphonophore. B = swimming bell; S = stem; C = cormidia.

B

S

C

1

2

3

The solitary coral *Heteropsammia* closely resembles an anemone at first glance. (Photo by Dr. L. P. Zann.)

Diagrammatic cross-section through a generalized sea anemone showing structures typical of the anthozoans. C = column; T = tentacle on the oral disc; S = sphincter; G = gullet; M = mesentery; A = acontia; P = pedal disc.

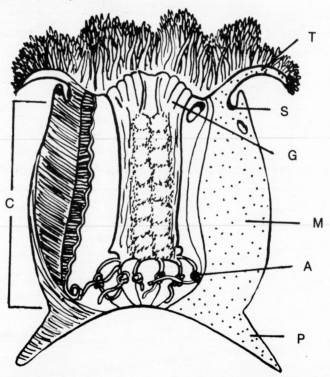

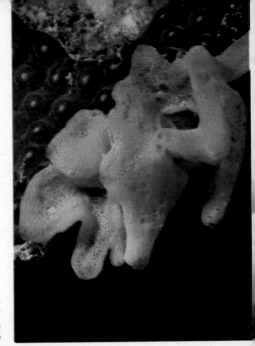

Clathrina coriacea (formerly *Leucosolenia canariensis*) (right), an abundant calcareous sponge (Atlantic; P. Colin photo).

An abundance of invertebrates use sponges as homes, as shown below by crabs, shrimp, and other animals removed from an Australian sponge (L. P. Zann photo).

All cnidarian animals have the same basic body plan. They are like two sacs, one placed inside the other. The mouth is the opening into the inner sac, while the closed end forms the base of a sea anemone or the umbrella of a jellyfish. The two sacs are comprised of cell layers, the outer layer being the epidermis, which provides protection and produces the stinging cells or nematocysts. It also has sense organs that detect food and other stimuli. The inner sac lines the body cavity or stomach. Between these two sacs is a layer called the mesogloea, a gelatinous substance that stiffens the animal. It is quite thin in sea anemones but is greatly thickened in jellyfish. Its primary function is that of support, but it is of little use to the animal out of the water.

The phylum Cnidaria is unique because all members possess stinging cells or nematocysts. Nematocysts are important as a defensive mechanism, and they aid these carnivorous marine animals in catching food. A unique feature of the Cnidaria, nematocysts are not really cells at all. They are actually structures formed inside special cells called nematoblasts, the small cells that contain the coiled nematocysts. When an organism touches a nematoblast, the nematocyst is forced out of the cell. This process takes only a fraction of a second. Once discharged, the shaft or thread portion of the nematocyst is about 100 times as long as the nematoblast cell.

Some nematocysts act to ensnare the prey organism and are fired in response to chemical and/or mechanical stimulation. Others have a small bulb filled with poison that is injected into the prey by means of a tube, much like a hypodermic syringe. Some are capable of cutting or slicing through the skin of the prey with tiny barbs or spines located on the shaft of the nematocyst. A nematocyst can be fired only once, so after it is discharged it is eventually cast off and discarded.

The exact cause of nematocyst firing is unknown, but it may be due to a change in cellular pressure brought about by a chemical reaction within the nematoblast cell. Because of this much research has been focused on nematocysts, their method of action, and the poisonous effects of their toxins.

The poison of some coelenterates is very potent, though human fatalities are rare. Because of the minute amount of poison in each nematocyst, a considerable number of them is required to kill the prey. The Portuguese man-of-war, *Physalia*, has a very toxic poison. Because of the long tentacles that trail below the animal in the water, large amounts of poison can be injected into the prey; its sting is very painful to humans and may kill. Although cnidarians die when removed from the water or washed onto a beach, the nematocysts may remain functional for several days.

Nematocysts can be seen with the aid of a microscope by squeezing a portion of a tentacle between a slide and a cover slip. There are many kinds of nematocysts, and the differences between them help biologists to identify many otherwise similar species of cnidarians.

By understanding the basic anatomy, behavior, and environmental requirements of the Cnidaria, the aquarist will be better able to duplicate suitable conditions for these fascinating animals in his aquaria. The phylum Cnidaria is divided into three large classes of animals, the Hydrozoa, the Scyphozoa, and the Anthozoa. This system of classification is based primarily on differences in body structure. The three classes are distinguished from one another by the degree of partitioning of the body cavity by longitudinal folds called septa.

In the Hydrozoa, the sole body cavity, the stomach, is a simple tube. In the Scyphozoa the stomach is partially divided into four chambers. The Anthozoa possess a body cavity that is partitioned into numerous small chambers by the septa.

Zoologists known as taxonomists classify the Cnidaria, as well as all animal life, so that the animals may be better understood. Within the three classes of the phylum Cnidaria, animals with similar characteristics and features are grouped into orders. Within the orders, similar and related animals are grouped into tribes, families, genera,

The presence of two or more growth forms in the same sponge genus is not uncommon. Shown are *Agelas clathrodes* (formerly *A. flabelliformis*) below and *Agelas* species to the left, both from the Caribbean (P. Colin photos).

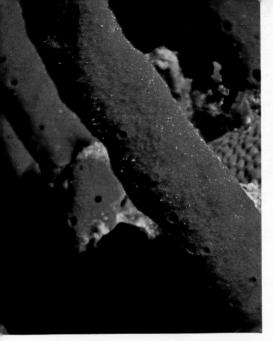

The red branches of *Haliclona compressa* (formerly *H. rubens*) (left) contrast strongly with the rather amorphous black mass of *Adocia carbonaria* (right), two common Caribbean sponges (P. Colin photos).

The boring and encrusting sponge *Cliona delitrix* is a common sight in the Caribbean (P. Colin photo).

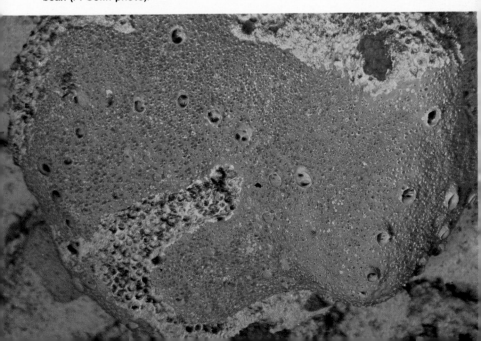

id species. The following classification scheme is one of the commonly accepted ways that cnidarian animals are classified, but it is by no means inflexible or accepted by everyone. The systematics of classifying animals has evolved slowly, and it will continue to change as we learn and understand more about them.

Phylum Cnidaria (Coelenterata)

Class Hydrozoa: *Cnidaria with a simple tube-like stomach. The medusa stage is usually small. The hydroid or polyp stage is small or large.*

Order Hydroida: Hydroids and hydromedusae; includes Hydra, Clava, Tubularia, Hydractina, Aequorea.

Order Milleporina: Stinging coral; hydrozoans with a heavy calcareous skeleton. Millepora.

Order Stylasterina: Hydrocorals; usually deep water, delicately branched. Stylaster, Allopora.

Order Siphonophora: Mostly planktonic colonial hydro-zoans. Physalia, Muggiaea.

Order Chondrophora: Planktonic hydrozoans similar to the siphonophores; adult stage usually found on the surface of the water. Velella, Porpita.

Class Scyphozoa: *Cnidaria with a four-chambered stomach. Generally the medusa stage is large while the polyp stage is small.*

Order Stauromedusae: Stalked medusae with a bottom-living polyp-like medusa stage. Haliclystus.

Order Cubomedusae: Sea wasps or box jellies; medusae with only four groups of tentacles. Chironex, Carybdea.

Order Semaeostomeae: Saucer-shaped medusae; common in bays along seashores. Aurelia, Chrysaora, Cyanea.

Order Rhizostomeae: Large tropical medusae that lack tentacles on the margin of the bell. Cassiopea, Stomolophus.

Class Anthozoa: *Cnidaria with a polyp stage only. No medusa stage ever present. The stomach is divided into numerous compartments.*

SUBCLASS OCTOCORALLIA: The octocorals; Anthozoa with eight tentacles.

Order Alcyonacea: The soft corals; colonies fleshy and usually lacking stiff skeletons. Tubipora, Anthomastus.

Order Gorgonacea: The gorgonians or sea fans; colonial anthozoans with a long central skeletal rod that supports the animal. Gorgonia, Corallium, Paragorgia.

Order Pennatulacea: The sea pens; usually long, colonial animals found anchored in a mud or sand bottom. Renilla, Pennatula, Cavernularia, Umbellula.

SUBCLASS ZOANTHARIA: Solitary or colonial anthozoans with twelve or more tentacles, usually in multiples of six. A calcareous skeleton may be present.

Order Actiniaria: The sea anemones; solitary anthozoans that lack a skeleton. Metridium, Actinia, Stoichactis.

Order Scleractinia: The true stony corals; solitary or colonial polyps with a calcareous skeleton. Astrangia, Acropora.

Order Ceriantharia: Solitary tube-dwelling, anemone-like polyps. Cerianthus.

Order Zoanthidea: Mostly colonial anemone-like polyps. Zoanthus.

Order Antipatharia: The black corals; colonial gorgonian-like polyps that secrete a heavy black skeleton.

Order Corallimorpharia: Solitary or colonial, flattened coral-like anemones. Corynactis, Ricordea.

The **Hydrozoa** are the most commonly found Cnidaria, for they occur in the ocean from shore to depths of well over a mile. There are also freshwater hydrozoan animals, such as hydras and the small freshwater jellyfish *Craspedacusta*. Because of the simple organization of the body, the hydrozoans are thought to be the most primitive group of Cnidaria.

The hydrozoans have a simple tube-like stomach with only a single opening, the mouth. This is in contrast to the other coelenterates, for in the Scyphozoa and Anthozoa the body cavity is divided into chambers or compartments by septa. Some species of hydrozoans are covered with a thin

Spinosella plicifera (above) and *Spinosella vaginalis* (below), both once placed in the genus *Callyspongia* (Caribbean; P. Colin photos).

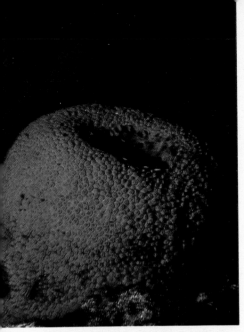

Pseudosuberites (left) and *Polymastia* (right), two East Pacific sponges of rather unusual appearance (A. Kerstitch photo on left; D. Gotshall photo on right).

Mycale sp., one of several bright red sponges often seen on Caribbean reefs (P. Colin photo).

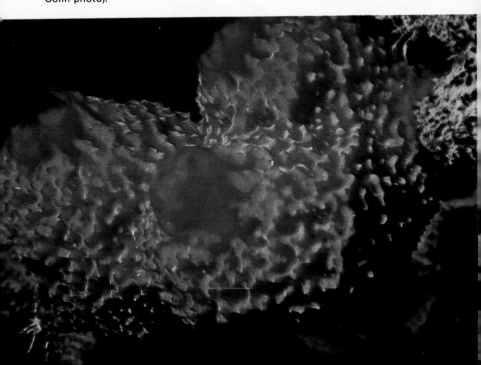

cuticle, while others may secrete a heavy skeleton much like that of the true corals.

Many hydrozoans exhibit alternation of generations, where part of the life cycle is spent attached to the sea bottom and part is spent as a drifting planktonic animal. The attached stage of the life cycle is the polyp or hydroid stage. The planktonic stage is the jellyfish or medusa stage. Usually the medusa stage is responsible for distributing the sex cells that unite to form new hydrozoan animals. Not all hydroids produce medusae, but those that do not have other structures that enable them to distribute the sex cells into the water so that fertilization will occur.

Most hydrozoans live in a marine environment and cannot tolerate fresh water (the hydras are of course common exceptions). *Clava* is a marine hydroid that is found along North American seashores and in protected bays. This half-inch tall hydroid is pink in color and forms fuzzy mats on algae. It is colonial and is classified as a naked hydroid because it lacks a cuticle.

Clava lacks a medusa stage, and during the summer and early fall small grape-like structures are found just below the ring of tentacles. Similar structures occur on another common hydroid, *Tubularia*. These are the reproductive structures that are responsible for releasing the sex cells at maturity, and they effectively take the place of the medusa stage.

Hydractina is another small naked hydroid similar to *Clava*. It is found on most sea coasts and lives attached to the shells occupied by hermit crabs, giving the shell a fuzzy appearance. Only when one examines a colony with a hand lens do the small, delicate polyps become visible. By living attached to the hermit crab shell, the hydroid is transported about and may obtain food particles when the hermit crab feeds. *Hydractina* does well in the aquarium and is quite attractive when the tiny polyps are fully expanded.

Most of the Hydrozoa have a planktonic or free-swimming medusa stage. The medusae of hydrozoans, called hydromedusae, are usually saucer- or cup-shaped. Often

The colonial hydroid *Stylactella* covering the shell of a mud whelk (Dr. L. P. Zann photo).

Millepora alcicornis, a fire coral, actually a stinging hydrozoan (G.v.d. Bossche photo).

An unidentified *Xestospongia* or *Cribrochalina* of vase-like form from the Caribbean (P. Colin photo).

Two cave-dwelling Caribbean sclerosponges. The small *Stromospongia norae* (left) is associated with serpulid worm tubes; the larger *Goreauiella auriculata* (right) may be over 150mm in diameter but only 2-3mm thick (P. Colin photos).

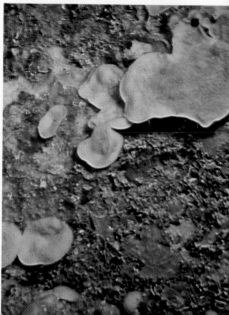

Halocordyle disticha, a large athecate or naked hydroid of tropical distribution (P. Colin photo).

Close-up of the polyp of *Tubularia* sp. (below), one of a large genus of common fouling hydroids (R. Larson photo).

there are a number of tentacles ringing the edge of the saucer or bell. The bell of the hydromedusa often has special sense organs. Simple eyes, called ocelli, are sensitive to light. Statocysts that are sensitive to gravity may be present and aid the medusa in general orientation. Besides the tentacles, a simple stomach and mouthtube hang from the center of the bell.

The hydromedusae are first formed on the hydroid. When they mature, the medusa is released and drifts about, distributing the sex cells, or others crawl about on the substrate. Medusae are usually either male or female. This distribution of sperm and egg cells by means of the medusae ensures that some animals will find a suitable location where they may successfully develop into new hydroids.

The hydromedusae are quite variable in size, some less than half an inch in diameter but others considerably larger. The medusa of *Aequorea* may grow to be four inches in diameter. It has a saucer-shaped bell or umbrella with a large number of short tentacles along the edge. Tubular canal structures, part of the stomach, radiate from the center of the bell on the underside to the edge of the animal. *Aequorea* is bioluminescent and can produce a brilliant flash at night.

Hydromedusae may be easily captured by dragging a fine mesh conical net through the water. Examination of the contents with a magnifying glass or hand lens will often reveal the medusa stage of many hydroids. The larger medusae may be raised in marine aquaria if fed live food such as freshly hatched brine shrimp. Care must be taken so that they are not caught in filter intakes.

Hydrozoan animals that cover themselves with a thick secreted skeleton are called the hydrocorals. Hydrocorals resemble hydroids, but because of their large external skeleton they are often mistaken for the true scleractinian corals. *Millepora*, or fire coral, secretes a heavy skeleton of calcium carbonate. The small polyps protrude through tiny pores in the skeleton to feed. *Millepora* is quite common in tropical seas and is found on coral reefs. Most skin and

SCUBA divers are familiar with this chocolate-brown hydrozoan because of its ability to irritate the skin when touched or handled.

Another group of skeleton-producing hydrozoans is the deepwater stylasterina corals. These delicately branched hydrocorals, such as *Stylaster* or *Allopora*, are brightly colored (usually pink or purple) because of the mineral salts within the skeleton.

The siphonophores are most unique hydrozoans. They are entirely planktonic and spend no part of their life cycle attached to the sea bottom. The siphonophore is made up of several types of specialized polyps. Instead of a free-swimming medusa stage, sex cells are released from an attached medusa-like structure into the water. This structure is a specialized polyp called a gonophore.

The Portuguese man-of-war, *Physalia*, stays afloat because of a pneumatophore or float. The polyp that forms the float secretes a gas, and the siphonophore can regulate the amount of gas within the float. Other siphonophores have a series of small swimming bells, called nectophores, that enable the animal to swim.

While some siphonophores drift with the prevailing currents, others can move about quite rapidly. *Muggiaea* is a streamlined siphonophore with a rocket-like nectophore. It is capable of moving about with considerable speed. Most siphonophores rely on thier nematocysts and transparency to avoid being preyed upon. Some siphonophores, such as *Agalma*, have both pneumatophores and nectophores. This arrangement allows the animal to move up and down in the water as well as to swim laterally in all directions.

Siphonophores also have polyps adapted for capturing food. Gastrozooids, or digestive polyps, swallow and digest food that is brought to them by the tentacles. The tentacles of siphonophores can be quite elastic and of considerable length. The tentacles of *Physalia* may be 25 feet or more in length.

Siphonophores occur throughout the oceans and occasionally may be washed ashore, often in a contracted state.

Sertularella speciosa, a stiff, feather-like hydroid (Caribbean; P. Colin photo).

The hydroid *Solanderia gracilis* (below) is so flattened and highly branched that it superficially resembles a sea fan (tropical; P. Colin photo).

A colony of the multitentacled *Clava*, a hydroid commonly found on seaweeds (R. Larson photo).

Species with nectophores can swim actively, and they are not as common on beaches as the siphonophores with floats only. The open ocean is the best place to observe the siphonophores, but not the most practical. Because they are so delicate, netting specimens often fragments them into a jelly-like soup. Many earlier described species later turned out to be only parts of much larger siphonophores.

A group of hydrozoans that resembles the siphonophores in many ways is the chondrophores. These are raft hydrozoans that lead a pelagic existence, but they do have a medusa stage. The chondrophores have no means of propulsion and rely upon the winds and currents for moving about.

Velella, or the ' by-the-wind-sailor, ' is a common chondrophore. It is native to all tropical seas and is often washed ashore on both coasts of North America. This beautiful blue animal is about one or two inches in length and is quite inconspicuous in the water. It consists of a raft of chitinous tubules and an erect, triangular sail. Hanging below the raft is a single gastrozooid or digestive polyp. The gastrozooid is ringed by gonophores which bud off the medusa stage. Stinging tentacle-like polyps, called dactylozooids, hang from the periphery of the raft.

Velella feeds on small crustaceans and planktonic animals that it may capture. Often storms will wash thousands of these chondrophores ashore. A predator of *Velella*, the pelagic snail *Janthina*, may also be washed ashore with them. *Janthina* feeds upon *Velella* and other planktonic organisms from its own raft of bubbles that keeps if afloat. Other predatory snails and nudibranchs are also found.

Another chondrophore, though not as common as *Velella*, is *Porpita*. This blue hydrozoan lacks the prominent sail of *Velella* and is shaped like a disc. It has a ring of tentacle polyps and feeds in the same manner as *Velella*.

Both *Velella* and *Porpita* will live in an aquarium for a short time, but they are not adapted for such an existence. Exactly why they do not fare well in captivity is not known, but they probably require pure ocean water of high salinity.

Because of their stinging nematocysts and the ease with which they may be damaged, siphonophores and chondrophores should be avoided as additions to a marine aquarium.

The **Scyphozoa**, or jellyfish, are a familiar group of coelenterate animals. They are often encountered in the summertime as a nuisance along swimming beaches. Scyphozoans exist as polyps too, and in one scyphozoan animal no medusa stage is present. Like the Hydrozoa, the Scyphozoa have an alternation of generations. The medusa stage, however, is considerably larger than that of the hydrozoans. The biologist Agassiz once encountered a ' lion's mane ' jellyfish that was over seven feet in diameter with tentacles over 100 feet in length.

The scyphozoans can be distinguished by the small finger-like processes of the stomach that aid in digestion. Scyphozoans often have long ribbony lips that hang from the mouth opening. Also, the typical scyphozoan stomach is partitioned into quarters, resulting in a stomach with four gastric pouches.

While the scyphozoans are characterized by their dominant medusa stage, the stauromedusae are thought to represent a combination of both polyp and medusa stages. *Haliclystus* is a common stauromedusa of both coasts of North America. It is greenish brown in color and has eight arms capped by knobbed tentacles. *Haliclystus* is often found attached to seaweeds or rocks and can move about by somersaulting in the fashion that *Hydra* does. It is about one inch in height.

The cubomedusae, or sea-wasps, as the name suggests, are the most dangerous of the coelenterates. The Australian cubomedusa *Chironex* has a potent toxin and has been responsible for several deaths. Cubomedusae in general, however, are not as dangerous as *Chironex*. They occur in tropical oceans and usually are not a hazard. Yet at night their well developed ocelli (eyes) enable them to swim toward the lights of divers, where they may become a nuisance.

Two common stinging hydroids of feather-like form: (above) *Aglaophenia* sp. (East Pacific; A. Kerstitch photo) and (below) *Lytocarpus philippinus* (Indo-Pacific).

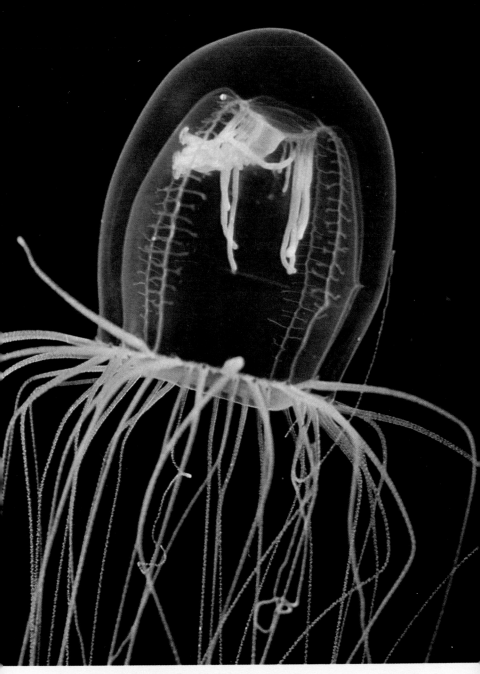

The medusa stages of many hydroids, such as this *Polyorchis* from the East Pacific, have a delicate beauty all their own (R. Larson photo).

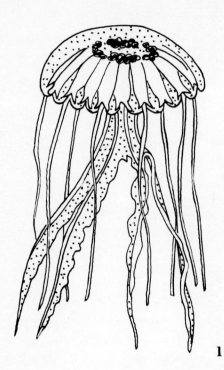

1) *Pelagia*, a rather typical jellyfish. 2) *Cassiopea*, a jellyfish modified for feeding while lying upside down.

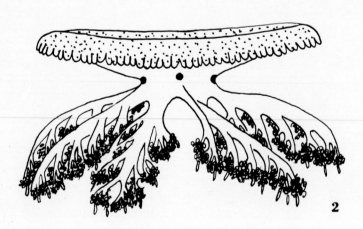

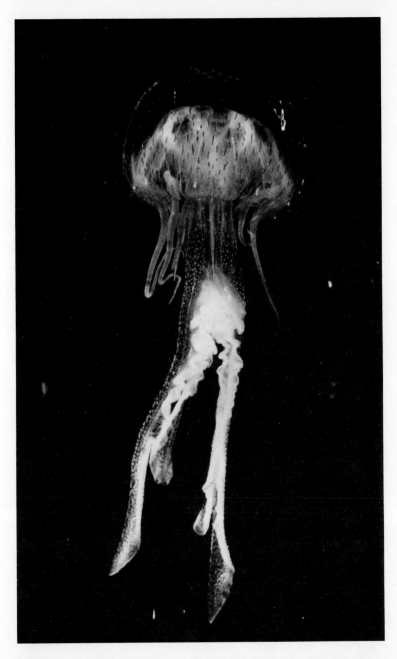

Pelagia noctiluca is one of the most familiar and beautiful of the jellyfishes (C. Arneson photo).

Two Caribbean fire "corals" (actually hydrozoans): *Millepora complanata* (left) and *Millepora alcicornis* (below) (P. Colin photos).

Allopora californica, a stylasterid hydrocoral (East Pacific; D. Gotshall photo).

Distichopora sp., a familiar type of stylasterid hydrocoral from the Indo-Pacific (S. Johnson photo).

The bell of the cubomedusae has a box-like appearance with tentacles hanging from the ' corners.' Usually the tentacles occur in groups of four. Cubomedusae are powerful swimmers, and with their strong poison they are capable of catching fish and shrimp. They are of interest not only because of their poisonous nature, but also because they may represent a possible link between the Hydrozoa and the Scyphozoa.

The coronate medusae are deepwater jellyfish that are seldom seen, but one species of *Linucte*, known as the ' sea thimble,' may occur in vast numbers in southern Florida waters and can be kept in marine aquaria on a diet of *Artemia* nauplii. They are usually captured by using special deepsea nets or trawls. The remaining groups of scyphozoan jellyfish are more familiar and can be quite abundant at times. The semaeostome medusae are the most common coastal jellyfish encountered. *Cyanea*, the ' lion's mane' jellyfish that gained fame in a Sherlock Holmes story, is common along both coasts of the United States. It is capable of giving a painful sting and is usually found in cooler waters.

A particularly bothersome jellyfish is *Chrysaora*, which is found along the Atlantic coast during the summer months. Large numbers of this medusa occasionally occur in Chesapeake Bay, where they present a hazard to sensitive swimmers. The name ' sea nettle' is well earned for *Chrysaora*, for jellyfish nets are erected around swimming beaches in this area during the summer months. *Chrysaora* is also found along the Pacific coast, though not as frequently as the large purple and white *Pelagia*, which often washes ashore on California beaches.

The large rhizostome jellyfish *Stomolophus* commonly occurs along southern shores of the United States. It has been called the ' cabbagehead' jellyfish, for it is quite large and has eight thick gelatinous lips that hang from the mouth. Because it lacks the long stinging tentacles that *Chrysaora* and *Cyanea* have, *Stomolophus* feeds on small planktonic animals.

Small scyphozoans can be maintained in the marine aquarium if fed live foods. Special care must be taken, however, to prevent them from becoming trapped or damaged in filter intakes. Air stones should be avoided in tanks with scyphomedusae, for they may become inflated with air and damaged. Because they can give a powerful sting, they should not be kept with other marine animals. The best medusa for the home aquarium is *Cassiopea*, or the ' upside down ' jellyfish. Because of its unusual habit of living upside down, it is often mistaken for a sea anemone. *Cassiopea* is quite common along the coast of southern Florida, in the Caribbean, and in Hawaii. It may be found in great numbers in calm, protected waters. *Cassiopea* will do quite well on a diet of freshly hatched brine shrimp, or even on dried fish food. Its sedentary nature and adaptability make it the most suitable jellyfish for the marine aquarium. There must be a source of artificial light to stimulate growth of the symbiotic algae (zooxanthellae) which live within the tissues of the medusa and give it its brownish yellow color.

The polyp stages of the Scyphozoa are quite small and often attached to old shells on the sea bottom. In the spring the polyps undergo a budding process as the water warms. Small medusae break off from the polyp and become mature by late summer. Once mature, the scyphomedusae release their sex cells before dying. Successful fertilization results in a free-swimming larva that eventually attaches to the bottom or to a solid substrate and develops into a polyp.

The polyps of scyphomedusae are easily kept in the aquarium, where the budding process can be observed. Larval medusae will readily eat larval brine shrimp, but it is difficult to raise them to maturity. Even scyphozoans that exist only as polyps, such as *Haliclystus*, are quite difficult to keep alive. Hopefully further experimentation will eventually result in successful techniques for raising and maintaining scyphozoans in the home aquarium.

Probably at one time or another everyone who has observed a sea anemone, gorgonian, or coral polyp has noticed the resemblance to some kind of exotic plant. For

Stylaster, a hydrocoral, has distinctive zigzag branches because of the arrangement of the polyps. To the left is a colony of *Stylaster roseus* from the Caribbean (P. Colin photo); below is a detail of the branches of a Pacific species (R. Larson photo).

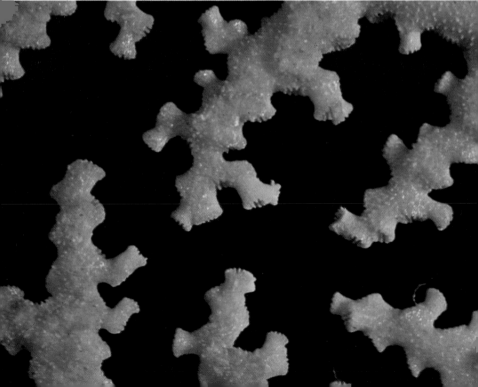

Parts of the colonies of two siphonophores: (above) *Stephanomia* (Caribbean; C. Arneson photo) and (below) *Hippopodius hippopus* (Atlantic; T.E. Thompson photo). These delicate animals fragment when collected and are seldom complete when photographed.

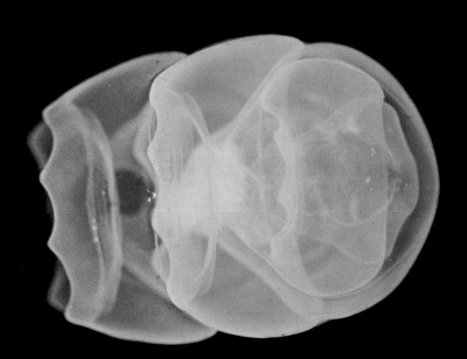

many years the **Anthoza** were thought to be plants or an intermediate life form between the plant and animal kingdoms.

In the Anthozoa the coelenteron or body cavity is divided into numerous compartments or alcoves by longitudinal septa. This division of the cnidarian body cavity by these septa has given the Anthozoa a considerable advantage over other cnidarian animals. The septa greatly increase the inside surface area of the digestive cavity, enabling more effective absorption of food by the surrounding tissues. As a result, the anthozoans have been able to grow much larger than most of their relatives in the Hydrozoa. Some tropical sea anemones reach several feet in diameter and some deepwater gorgonians such as *Paragorgia* may grow to be six feet in height and half a foot in diameter.

Nearly all anthozoans are bottom-living forms, and only a few species are planktonic during their adult life. The Anthozoa are also characterized by not having a medusa stage at any time during the life cycle.

Reproduction in the anthozoans is basically similar in all the groups. Sex cells are produced along the edges of the longitudinal septa and are released into the body cavity, from which they are eventually shed into the ocean. Fertilization results in a free-swimming larval stage that eventually settles to the bottom to become a new polyp. Asexual reproduction also occurs, particularly in colonial anthozoans where new polyps bud off to increase the size of the colony.

The **octocorals** (subclass Octocorallia) are the more primitive of the anthozoans. They derive their name from the eight pinnate tentacles that surround the mouth of the polyp. The remaining anthozoans possess multiples of six tentacles and belong to the subclass Zoantharia.

The tentacles of the octocorals are delicately branched, making them look like feathers. The octocorals embody a diverse assemblage of animals characterized by an internal skeleton made of a horny, wood-like substance. They lack the calcium carbonate skeleton that the true scleractinian corals have. Instead, small secretions of calcium carbonate are present as spicules within the tissue of the animal. These spicules are quite small and very diverse in shape. They are one of the means by which specialists can identify the octocorals.

Octocorals also tend to be colonial, and in the sea pens the polyps function somewhat like the polyps of the siphonophores. Octocorals are found living in shallow tropical seas as well as in very deep water.

Many of the octocorals are most unusual looking and oddly shaped animals. The gorgonians or sea fans, however, are aptly described by their name. They are fan-shaped animals that grow in shallow tropical waters and are common on most coral reefs. Gorgonians have a horny skeleton that supports the colony of polyps and attaches the colony to rocks or corals. Gorgonians will orient themselves to the prevailing current as they grow, in this way capturing drifting food organisms. The skeleton of gorgonians is covered with a thick skin called the rind because of its resemblance to the rind of a melon.

Gorgonia, the purple sea fan, is a common Caribbean octocoral. It is familiar to almost everyone as the lacy sea fan found dried in souvenir shops and in marine exhibits.

Heliopora is an unusual octocoral because it has a heavy skeleton similar to that of the hydrozoan *Millepora*. *Heliopora* is often called the blue coral because of its coloration, despite the fact that it is not a true scleractinian coral. The alcyonacean *Tubipora* or pipe organ coral also has a skeleton of calcium carbonate. Its ornate appearance makes it a highly desirable display item for the marine aquarium, although it is only found in the tropical southern Pacific Ocean.

The octocoral *Corallium* is also prized for decorative purposes because of its reddish skeleton. It is utilized for jewelry and is collected for this purpose in the Mediterranean Sea and near Hawaii and Japan.

The pennatulids or sea pens are another group of octocorals. Unlike most other octocorals, sea pens are adapted for living in sandy or muddy areas and are usually found in

Physalia physalis, the man-of-war, is the most commonly seen siphon-ophore and the only one at all familiar to most aquarists. It has actually caused human deaths and each year causes hundreds of severe stingings. On the facing page is a nearly intact colony (and the symbiotic man-of-war fish, *Nomeus*); the tentacles can extend for several meters when fully relax-ed (C. Arneson photo). Above is a detail of the smaller polyps of the colony.

deep water. The sea pens are colonial animals comprised of specialized polyps. One polyp serves to anchor the sea pen in the sea floor or bottom and is called the stalk or peduncle. The remaining polyps are attached to the upper portion of the peduncle, which is called the rachis. The polyps that attach to the rachis branch outward and give the animal the appearance of a feather.

Pennatula is a typical sea pen that occurs in the western Atlantic Ocean. It has a horny rachis bearing special polyps which pump water into the colony to keep it erect. Normally sea pens grow to a foot in height, but some species reach a height of six feet.

The deep-water pennatulid *Umbellula* resembles a long-handled umbrella because of the long peduncle that is inserted into the bottom sediment. The upper end is topped with a cluster of polyps that resemble feathers and enable the animal to obtain food.

A more common but unusual appearing pennatulid is the sea pansy, *Renilla*. *Renilla* has a reduced peduncle and a rachis flattened and shaped like a leaf. This sea pen rests on the bottom, the polyps attached only to the upper surface of the rachis. *Renilla* is found on sandy bottoms along the southern coast of the United States. Many pennatulids are bioluminescent and glow or give off a bright bluish light when disturbed. In *Renilla*, this bioluminescence passes over the rachis in the form of a wave.

The remaining groups of anthozoan animals belong to the subclass **Zoantharia**. While the octocorals are exclusively colonial, many of the Zoantharia are solitary, such as the sea anemones. Other zoantharians are colonial, such as the scleractinian corals. The Zoantharia have more than eight tentacles, which usually occur in multiples of six. These anthozoan animals are also the most familiar of the Cnidaria. Besides the sea anemones and the true corals, the zoanthids, cerianthids, and the antipatharians are all members of the subclass Zoantharia.

Perhaps of all the Cnidaria kept in saltwater aquaria, none are as fascinating as the sea anemones or Actiniaria.

There are upwards of 1,000 species of these graceful and colorful animals inhabiting the oceans of the world. Of the anthozoans only the Ceriantharia (burrowing anemone-like anthozoans) and the Zoanthidea (colonial anemone-like anthozoans) resemble the sea anemones.

Although plant-like in appearance, sea anemones are true animals. Even the common names of sea anemones reflect their variety of shapes and appearances. During the 1800's, British naturalists were collecting corklet, beadlet, pearlet, pimplet, and pufflet anemones. Unfortunately the identification of sea anemones is most difficult without examining the internal features of these animals. Therefore the aquarist who wishes to attempt identifying his sea anemones will find it helpful to become familiar with their anatomy.

Sea anemones appear simply as an extendable cylinder, the column capped at the top by an oral disc and closed at the bottom by the pedal disc. Most sea anemones are immediately recognizable by the tentacles found on the oral disc, which usually obscure the mouth. Like all other cnidarians, the mouth is the only opening to the body cavity.

The tentacles are most conspicuous on a fully extended sea anemone. They function in feeding, defense, holding prey, and even in locomotion. They are hollow, the cavity within being continuous with that of the body cavity. Nematocysts cover the surface of the tentacles and provide the animal with a formidable weapon for defense.

Tentacle number is quite variable, even among anemones of the same species. An individual ring of tentacles is called a cycle, the primary or innermost cycle containing the oldest and largest tentacles. Subsequent cycles (secondary, tertiary, etc.) are comprised of younger and smaller tentacles. The arrangement of the tentacles on the oral disc is determined by the internal paired septa and their manner of attachment to the oral disc.

The oral disc closes off the body at the top and bears the tentacles. A smooth unfeatured area usually surrounds the mouth in most species. From the inside, paired septa attach to the oral disc. These septa are radially arranged and effec-

The pelagic *Porpita* and *Velella* were once considered siphonophores, but current belief is that they are highly modified hydroids. *Porpita porpita* (above) is fed upon by the sea slug *Glaucus*, which can re-use the nematocysts of its prey (C. Arneson photo). *Velella velella* (below) is preyed upon by *Janthina*, a pelagic snail (T.E. Thompson photo).

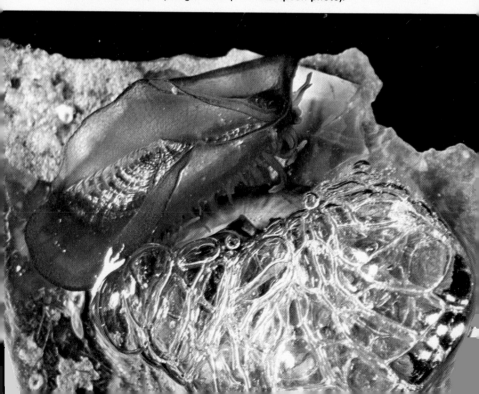

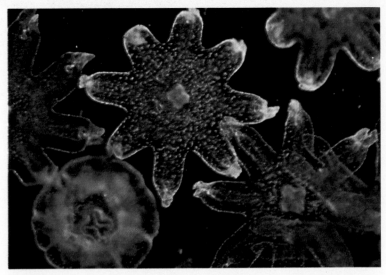

Ephyra larvae of the jellyfish *Aurelia* (Y. Takemura and K. Suzuki photo).

Haliclystus, a small attached jellyfish of cool waters. The animal can move about by using the "anchors" (dark ovals between the tentacle bundles) to attach while moving the pedal disc (R. Larson photo).

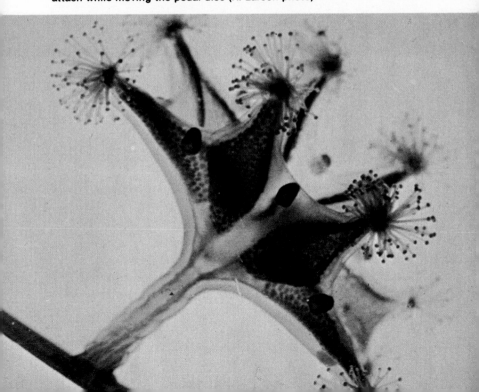

tively increase the surface area of the sea anemone's digestive cavity. The septa also bear longitudinal muscles important for the withdrawal of the mouth, oral disc, and tentacles. The septa also support the mouth-pharynx structure and bear other structures that will be discussed in conjunction with the internal anatomy.

The column of most sea anemones is capable of considerable expansion and contraction. The region where the column and the oral disc join is called the margin. Immediately below the margin, the thin-walled portion of the column is termed the capitulum. A marginal sphincter muscle separates the capitulum from the scapus, which is the stout lower portion of the column.

The marginal sphincter muscle aids in contraction by pulling the oral disc inward while the tentacles and column are contracting. The ability to close up in this manner is of considerable importance to sea anemones that are exposed by low tides.

The column of many species of sea anemones bears various formations and specializations. Columnar warts or verrucae are present on the anemones *Tealia* and *Anthopleura*. By attaching shell fragments, sand grains, and pebbles to these protuberances, the anemone is provided with an excellent means of camouflage and protection. Vesicular formations or blisters frequently occur in tropical sea anemones. The tropical anemone *Lebrunia* is covered with ornate evaginations of the column.

Various species of sea anemones have openings called cinclides in the column, providing a direct connection between the digestive cavity and the outside environment. In the event of danger the cinclides allow rapid purging of internal fluids during contraction of the anemone. Filaments laden with nematocysts and called acontia often protrude from cinclides of some genera such as *Metridium*. The cinclides of *Calliactis* are arranged near the pedal disc and provide protection for the hermit crab upon which this anemone dwells.

The pedal disc or base functions in attaching the sea

Stomphia coccinea, a swimming sea anemone from the northeastern Pacific (U. E. Friese photo).

Two separate mouths are readily visible in this rare photo of an anemone reproducing by division (E. Temke photo).

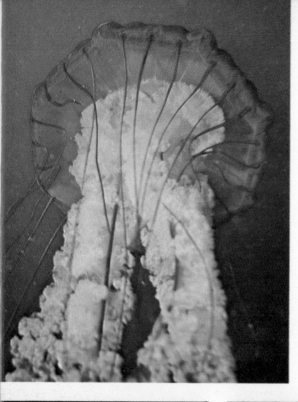

Three familiar jellyfishes: *Chrysaora* (upper left; D. Gotshall photo); *Pelagia* (lower left; D. Gotshall photo); and *Cyanae* (lower right; R. Abrams photo).

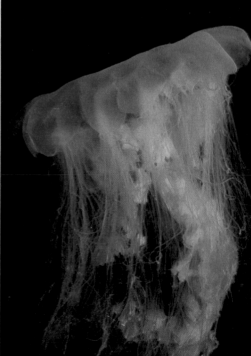

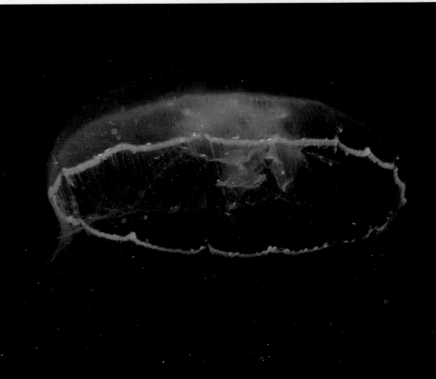

Chiropsalmus (at right), a stinging jellyfish or sea wasp, one of the cubomedusids (Caribbean; P. Colin photo). *Aurelia aurita,* the moon jelly (below), a widely distributed jellyfish that feeds on plankton rather than larger animals (P. Colin photo).

anemone to the bottom or to other objects. A constriction, the limbus, differentiates the pedal disc from the column. Although most anemones, such as *Actinia*, *Metridium*, and *Tealia*, have a base adapted for attachment to the bottom or to rocks, many have most unique pedal disc modifications.

Burrowing sea anemones have a pedal disc that may be of reduced size and is termed a physa. A pedal disc of this type may enclose a portion of the bottom sediment so as to secure the animal to one spot. *Edwardsia* and *Halcampoides* are anemones that have physa.

The planktonic anemone *Minyas* has a most curious modification. Here the pedal disc is adapted for floating, the anemone drifting about just below the surface of the water. In addition to these various modifications, many sea anemones live in association with other marine animals. The modifications of the pedal disc of these species will be discussed later.

The internal anatomy of most cnidarian polyps is quite similar, with few exceptions. The mouth and digestive cavity are connected by a pharynx or gullet of variable length. Ciliated grooves called siphonoglyphs are the most distinctive feature of the pharynx. The number of siphonoglyphs is variable, and they function to maintain a slight internal fluid pressure by propelling water into the body cavity.

The pharynx is supported by the surrounding paired septa that radiate outward and connect to the column. These are the primary or complete septa. Septa that are attached to the column but do not connect to the pharynx are secondary or incomplete septa.

Below the gullet or pharyngeal region, the remainder of the body cavity is dominated by the digestive cavity. In this region the septa are characterized by having filaments along their free edges. These septal filaments aid in water circulation and digestion and terminate in filamentous threads, the acontia. The acontia are often visible as a coiled mass in the digestive cavity, or they may be seen protruding from the mouth or cinclides. The reproductive organs, the gonads, are also located on the septa. They are responsible for releasing the sex cells, which are then expelled into the water.

Variations in the anthozoan reproductive scheme do occur, however. Some anemones actually "brood" the young within the alcoves of the body cavity or in specialized pouches of the body wall. The young of *Epiactis* can be found attached around the base of the parent anemone. In some species the larval stage is parasitic upon other animals. Larval *Edwardsia* parasitize ctenophores, and some sea anemone larvae are parasitic upon both hydromedusae and scyphomedusae.

Asexual reproduction by budding or fission takes place, too. Pedal laceration, where a new individual is produced by a constriction of the base of the parent anemone, is common, as in *Metridium*. In many cases a ring or trail of miniature sea anemones may be found in the vicinity of the larger parent anemone. Miniature anemones may bud off the column of an anemone, but this is rarely observed. The splitting of a sea anemone into two individuals also occurs, initiated by a furrow that eventually separates the two resultant anemones. In this manner large clonal aggregations of *Anthopleura elegantissima* are produced.

The process of regeneration of whole organisms from fragments or portions is one that fascinates scientists and laymen alike. The echinoderms and the cnidarians exhibit considerable ability to produce new animals from only parts. The anemone *Metridium* can regenerate a new oral disc if cut across the column at any point. The anemone *Boloceroides* can even regenerate a whole new anemone from a piece of the column alone. A gastrodermal sphincter at the base of each tentacle in *Boloceroides* even allows the anemone to cast them off, each of these cast-off tentacles being able to regenerate into a complete anemone.

Unfortunately for taxonomists, regeneration and asexual reproduction make it even more difficult to identify such individuals. The arrangement of the septa and the number of siphonoglyphs and tentacles can become quite variable and cannot be used as an accurate guide to identification. More

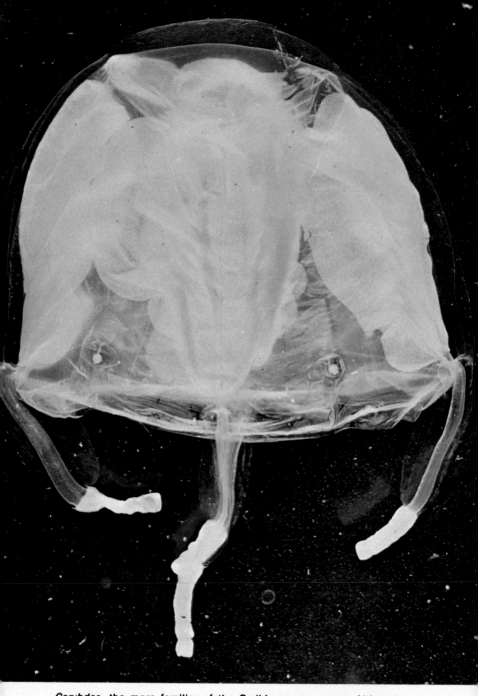

Carybdea, the more familiar of the Caribbean sea wasps. Although not seriously dangerous, the stings can be very painful (R. Larson photo).

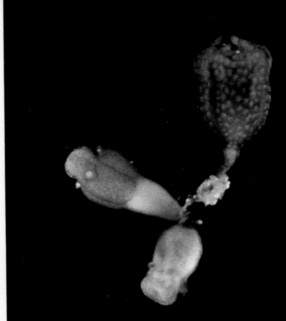

Metamorphosing polyps of *Carybdea alata* (Caribbean; C. Arneson photo).

The bottom-dwelling, plankton-feding *Cassiopea* is virtually the only jellyfish that can adapt to life in a marine aquarium, and even it is not the best of pets (Caribbean; D.L. Ballantine photo).

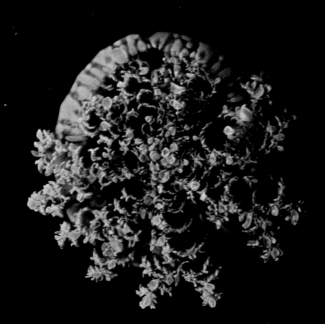

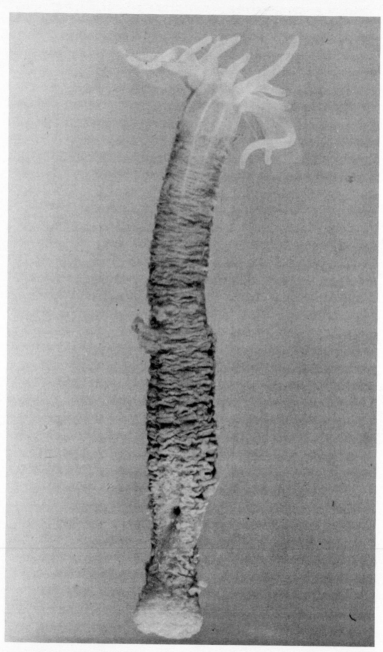

Edwardsia, a burrowing anemone (C. Arneson photo). The pedal disc is reduced to a physa that aids in burrowing.

Alicia is a nocturnal anemone characterized by the nematocyst-rich warts on the column; it is reputed to be venomous to divers (C. Arneson photo).

Young medusa of *Cassiopea* budding from the polyp stage (R. Larson photo).

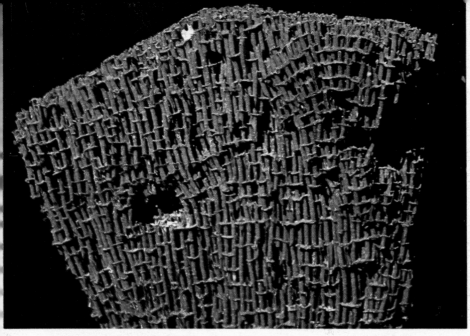

Organ-pipe coral, *Tubipora musica,* is actually an Indo-Pacific soft coral living in rows of parallel tubes made of fused spicules. When expanded (below) the green polyps hide the red tubes; the tubular skeleton (above) is often seen in pet shops (H.R. Axelrod photo above; B. Carlson photo below).

reliable characters are tentacle and column morphology, sphincter muscles, and the cnidom (kinds of nematocysts).

Sea anemones lack some of the complex life-support systems found in higher forms of marine organisms. A definitive respiratory system is lacking, therefore clean water with a high oxygen content is a necessity for life. Anemones do not have a centralized nervous system with a brain, but they do show remarkable muscular coordination using a nerve net system.

Sea anemones occur in a wide variety of habitats and often may be found in great numbers. It is unfortunate that as man encroaches upon the coastlines, sea anemones will undoubtedly succumb to the pollution being introduced into their natural environment. In spite of this, sea anemones are remarkably hardy animals. A visit to a rocky coast to observe anemones securely perched on, under, and between rocks in the surf zone gives some indication of their adaptive ability. There is strong evidence that sea anemones live to ripe old ages, even in the aquarium.

A fascinating account of an anemone living to sixty-six years of age was reported in the late 1800's. Sir John Dalyell collected an *Actinia* that eventually outlived him. During its productive lifetime it bore over 750 young, 150 of them born at the age of 50! With this in mind, it would be hard to guess the age of sea anemones that live peacefully in the uninhabited regions of the world's oceans.

Size is no indication of the age of a sea anemone either, for with lack of food they dwindle in size instead of dying. Therefore the aquarist who travels frequently need not worry about the health of his sea anemones in his absence.

While some sea anemones prefer to live in the surging tide zone, others are fond of making their existence under less rigorous conditions. Many prefer calmer water and are not profoundly influenced by tidal cycles. *Condylactis passiflora* may be found in shallow water grass beds. Here purple, green, and white colored varieties may live side by side in the warm tropical waters. Another shallow-water dweller is *Anemonia sulcata*, which is found in the eastern

Altantic and the Mediterranean Sea. Other sea anemones that can be found in shallow water are those adapted to a burrowing existence. These anemones secure themselves with the aid of their physa, and usually the oral disc is the only portion of the body that protrudes above the bottom sediment. Many of these burrowing sea anemones lack basilar musculature, and they will retract into the sand when disturbed by a predator or other stimulus. The Edwardsiidae anemones are the most typical of the burrowing forms. They are often mistaken for cerianthids by the novice collector.

While the burrowing sea anemones secure themselves with their bulbous physa and anemones of rocky coasts utilize their well developed pedal discs, the planktonic *Minyas* floats about freely. With a pedal disc modified into a float, it would appear best adapted for survival in the open ocean. *Minyas*, however, has a predator equally as unique. The Recluzia snail, *Recluzia*, preys upon this ocean-going anemone from its own float of air bubbles.

Various species of sea anemones seek habitats where they are associated with other marine animals. Conversely, other animals utilize sea anemones for protection and for a place to live. The term symbiosis is used for an association where two animals live together. Three of the most familiar symbiotic relationships are mutualism, commensalism, and parasitism. Many classical biological examples of symbiosis involve members of the phylum Cnidaria, particularly the sea anemones.

Parasitism is the most obvious relationship, where one partner in the relationship benefits or gains a definite advantage to the disadvantage of the other. The early larval stages of many burrowing sea anemones are parasitic upon the medusae of scyphozoan jellyfish. The larvae attach to the medusae and gain nourishment and a protected place to live until they are ready to attach to the ocean bottom and lead a life of their own.

Commensal relationships exist between many crabs and sea anemones. Here one partner, the crab, derives an ad-

Telesto smithii, a rather primitive octocoral in which the terminal polyp is larger than the lateral ones (Pacific; W. Deas photo).

The tentacles of *Xenia* arise in clumps from a rather stout base. Colonies of some species are bright blue in color, others tan (Pacific; W. Deas photo).

Often the spicules of soft corals are more conspicuous than the polyps. In this Pacific specimen the spicules are obvious and projecting (G.R. Allen photo).

vantage from the sea anemone, which is not affected by the relationship. In a commensal relationship it is not unusual to see the two partners, or symbionts, living independently of one another.

The Hawaiian crab *Lybia tessellata* is often found using the sea anemone *Triactis* for protective purposes. This inhabitant of Pacific coral reefs will remove sea anemones from the bottom and carry them about in the pincers. Besides threatening predators and enemies with these anemones, *Lybia* will steal food from them. Often when both pincers are laden with sea anemones, the crab will use its first walking leg to secure food, even if it must remove the food from the mouth and gullet of the anemone. Here the crab receives considerable benefit, while the sea anemone gains nothing from the association. The Hawaiian anemone *Telmatactis* can be found associated with the crab *Polydectus*.

Mutualism is the last and most studied form of symbiosis, as far as the sea anemones are concerned. In a mutualistic relationship, both symbionts benefit from the association, with no apparent disadvantage to either one. The most extreme form is exhibited by the anemone *Adamsia palliata* and the hermit crab *Pagurus prideauxi*. Mature individuals of either animal are never found without the other being present. Separation of the hermit crab and the anemone results in the death of the sea anemone. In this relationship the anemone is situated on the shell of the hermit crab so that it may obtain scraps of food. In return, the crab benefits from the protection afforded by the nematocysts of the anemone when it is expanded, and from the acontia when it is contracted.

A similar relationship exists between the anemone *Calliactis parasitica* and several species of *Pagurus* hermit crabs. In this association the anemone is not essential for the existence of the crab, or vice versa. Here *Calliactis* is found attached to shells occupied by at least three species of crab.

It is of interest to note that *Calliactis* also seeks out uninhabited shells and can make an unassisted transfer from shell to shell in about thirty minutes. Scientists believe that an unidentified 'shell factor' initially excites the sea anemone, for *Calliactis* will not attach itself to shells boiled in lye. If a suitable shell is within reach, *Calliactis* will adhere its tentacles to the shell, detach its pedal disc, and flex its column in order to complete this acrobatic routine by reattaching to its new home.

The benefits for both crab and sea anemone are the same in the *Calliactis-Pagurus* association as they are for the *Adamsia-Pagurus* association. In both situations the occupant anemone will be transferred to a new shell by the crab when it outgrows its smaller shell. The *Pagurus* will coax the sea anemone to loosen itself by gently prodding, stroking, and nipping at the column of the animal. Once freed, the crab will position the anemone, like a new hat, upon the new shell and go about its business. This transfer process occurs less frequently with *Adamsia* and *Pagurus*, for here the sea anemone secretes a substance from the pedal disc that effectively increases the shelter afforded the hermit crab.

Of all the symbiotic relationships that involve sea anemones, the one of most interest to aquarists is probably that between the Indo-Pacific damselfish *Amphiprion* and the sea anemones of the family Stoichactiidae. Many aquarists first become familiar with sea anemones by observing this fish-anemone association in a marine aquarium. In 1868, the biologist Collingwood first reported that fish lived among the tentacles of tropical sea anemones. The relationship between *Amphiprion* and the stoichactiid anemones is again a mutualistic one where both of the symbionts benefit to a certain degree.

An unprotected *Amphiprion*, or one that has been isolated from its anemone for a day or two, must acclimate to the anemone. The fish will prepare for a 'bath' among the poisonous tentacles in the following manner. The acclimation process is initiated by a nipping or nudging of the tentacles at various locations. This is followed by repeated passes over the sea anemone, where the fish will graze its fins against the tentacles. The *Amphiprion* will make repeated contact with its fins and finally with portions of

Dendronephthya, a common tree-like soft coral (Pacific; Conde photo courtesy Nancy Aquarium).

The spreading flat lobes of *Sarcophyton* may occupy large areas of reef (Pacific; A. Power photo).

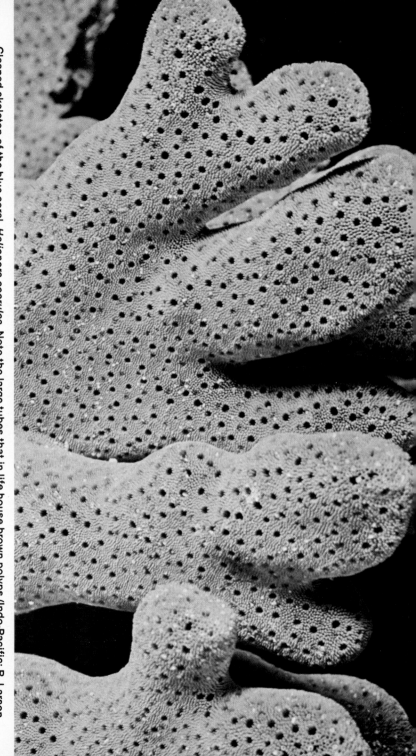

Cleaned skeleton of the blue coral, *Heliopora coerulea*. Note the large tubes that in life house brown polyps (Indo-Pacific; R. Larson photo).

the body, where the time of contact with the sea anemone is increased. These periods of contact are only interrupted by brief intervals of hovering above or around the anemone by the *Amphiprion*. The acclimation is completed by either a head- or tail-first settling among the tentacles by the fish.

The process of acclimation has been observed by numerous biologists and aquarists alike. The time for the process to be completed varies from ten minutes to an hour or more, depending upon the symbionts involved. It would appear that the fish develops some protection against the stinging nematocysts, possibly by thickening the secreted mucous coating that normally covers the skin of fish. The scraping away of this mucus results in the *Amphiprion* being readily stung.

In the wild, approximately two dozen species of *Amphiprion* have been observed in association with various species of sea anemones. *Stoichactis* and *Radianthus* are the most common sea anemones that have fishes associated with them. Those anemones from the Indian Ocean such as *Radianthus ritteri* and *R. malu* can be distinguished from Indian Ocean *Stoichactis* by their longer tentacles. *Stoichactis* is characterized by shorter, wart-like tentacles. The anemones of this family are poorly described and unstudied, and all the names used here are simply those familiar in the literature. More detailed studies may result in the change of familiar names.

These underwater giants often have oral discs that measure more than a meter across. Both *Stoichactis* and *Radianthus* are inhabitants of tropical coral reefs and are found in protected lagoons or on patch reefs. Here they may be found living in exposed areas, often living close to one another. These two genera of anemones are particularly responsive to light, and they droop their tentacles in the absence of daylight. The stoichactiid anemones are quite variable in coloration, and it appears doubtful whether *Amphiprion* choose their anemones on the basis of color. Visual contact, however, is important to the fish in order that they may find their way back to the home anemone.

Other sea anemones with fish in residence have been noted by biologists. In particular, two actiniid anemones, *Physobrachia ramsayi* and *Macrodactyla gelam*, one actinostolid anemone, *Parasicyonis actinostoloides*, and one thalassianthid anemone, *Cryptodendrum adhesivum*, have been observed in the Indo-Pacific. *Anthopleura elegantissima* and *A. xanthogrammica* are two Pacific coast anemones that anemonefish will acclimate to in the marine aquarium.

Physobrachia is found on tropical reefs, but it prefers to dwell in cracks, crevices, and protected areas away from light. *Physobrachia* also occurs in clusters, and what may appear to be a single anemone may actually be several smaller animals grouped together. The tentacles of this anemone are tipped by bulbous formations.

Other fish take advantage of these tropical sea anemones for shelter and protection in addition to *Amphiprion*. The damselfish *Dascyllus* (*D. trimaculatus* and *D. albisella*) associate with stoichactiid anemones as juveniles but not as adults.

The benefits derived by anemonefish from this relationship are fairly obvious, for they receive protection and shelter among the tentacles of their host anemone. Without an anemone to associate with in the wild, anemonefish are preyed upon without any hesitation on the part of larger fish. Those *Amphiprion* that have laid claim to an anemone exhibit strong territorial behavior and will ward off any intruders, be they other *Amphiprion*, larger fish, or even divers.

The *Amphiprion* will also utilize the vicinity of a host anemone to lay their eggs, which are cared for by both parents. After a week, the young *Amphiprion* hatch, usually at night. Within six to ten days they begin to seek out anemones of their own.

Besides these benefits derived from associating with stoichactiid anemones, anemonefish have been observed to feed on waste material that is expelled from the coelenteron of the anemone. Thus the *Amphiprion* may benefit from this

Iciligorgia schrammi differs from most gorgonians in having the polyps in only two marginal rows (Caribbean; P. Colin photo).

Corallium, the precious red gorgonian "coral" found in commercial quantities in only a few areas (Indo-Pacific; G.R. Allen photo).

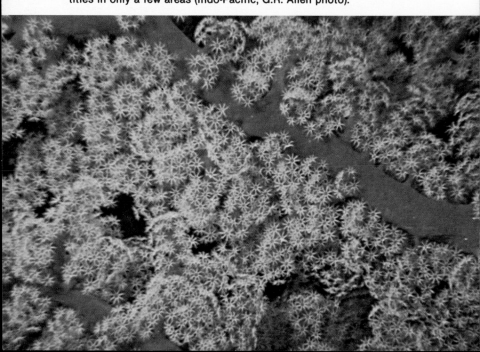

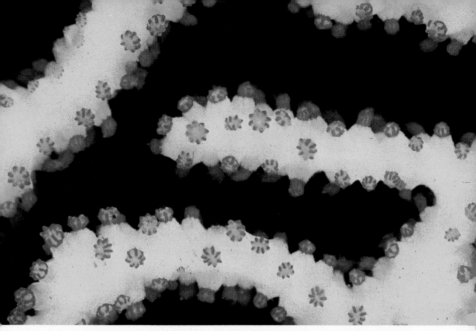

The gorgonian *Mopsella ellisi* is one of the more attractive species, but the colorful branches cannot be placed in the aquarium (Indo-Pacific; K. Gillett photo).

Gorgonians provide shelter for many animals. The resemblance of these *Hippocampus bargibanti* to the *Anthogorgia* is remarkable (New Hebrides; A. Power photo).

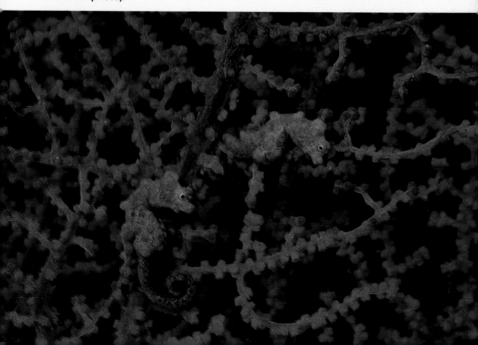

supplemental source of food, and the sea anemone may have its tentacles and oral disc cleaned on occasion by the fish. Stomach analyses of *Amphiprion* reveal that they feed primarily on algae and planktonic copepods, so it is doubtful that this source of food from the anemone is a major one.

Some *Amphiprion* have been observed to "feed" their host anemone with pieces of food. Instances have been recorded where these fish have returned to the anemone prey (small fish) injured by the nematocysts. This behavior, however, has been observed more often in the aquarium than in the wild. In general, experiments have shown that some species of *Amphiprion* will return organic and inorganic material to their anemone, as long as it is too big for the fish to swallow. *Amphiprion* will also exhibit 'feeding' behavior to substitute anemones, as well as to depressions in the sand, clumps of algae, and the bubble stream of an airstone. Anemonefish will also consume bits of anemone tentacles, but what benefit this may have, if any, for either fish or anemone remains to be seen.

Other marine animals besides fish utilize sea anemones for protection and shelter. Various crabs can often be found among the tentacles of anemones. *Petrolisthes maculatus*, a porcelain crab, *Periclimenes brevicarpalis* and *Periclimenes inornatus*, small shrimp, may live in *Radianthus* and *Stoichactis* anemones. Other shrimp such as *Thor amboinensis* can be found on the oral disc of these anemones. These relationships between crustaceans and anemones are not restricted to tropical species alone. On the eastern Pacific coast the shrimp *Lebbeus grandimanus* is found associated with anemones such as *Tealia crassicornis*.

Although sea anemones discourage many would-be predators with their nematocysts, they still have a fair number of enemies. Various species of fish, such as cod and flounder, feed on sea anemones with little or no harm. In fact, sea anemones were once used for bait along the coast of Scotland. Various gastropods, such as wentletraps of the family Epitoniidae, are known to suck the body fluids of sea anemones. The pycnogonids or sea spiders also have been observed to feed on sea anemones in the same manner. Other known predators include crabs and starfish. Several nudibranchs, such as *Aeolidia papillosa, Spurilla oliviae,* and *S. chromosoma,* are known to prey on anemones. Of course, man and his pollution of coastal waters must also be added to the list of dangers.

While most sea anemones have no means of escape from predators, the deep-water anemone *Stomphia coccinea* exhibits a most unique response to certain predators. When approached by an enemy, *Stomphia* ' swims ' by freeing the pedal disc and flexing the muscles of its column so as to be propelled away from danger. This response is thought to be brought about by chemical stimuli that the anemone can detect. This so-called ' swimming' response is most pronounced in the presence of the starfishes *Dermasterias* and *Hippeasteria.* The nudibranch *Aeolidia papillosa* also causes this response to occur.

The identification of actiniarians is difficult for both biologists and aquarists. For many anemones a microscope is required to examine the septal arrangement, the musculature, and the kinds of nematocysts present. Identification is further complicated by the contracted state that most sea anemones assume when preserved for study and by the anatomical variations that result from asexual reproduction.

With this in mind, the following classification scheme is presented so that the aquarist may begin to understand the basic systematics of the Actiniaria. It is also hoped that he may make tentative identifications of sea anemones that may be collected or kept in the marine aquarium. This simplified scheme is based primarily upon the survey of the Actiniaria completed by Dr. Oskar Carlgren in 1949.

Order Actiniaria

Suborder Nynantheae

This group contains the majority of the sea anemones that are commonly observed and most often collected by biologists and aquarists.

Muricea pinnata (Caribbean; P. Colin photo) at upper left; *Plexura homomalla* (Caribbean; P. Colin photo) at upper right; and *Muricea californica* (East Pacific; D. Gotshall photo) below, three rather typical gorgonians.

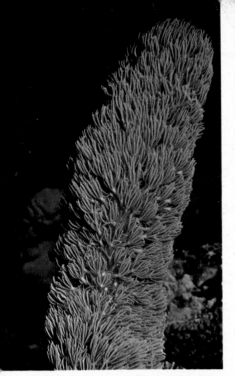

Some Caribbean gorgonians of more unusual form: *Plexaurella* (upper left); *Pseudopterogorgia* (upper right); and *Eunicea* (below) (P. Colin photos).

Tribe Athenaria

These elongate, worm-like anemones are distinguished by being adapted for a burrowing existence. They lack basilar musculature, and the pedal disc (in this group called a physa) may be tapered or rounded so as to aid in digging.

Nine families, including:
    Family Edwardsiidae
    Family Halcampoididae

Tribe Boloceroidaria

This small group of anemones is similar to the Athenaria, for they lack basilar muscles. The pedal disc is not physa-like.

Two families, including:
    Family Boloceroididae
        Boloceroides

Tribe Thenaria

These sea anemones are the more typical and representative group of actiniarians. They possess a differentiated pedal disc and have basilar muscles. They attach to solid objects and are further distinguished by their acontia or by the absence of them.

Sea Anemones Possessing Acontia:
Eleven families, including:
    Family Hormathiidae
        Adamsia palliata
        Calliactis parasitica
    Family Metridiidae
        Metridium senile
    Family Aiptasiidae
        Aiptasia spp.
        Bartholomea annulata
    Family Isophellidae
        Telmatactis decora

Sea Anemones Lacking Acontia:
Ten families, including:

Family Actiniidae
  Actinia equina, A. tenebrosa
  Anemonia sulcata
  Oulactis muscosa
  Anthopleura xanthogrammica, A. elegantissima
  Condylactis passiflora
  Tealia columbiana, T. coriacea, T. crassicornis
  Physobrachia ramsayi
  Macrodactyla gelam
  Epiactis prolifera
Family Thalassianthidae
  Cryptodendrum adhesivum
Family Minyadidae
  Minyas spp.
Family Stoichactiidae
  Stoichactis kenti, S. giganteum
  Radianthus ritteri, R. malu
Family Actinostolidae
  Parasicyonis actinostoloides
  Stomphia coccinea
Family Aliciidae
  Triactis producta
  Lebrunia danae

The **true scleractinian** corals, or hexacorals, are members of the Anthozoa. The term hexacoral aptly describes the arrangement of the six septa that divide the gastric cavity of the coral polyp. The corals are quite similar to the sea anemones, but they have the ability to secrete a calcareous skeleton. This skeleton provides the polyp with protection and serves as a substrate or place of attachment for other corals to grow. The corals are usually colonial.

The skeleton of corals is produced by the outer layer of skin and is actually a mold of the coral polyp. As the polyp continues to grow and expand, more skeletal material is secreted. The budding of polyps from the parent adds new individuals to the colony, and the coral formation can grow to immense size. The skeleton of corals can be used for identification purposes, but sometimes environmental factors

Gorgonia mariae, the smallest and probably rarest of the Caribbean sea fans (P. Colin photo).

Eugorgia rubens, a close relative of the sea fans but of very different shape (East Pacific; D. Gotshall photo).

Ellisella sp., a nearly unbranched gorgonian of bright red color (Caribbean; P. Colin photo).

Pterogorgia citrina, a gorgonian with greatly flattened branches (Caribbean; P. Colin photo).

can influence their growth and cause variations in their structure.

Living reef-building (hermatypic) corals are usually a greenish brown or yellow color. This is due to the microscopic yellow-green algae that live within the tissue of the coral. These single celled algae, called zooxanthellae, are also present in a wide variety of other marine animals, such as sea anemones, molluscs, bryozoans, worms, and tunicates. These algae have been found to be beneficial in the building of the skeleton by the polyp.

The famous naturalist Charles Darwin discovered that corals only flourish near the surface and in warm climates. During the voyage of the *Beagle*, Darwin studied coral atolls, which only occur in the middle of tropical oceans. He knew that corals could not grow unaided from the ocean floor to the surface to form these coral islands, so from his knowledge of geology he postulated that the ocean floor must have once been near or at the surface of the ocean at one time. He thought that corals must have become established and continued to grow at such a rate as to produce an island when the sea floor underwent the slow geological process of sinking. Today we know that Darwin was partially correct, for core samples taken from numerous coral atolls indicate that coral skeletal remains may be more than a mile thick.

Some species of corals do not form reefs. Deepwater corals live at depths where there is no available light. They may also be found where the water temperature is very close to freezing. These deepwater corals are usually solitary polyps, and they are adapted for living on a mud or silty bottom.

Some non-reef-building (ahermatypic) corals occur along the coasts of the United States. The star coral, *Astrangia*, is found in the western Atlantic Ocean and attaches to hard objects such as shells and rocks. *Astrangia* can be kept in the aquarium, where it will feed on bits of pulverized clam or shrimp. *Balanophyllia* is a solitary coral found along the Pacific coast. It is bright orange in color and is often found below the low tide line under rocks.

In the United States, reef-building corals occur only along the southernmost coast of Florida. Most of the state of Florida is the remains of a fossil coral reef that once was covered by a shallow sea. The coral reefs of Florida are quite extensive, but today development threatens their existence, as dredging activities in the Florida Keys have introduced large amounts of suspended sediment that is detrimental to corals.

Corals secrete a mucus that is utilized in catching tiny food particles, as well as for keeping the polyp free of settling sediment. Too much sand or sediment interferes with the cleansing and feeding processes, and the coral polyp eventually dies. The sediment also prevents the larval stage of corals from finding a suitable location to settle and begin growth.

To successfully raise or maintain any marine organism in an aquarium, the beneficial factors that nature provides must be supplied, while harmful factors must be minimized or eliminated. Coral reefs only occur where certain environmental requirements are satisfied, and an aquarist must strive to duplicate these conditions to keep corals alive in the marine aquarium.

Reef corals are found where the ocean temperature ranges from 78-85°F. and the specific gravity of the sea water is about 1.025. The water surrounding a coral reef is usually saturated with dissolved oxygen and is free of suspended sediments. The careful aquarist can duplicate most if not all of these conditions in his tank. A pH of about 8 (slightly alkaline) is beneficial, too. Natural sunlight can be approximated by the use of fluorescent bulbs for about twelve hours a day. Since many species feed at night, it may be best to carefully control the amount of light that your corals receive during the evening hours. Since reef-building corals contain symbiotic algae, light is essential for their growth.

Feeding corals does present a formidable problem, for many kinds of corals cannot obtain the food that they require in the home aquarium. Large coral polyps, such as *Fungia*, are known to capture small fish, but most corals are

Close-up of the fully expanded polyps of the sea pen *Cavernularia obesa* (Pacific; W. Deas photo).

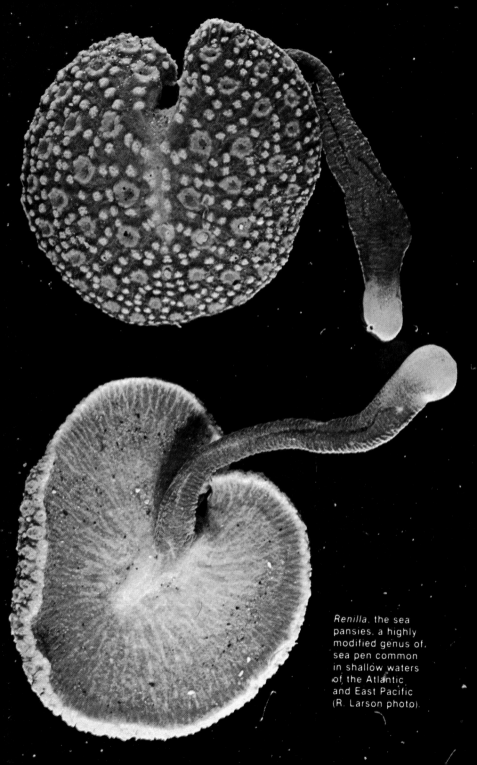

Renilla, the sea
pansies, a highly
modified genus of
sea pen common
in shallow waters
of the Atlantic
and East Pacific
(R. Larson photo).

particle feeders. In this case finely chopped and pulverized clam, shrimp, fish, or brine shrimp nauplii that are suspended in sea water may be used. This mixture may then be sprayed into the tank and onto the expanded coral polyps to allow them to feed. Success in maintaining corals alive in the aquarium is limited, for their food requirements mean that they demand constant attention. Additionally, corals may be readily preyed upon by other tank inhabitants.

Continued experimentation with various types of foods for corals in the aquarium may produce satisfactory results. Those species of corals found in shallow inshore waters can withstand changes in their environment better than deepwater, delicate species. Because they are colonial animals, small pieces should be selected for the aquarium. Dead polyps and invertebrates that are attached or living within the colony should be removed to prevent any fouling of the water. Maintaining delicate animals such as corals requires considerable patience and dedication, but the results can be quite rewarding for the aquarist who succeeds. By being careful and by using common sense and ingenuity, reef corals can be the star attraction of your marine aquarium.

The dead skeletons of reef corals are suitable for use in the aquarium, but care should be exercised so as not to contaminate the water in the tank. The skeleton itself can be cleaned of dead tissue by immersing it in a solution of bleach and water. If too strong a solution is used, it may dissolve delicate parts of the skeleton, so the coral should be checked frequently. After all the organic material has been removed from the skeleton, it should be hosed off and soaked in fresh water to remove excess tissue and any of the bleach solution. The skeleton can then be dried for several days in the sun and may have to be soaked again to remove any tissue that may remain. It is important that the coral skeleton be free of anything that may contaminate the water in the aquarium. A clean, intact coral skeleton can be a most attractive addition to the marine community tank and can serve as an excellent habitat for many fish and invertebrates.

It should be emphasized that collecting living or even

Cleaned skeletons of two common reef-forming corals, *Pectinia* above and *Acropora* below (Dr. Herbert R. Axelrod photos).

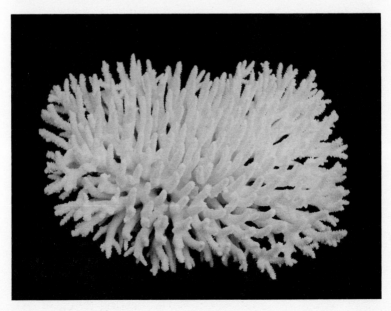

This red plume-like colony of *Pennatula aculeata* shows the typical sea pen organization of two rows of polyps (autozoids) on each side of a stalk (rachis) (Atlantic: R. Larson photo).

Expanded colonies of two East Pacific sea pens, *Stylatula elongata* (left) and *Ptilosarcus guernii* (right) (D. Gotshall photos).

Uprooted colony of *Ptilosarcus.* Note the arrangement of the polyps on the stalk and the bulbous "root" that anchors the colony in the substrate (East Pacific; T.E. Thompson photo).

dead coral is considered 'ecologically unsound' in many areas. Shallow-water reefs become fewer each year from pollution and commercial collecting, and new reefs take many years to establish. Excellent plastic corals are now available.

The Ceriantharia, **cerianthids**, closely resemble the sea anemones and may be mistaken for them. Cerianthids are adapted for a burrowing existence, and they lack the pedal disc that most sea anemones possess. They live within tubes of secreted material that are coated with sand grains, shell fragments, and other material.

Cerianthids also lack the marginal sphincter that many anemones have. Because of this they cannot withdraw the oral end of their body into the column. Instead, when they are disturbed they retreat into their tube by contracting the column musculature.

The tentacles of these animals are arranged in two cycles, with one smaller set surrounding the mouth and the other, larger, set ringing the margin of the oral end of the body. The cerianthids possess a single siphonoglyph, which aids the animal in inflating itself once it has contracted within its tube.

Unlike the Actiniaria and the Zoanthidea, the cerianthids have numerous unpaired septa that divide the body cavity. The cerianthids are capable of considerable extension of their bodies, so they may search for food on the sea bottom without having to leave their tube-dwelling. They may be found in shallow tropical waters and can be collected with a shovel at night. Only small specimens should be taken for the aquarium, for they can expand to two feet or more in length and may live in tubes that are three feet in length. Care should be used when collecting these animals, for they will often throw off their tentacles if roughly handled. *Pachycerianthus fimbriatus* occurs in muddy-sandy areas of bays and estuaries along the Pacific coast.

Cerianthids are quite popular marine invertebrates for the aquarium. They adapt quite well to living in an aquarium and reportedly live for a considerable number of years under favorable conditions. Some species are reported to secrete a noxious mucus when disturbed, however.

The Zoanthidea, or **zoanthids**, look like miniature sea anemones but are colonial. They usually are small and are found attached to coral, sponges, worm tubes, bryozoans, gorgonians, and rocks. Most often they are found in shallow tropical waters. They are difficult to identify without examination of their nematocysts and internal anatomy. Representative genera are *Palythoa* and *Zoanthus*.

These anemone-like coelenterates have a single ring of tentacles arranged around the oral disc. The mouth has a single siphonoglyph, and the pharynx is supported by paired septa, as in the sea anemones. The individual polyps of the colony do not have a pedal disc but are united at the bottom.

Some zoanthids exhibit an unusual association with hermit crabs. Once attached to the shell of the crab, the united portion of the colony, called the basal mass, begins to dissolve the shell. Eventually this mass encloses the hermit crab just as the shell originally did. The crab continues to live within the colony of zoanthids, and the resultant structure is called a carcinoecium.

Zoanthids make a good addition to the marine aquarium, but care should be taken to remove dead polyps from the colony to prevent fouling of the water. Because some have symbiotic zooxanthellae within their tissues, they do require a good deal of light. They may be fed planktonic organisms, brine shrimp, or substitutes. Clam, shrimp, or crab is satisfactory if administered in solution, but some are very difficult to feed. However, these may get their nutrition mostly from zooxanthellae.

The Antipatharia, black corals, are deepwater Anthozoa that are seldom seen except by SCUBA divers. They are often found in deep water around tropical coral reefs and seldom occur in depths of less than 100 feet. Very little is known about them because of the habitat in which they live.

Some species of Antipatharia are very long and whip-like, often reaching lengths of ten feet or more. Others branch

Expanded colony
of *Leioptilus* sp.
from the East Pa-
cific (A. Kerstitch
photo).

Close-up of the "root" of the sea pen *Virgularia* (Atlantic; T.E. Thompson photo).

The several species of the anemone genus *Calliactis* are associated with hermit crabs as commensals. The tentacles are numerous and may be banded or solid colors (U.E. Friese photo).

Peachia hastata, an anemone of cool North Atlantic waters; in some species of *Peachia* the larvae are parasitic on jellyfishes (T.E. Thompson photo).

like gorgonians. The black horny skeleton is highly prized as jewelry and is becoming depleted because of the heavy collecting being done in certain areas. The aquarist would rarely have the opportunity to keep antipatharians in the marine aquarium, and nothing is known of their feeding habits or their manner of reproduction.

The Corallimorpharia, as their name suggests, are similar to the stony corals. Some species have unusual, short or warty club-like tentacles and their septal arrangement resembles the true corals. But since they lack a calcareous skeleton, they can easily be mistaken for anemones. Most species are only a few centimeters in diameter, yet they are conspicuous animals underwater, usually being found in large numbers as brightly pigmented forms on hard substrates. *Corynactis californica* from the rocky subtidal of California has an intensely red column and white-tipped tentacles. *Ricordea* and *Rhodactis* are two very noticeable corallimorphs from southern Florida and the Caribbean which are red, green, or other colors and often cover several square meters of reef. A species of *Corynactis* is also known from Hawaiian waters.

Very little is known about the biology of corallimorphs. However, they can be kept in aquaria. *Corynactis*, *Ricordea*, and *Rhodactis* will accept a variety of foods, but the latter two genera require supplemental light since they have zooxanthellae in their tissues. Because of their bright colors and unusual appearance, they would be attractive and exotic additions to the marine aquarium.

With an ever-increasing interest in the sea and its life, more and more people are collecting, observing and keeping marine animals. With the aid of snorkeling and SCUBA diving equipment, the habitats of many marine animals are within reach of those who desire to learn more about the sea. Coelenterates are found along every sea coast, especially preferring rocky coasts and coral reefs. The practice of keeping cnidarians in aquariums is well over one hundred years old, and with the recent perfection of synthetic sea salts, even the novice can keep marine animals in the home. There are few hobbies as demanding as successfully maintaining marine animals in the home aquarium. Because of the sedentary nature of most coelenterates, they make a suitable addition to any marine aquarium and can be a continous source of enjoyment for the aquarist.

Cnidaria for marine aquaria can be obtained either by purchasing them from a reputable dealer that specializes in marine life or by collecting them yourself. Collecting your own specimens can be fun and rewarding, as well as educational. Before collecting, however, it would be advisable to inquire about laws with regard to disturbing the marine fauna. Collecting is prohibited in all national parks and seashores. In addition, some states require permits, particularly along the western coast of the United States. A preliminary inquiry may prevent the enthusiastic collector from being fined for his efforts.

There are a few simple rules that every good collector should follow. The most important thing to remember when collecting your own specimens is not to overcollect. Many popular areas have been seriously damaged by overzealous collectors who have taken more specimens than necessary. Collect only the animals that you need or that your aquarium can accommodate. Also take adequate precautions so that collected specimens are not crowded for the return trip to the aquarium.

For the aquarist who wishes to obtain coelenterates native to his area, a trip to the nearest seashore can be rewarding, especially for those who know where and when to look. A rocky coast, with its pools, crevices, and cracks, is an ideal location. Alexander Agassiz, the famous nineteenth-century biologist, describes one of his fruitful searches for sea anemones:

> *"it (a grotto) can only be reached at low tide, and then one is obliged to creep on hands and knees to its entrance, in order to see through its entire length; but its whole interior is studded with these animals, and they are of various hues—pink, brown, orange, purple, or pure white, the effect is like that of brightly colored mosaics set in the roof and walls."*

Aiptasia tagetes, a small non-descript Caribbean anemone (P. Colin photo).

Because it must be fed on clouds of fine food, *Metridium senile* is not suitable for the beginner's aquarium; stick with anemones that feed on chunks of fish or shrimp (East Pacific; K. Lucas photo at Steinhart Aquarium).

The large Caribbean *Bartholomea annulata* keeps its column hidden, only the numerous tentacles being visible; notice the circles of nematocysts on the tentacles (P. Colin photo).

The attractive *Telmatactis decora* is carried as a juvenile by the teddy-bear crab, *Polydectus* (Indo-Pacific; S. Johnson photo).

Tools helpful for collecting in this type of habitat include a good pair of boots or tennis shoes, gloves, a bucket, a thermometer, a knife, and perhaps a small geology pick for loosening rocks. Other instruments such as oyster knives, putty knives, and even letter openers have been found to be effective, too. In England during the 1860's, even blunted hairpins were utilized by an enterprising young woman to loosen anemones from crevices without injury. Sea anemones are safely removed intact from flat substrates without injury by hand and with a modicum of patience.

A small hammer or pick is helpful when removing rocks with specimens attached. Corals may be broken loose in this manner, or small pieces may be picked up loose from shallow water areas. When working around corals or handling jellyfish, it is advisable to wear gloves for protection. If stung by a jellyfish, it is important to carefully remove any remaining tentacles. Flush the stung area with water and apply medication such as a topical antihistamine to eliminate the swelling or burning sensation that may result. A doctor should be consulted if serious complications arise.

Several attempts at removing attached hydroids and sea anemones will allow the collector to perfect his techniques, but care should be exercised to replace overturned rocks and stones. Surplus animals should be released in safe spots rather than risk contamination from several ill-cared-for-specimens.

Consulting tide tables will also aid the collector in his search, as well as prevent any hasty retreat to safety in the event of an incoming tide. Low tides, and in particular spring tides, make for the best collecting.

Rocky shores are good collecting areas, but the calm waters of harbors, coves, and bays provide shelter for many cnidarians, especially jellyfish. They should be dipped from the water by using a small bucket attached to a pole. Nets will often tear delicate specimens, so they should not be used. Jellyfish and the floating siphonophores may be collected offshore, too.

Collecting at night can also produce results, particularly in sandy lagoons or on mud flats. Here burrowing sea anemone and tube-dwelling cerianthids can be collected with the aid of a shovel. Because the cerianthids can extend their bodies a foot or more in length, it is wise to collect only small specimens. When collecting these burrowers, don't worry about the tube dwelling they make, because they will immediately secrete mucus to form a new tube when they are placed in the sand.

Other areas where cnidarians may be found only require that the collector be observant. Dock pilings, boat hulls, rock jetties, and the roots of mangrove trees are often covered with marine life. The use of mask and flippers can extend your collecting range underwater, and proper instruction with SCUBA gear can provide access to reefs and deepwater areas.

A good collector will make sure that each specimen is isolated upon capture for the trip home. Plastic freezer containers work well for smaller cnidarians if they are provided with adequate aeration. Insulated containers such as ice chests help to keep the water temperature from rising, and battery operated pumps (often sold in fishing tackle shops) will supply sufficient aeration during transportation.

In many cases the aquarist has neither the opportunity nor the time to undertake collecting trips to obtain cnidarians for his marine aquarium. Therefore, the most convenient way to obtain marine animals is from a dealer. Care should be exercised in the selection of healthy marine invertebrates, for they are very sensitive to changes in their environment. A preliminary examination of a purchase can save the buyer considerable time and money in the long run.

While the following guidelines apply primarily to sea anemones, they also apply to almost any of the common invertebrates found in marine aquarium shops. The tanks of any dealer should be examined as critically as your own for any faulty maintenance. Clean tanks with adequate filtration and aeration are a must.

When observing sea anemones for only a brief time, it is often most difficult to judge the health of any individual animal. Because of the expansion and contraction abilities

A brilliantly colored Australian form of the anemone *Actinia tenebrosa* (K. Gillett photo).

Two types of long-tentacled or snakelocks anemones: (above) *Anemonia sulcata* (Atlantic; U.E. Friese photo) and (below) *Cnidopus verator* (Pacific; U.E. Friese photo).

of these animals, not all sea anemones can be assumed to be ill if contracted. However, a check should be made for a damaged column or torn pedal disc of the animal. Also check with your dealer about what he feeds his sea anemones, for knowing this information at the time of purchase can be of considerable help to the buyer.

Examine colonial cnidarian animals, such as corals and zoanthids, carefully because dead or infected polyps can endanger the health of the entire colony. Such polyps should be removed at once.

After purchasing new animals, it is wise to isolate new arrivals before introducing them into a community tank. The use of an antibiotic in the water of a holding tank, coupled with rapid filtration, will help keep bacterial growths to a minimum. The use of antibiotics in a community tank should be avoided, for certain bacteria function to break down toxic organic compounds in the waste products.

Once acclimated, feeding and healthy looking specimens may be introduced to the community tank with caution. Many marine animals eat coelenterates as part of their natural diets, therefore crabs, shrimp, molluscs, and many species of fish may not make the best tank-mates. With this in mind, it may be wise to keep sea anemones, cerianthids, corals, and zoanthids with tried and proven neighbors. To facilitate the easy removal of sea anemones from the aquarium for any reason, it is helpful to position them on small rocks or pieces of coral.

Once established in the aquarium, the fundamental needs of cnidarian animals may call for some modification in tank maintenance. Cnidarians require a good deal of oxygen, so aeration and filtration must be of a high degree. Sea anemones and many other coelenterate animals such as corals and zoanthids can withstand considerable crowding in their natural habitat, but only if their oxygen requirements can be satisfied. As with fish, it is better to understock an aquarium than to risk the chance of disease from overcrowding. Lack of oxygen will be apparent if sea anemones appear flaccid or highly distended.

In addition to oxygen requirements, water temperature constancy is a must. Most cnidarians handled by dealers are tropical varieties that are suitable in most marine aquariums. Many cnidarians, especially those from cooler temperate waters, cannot survive in tropical marine aquariums.

Light is an important factor for assuring the survival and optimum appearance of corals, sea anemones, and colonial zoanthids. Some tropical sea anemones, such as *Anemonia sulcata* and *Radianthus ritteri*, are adapted for exposure to the sun for long periods. However, many temperate water anemones require more subdued light. This is true for other cnidarian animals, too. Those cnidarians with symbiotic algae in their tissues will require a good deal of light if the algae are to survive.

Feeding coelenterates involves a good cleaning up after mealtime. Sea anemones and cerianthids expel digested material from the stomach or coelenteron about a day after being fed. Therefore the siphoning-off of wastes and excess food is recommended to prevent fouling and bacterial growths.

Because cnidarians are carnivorous, live prey is without a doubt the best food. For the average aquarist, this may present a problem. Most sea anemones will accept substitutes. Pieces of fresh or frozen seafoods, such as shrimp, clam, crab, or fish, are usually satisfactory. Freshly hatched brine shrimp are accepted as well and will usually satisfy the needs of most cnidarians, including corals. An eyedropper or a length of glass tubing can be utilized to assure that some brine shrimp will be captured by your animals.

For the novice, it would be best to consult your dealer as to the amount of food that should be given to your particular sea anemone or cerianthid. In general, if the tentacles remain contracted during feeding for any great length of time, it is doubtful that the animal will accept any more food. As a rule, it may be best to underfeed your specimens to avoid the task of removing excess food from the aquarium. Although the care and maintenance of cnidarian animals call for considerable patience and many hours of

Partially contracted speckled anemone, *Oulactis muscosa,* a species that covers the column with debris (Pacific; U.E. Friese photo).

Stinging animals occur in many groups of marine organisms, and anemones are no exception. This small *Anthothoe* species from Australia can cause painful rashes if it touches delicate skin (W. Deas photo).

Two of the most attractive species of anemones, both from the East Pacific: (above) *Anthopleura artemisia* and (below) *Anthopleura xanthogrammica* (K. Lucas photos at Steinhart Aquarium).

work, there are few sights as beautiful as a well-balanced marine aquarium with a complement of sea anemones, corals, and cerianthids.

Many of the animals of the phylum Cnidaria have been known since ancient times because of their delicate and intricate beauty. This is still true today, for more and more people are expressing their interest in marine life by keeping marine aquariums in their homes.

Besides their esthetic appeal, cnidarian animals are kept, observed, and studied for other reasons. Studies of nematocyst toxins and their manner of action and the effect they have on other animals are of particular interest to scientists. Also the unique reproductive cycle of the Cnidaria, with both a polyp and a free-swimming stage, is of interest, as well as the evolutionary importance of the Cnidaria in the development of higher life forms. But no matter who keeps these animals in the aquarium, whether scientist or hobbyist, for either pleasure or research, it is only by continued observation that we will discover the role of the Cnidaria in the marine communities of our waters.

Phylum Ctenophora

This small phylum (only about 80 species known) is strictly oceanic in distribution, its members seldom being found even in brackish waters. They have often been included within the Cnidaria and do superficially resemble jellyfish, but since they differ in several important aspects most workers now recognize the comb-jellies or ctenophorans as a full phylum. The most obvious distinguishing feature is the absence in all comb-jellies of nematocysts; instead, specialized cells known as colloblasts are present. These produce a sticky substance and are not replaced after being used. Most genera have specialized plates of cilia in eight rows (the combs) which are in constant motion. Usually two tentacles are present and can be withdrawn into sheaths.

Class Tentaculata: Tentacles present; body form oval to elongate. Sea walnuts, Venus' girdle, creeping comb-jellies.
Class Nuda: Tentacles lacking; body rather conical or thimble-like. Beroe.

The comb-jellies, although exceptionally gorgeous to look at in the sea, are also exceptionally delicate. For this reason they are almost impossible to collect uninjured and are never seen in aquaria. They are even difficult to preserve for scientific identification as they have a disturbing tendency to dissolve or fragment into globs of jelly.

CTENOPHORES

Ctenophores are in the main delicate, transparent planktonic carnivores between a centimeter and 20 centimeters in length. The common coastal *Pleurobrachia*, often called the sea walnut or sea gooseberry, is a good example of the

Phlyctenactis tuber-culosa, an anemone in which the column is covered with large vesicles (Australian; U.E. Friese photo).

Condylactis passiflora is one of the largest Caribbean anemones, its purple-tipped tentacles often reaching 100mm in length (U.E. Friese photo).

Tealia colum-biana, an interesting East Pacific anemone (D. Gotshall photo).

Physobrachia, one of the giant anemones in which anemonefish are usually found; note the bulbous tentacles (partially retracted)(Indo-Pacific; U.E. Friese photo).

typical comb-jelly organization. In size and shape it is much like a walnut. The bulk of the body consists of a jelly-like substance, with the mouth at one pole of the body and leading via the pharynx to the gastric cavity. From the gastric cavity (stomach) arise four short horizontal tubes which each branch, leading to blind tubes under the surface which resemble lines of longitude; there are eight such meridional tubes. The gut is a simple tube leading to one or more anal pores at the opposite pole from the mouth.

At this pole is also to be found the apical sense organ, consisting of a small solid body (the statolith) supported on four bristle-like tufts of fused cilia which bend with the pull of gravity. In a way not fully understood, nervous impulses are thought to pass outward from the sense organ to each of the eight rows of comb plates, which are stimulated to beat like tiny oars and propel the animal. Different 'messages' between the comb plates of one side and another will cause repositioning of the ctenophore.

A row of comb plates (comb row) comprises several dozen plates each consisting of 100,000 or more cilia fused together symmetrically. Each plate beats just after its neighbor and in turn is thought to induce beating in the next, so a ripple of beating passes down the comb row. The plates act in combination and are evidently very effective since the animal can move through its own length, from rest, in a fraction of a second. They are unique to the ctenophores.

Also uniquely ctenophoran are the two long, readily contracted tentacles, each housed in a sheath. They impart a bilateral symmetry to the otherwise largely radial plan of the animal. Each is a long solid filament bearing short lateral branches and has a muscular core which loosely controls its movements. The tentacle ectoderm is studded with special cells called colloblasts which are sticky and adhere to prey. Although apparently somewhat similar in function to nematocysts, they are different in structure and need not be replaced after each use.

Comb-jellies swim about in the plankton, no doubt aided by their almost neutrally buoyant gelatinous bulk. The ten-

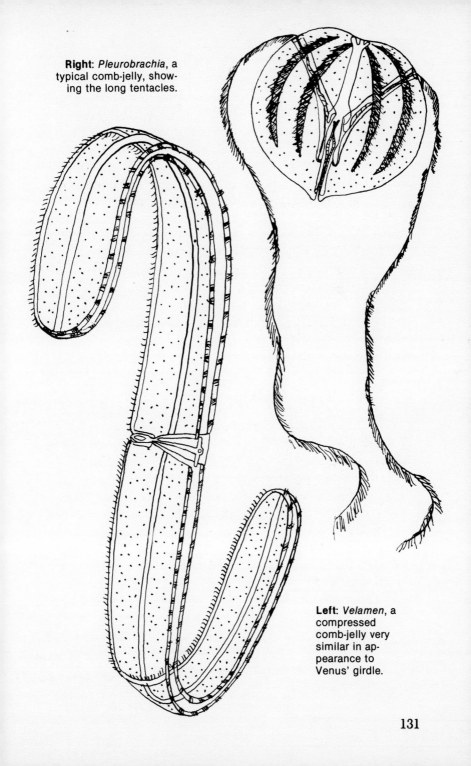

Right: *Pleurobrachia*, a typical comb-jelly, showing the long tentacles.

Left: *Velamen*, a compressed comb-jelly very similar in appearance to Venus' girdle.

The giant anemones are difficult to distinguish as they vary considerably with species, locality, age and size, and degree of retraction. Above is *Stoichactis helianthus;* below is *Radianthus malu* (Indo-Pacific; U.E. Friese photos).

The branched pseudotentacles of *Lebrunia danae* are large and distinctive, but the true tentacles are hidden and seldom seen (Caribbean; Savitt and Silver photo).

As in several other anemones, the column of *Bundosoma granulifera* is usually hidden in a crevice or hole (Caribbean; P. Colin photo).

tacles are used to capture such prey as small fish, copepods, jellyfish, and similar small animals. They are in turn preyed upon by larger fish, crustaceans, and jellyfish. In such types as *Pleurobrachia* the adults are hermaphrodites and eggs and sperm are released into the sea for fertilization. Commonly there is only one spawning a year and the adults live for only a year. The larvae closely resemble the adults in many features.

Several comb-jellies are very distinctive in form. The Venus' girdle, *Cestum*, is much compressed in the plane of the tentacles, which are rudimentary and have no sheaths. This results in an extremely wide body (up to 1.5 meters) which is narrow and thin. The animal swims by sinuous body movements caused by musculature, the comb plates being used only for repositioning the body. Venus' girdle is typically a tropical species found in deeper water.

The creeping comb-jellies are slug-like in form and creep over the substrate. Some are able to swim weakly and a few are even planktonic. Although tentacles and sheaths are present, the comb plates are lost in the larval stages. Once the creeping comb-jellies were thought to be possible relatives of the flatworms, though this is now known to be because of the superficial resemblance of shape instead of important structural and development characters.

The tentacle-less *Beroe* and allies are even more delicate than most other comb-jellies and resemble a transparent to pale pink upside down plastic bag. Many species are able to produce light. These ctenophores may fragment at a touch.

Phylum Platyhelminthes

Of the worm-like animal groups, the platyhelminths or flatworms are easily recognized in most cases by their flattened shape and the lack of an anus. With relatively few exceptions these are small worms of fragile appearance; since there is no body cavity, the body is nearly solid except for the gut and is easily torn. Of the about 13,000 flatworm species, over 10,000 are parasitic in vertebrates and other animals, forming the well-known flukes and tapeworms.

Class Turbellaria: *Generally free-living flatworms with simple life cycles not involving intermediate hosts and resting stages; the body is covered with minute cilia used for locomotion.*

Order Acoela: *Very small marine worms lacking an intestine and often without a pharynx and gonads.*

Order Rhabdocoela: *Small marine or freshwater worms with a sac-like intestine without branches and two simple nerve trunks.*

Order Alloeocoela: *Small mostly marine worms with a relatively simple intestine with few or no branches; there are three or four pairs of nerve trunks.*

Order Tricladida: *Larger worms living in marine, freshwater, or even terrestrial habitats; the intestine has three well developed branches, one forward and two backward; the eyes are usually few in number; most species are rather elongate.*

Order Polycladida: *Larger worms that are almost strictly marine; the intestine has numerous lateral branches or diverticula often extending to the margins of the body; eyes usually very numerous; body usually broadly oval.*

The staghorn corals *(Acropora)* are common reef-builders in warm seas and oc-cur in a bewildering array of sizes and shapes. Shown above is a rather typical species with slender branches (Pacific; G.R. Allen photo), while below is the bizarre elkhorn coral, *Acropora palmata,* of the Caribbean (P. Colin photo).

Corals can compete for access to open space just like more motile animals. Here the larger *Euphyllia turgida* is attacking and destroying part of a small colony of *Fungia* (Pacific; B. Carlson photo).

Plate corals, *Agaricia*, occur in numerous species in the Caribbean. Some, like this *A. tenuifolia*, have polyps on both sides of the plates, while other species have polyps on only one side (P. Colin photo).

Class Trematoda: *Parasitic; the flukes. Body without cilia but with one or more suckers; often a complex life cycle.*

Class Cestoda: *Parasitic; the tapeworms. Body without cilia; specialized reproductive segments (proglottids) usually present; attached end with suckers; life cycles complex and involving intermediate hosts.*

Few of the flatworms are of more than casual interest to aquarists. Some triclads and rhabdocoels are likely to be introduced into the aquarium with shells or other freshly collected material, but these seldom survive long; some are good food for fish in a community tank and none are likely to do any harm. A few of the larger and more attractively patterned polyclads are sometimes available in shops. The margins of the body contain extensions of the gut, so even minor tears can cause considerable damage and eventual death. Many polyclads will feed on small polychaetes or on the flesh of clams. Their elegant swimming motions combined with attractive and often bizarre color patterns make them the only rivals of nudibranchs in the aquarium, but they are definitely hard to maintain.

The parasitic flatworms are common inhabitants of fish and are sometimes also seen in crustaceans. Except for some flukes which are found on the gills or skin of fish (ectoparasites), the others are found only in the digestive or circulatory systems and are not likely to be seen in a living host. These worms are more properly the subject of books on parasitology and diseases.

The flatworms or Platyhelminthes are the most primitive worms, having a relatively simple organization of parts. Unlike the more advanced types of worm, flatworms do not possess a coelom or body cavity, and their anatomy is therefore arranged somewhat differently.

There are three classes of flatworm, of which the trematodes and cestodes are highly specialized parasitic forms probably evolved from an ancestor very similar to representatives of the third class, the turbellarians. This class contains a few parasitic species, but most are free-living worms occurring in a wide variety of habitats.

The turbellarians are divided by some authorities into five orders, all of which contain marine representatives or may even be exclusively marine as in the case of the polyclads.

Whatever the order, all turbellarians have certain underlying features uniting them as a group. As the common name suggests, they are flattened dorsoventrally so that they are virtually two-dimensional in appearance. Unlike other worms they do not show the typical long, thin, rounded shape, but are somewhat elongated in form. In some instances the head region is marked off by wings or auricles. Eyespots are generally apparent anteriorly on the dorsal surface. The mouth is situated on the ventral side and, as it is the only opening, it also serves as an anus in some species.

The name Turbellaria was given to these worms by a German scientist who noticed that the body surface is covered with cilia which, by their continual beating, cause turbulence in the surrounding water. It is now known that these cilia play an important part in the locomotion of flatworms.

The most primitive of the turbellarians are probably the Acoela, although it has been suggested that some of their so-called primitive attributes are in reality due to regressive evolution. Nevertheless their simple construction makes it easier to discuss them first before turning to their more complex relations.

Acoeles are very small marine turbellarians characterized by the complete absence of a separate digestive tract. Like most other marine turbellarians, they are found in the littoral zone, although their apparent absence from deeper waters might reflect difficulties in collection rather than a true abhorrence of subtidal conditions. Because of the nature of the body wall, acoeles (and other turbellarians) are in great danger of desiccation at low tide and all species show adaptations to avoid drying out. They respond negatively to light and positively to gravity and by this means tend to stay in sheltered damp environments. In addition some species are found among weeds while many others are restricted to rock pools on the upper shore.

Although adult fungus corals, *Fungia,* are solitary and unattached, juveniles (shown here) are colonial (Pacific; J. Jaubert photo courtesy Nancy Aquarium).

Many coral genera have characteristic shapes, but the familiar *Porites porites* may be branched, lobed, head-like, or plate-like (Caribbean; P. Colin photo).

The brain corals comprise several genera in which the polyps occur single-file in valleys between long ridges. Above is *Diploria clivosa;* below is *Colpophyllia natans* (both Caribbean; P. Colin photos).

A few species of acoeles show an interesting feeding modification. *Convoluta paradoxa*, for example, is a common brown flatworm about one centimeter in length. Close examination shows that the brown coloration is caused by the presence of algae living symbiotically in the worm's

Convoluta, a small flatworm that usually contains the symbiotic alga *Platymonas* (Dr. L. P. Zann photo).

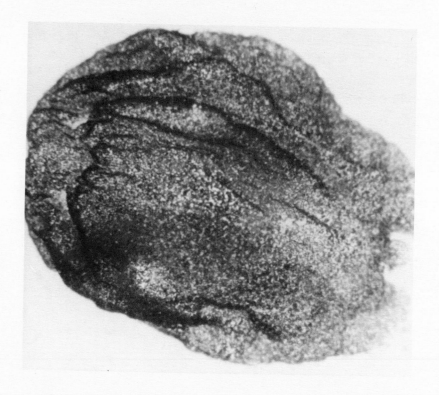

tissues. A similar phenomenon is found in *Convoluta roscoffensis*, a closely related species from northern France. This is colored green by the chlorophyll in its algae and has been studied in detail by workers interested in symbiosis. This species is found in great numbers in sandy pools; when the sun is shining the worms cluster near the surface, giving the appearance of a mass of seaweed. This enables the algae to photosynthesize. It is believed that the worm can somehow make use of this assimilated energy itself. The relationship is probably one of very long standing since both the plant and the animal have undergone extensive modifications. Thus the alga no longer has the typical plant cell wall and cannot survive outside of its host, and the flatworm has lost any carnivorous tendencies and, more remarkably, has developed a positive response to light.

Such species are the exception however, most acoeles being carnivores commonly feeding on small invertebrates like protozoans and crustaceans. Observations on the feeding behavior show that the prey is captured in a mass of digestive cells which are everted through the mouth. Digestion occurs intracellularly, small particles being captured phagocytically.

Typical of the phylum, acoeles are hermaphroditic, cross-fertilization being the rule. Because of the absence of a coelom, specialized genital ducts are often wanting for the female products at least, although many acoeles possess a muscular penis. Fertilization is internal; when there is no female opening the penis is inserted hypodermically and is therefore armed with barbs to help penetrate the body wall. Fertilization usually results in the production of hard-shelled eggs which are deposited in gelatinous capsules. The process has been observed in *Polychoerus carmelensis*, a tiny red acoele, and is known to occur at night when there is presumably less danger of predation.

All the higher turbellarians possess a definite digestive tract but are otherwise fairly similar to the acoeles. The presence of a gut improves the transportation of nutrients around the body and allows these flatworms to attain a greater size. None are large, however, the largest marine polyclads reaching only about fifteen centimeters in length.

The order Rhabdocoela, however, consists of very small worms, some only one or two millimeters long. There are many marine species, and **rhabdocoeles** are an important element in the fauna of muddy sediments in estuaries where they feed on decaying organic matter. Because of their small

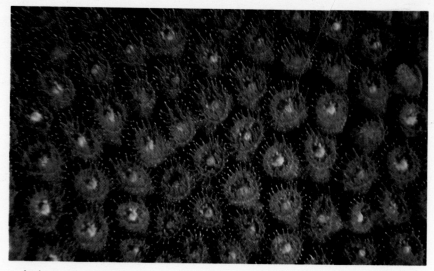

As in most corals, the polyps of *Montastrea cavernosa* expand at night (Caribbean; P. Colin photo).

In *Cladocora* the polyps are terminal on the branches, perhaps an adaptation to a slity environment (Caribbean; P. Colin photo).

Detail of the polyp of the solitary coral *Astrangia* (K. Gillett photo).

size they are seldom encountered and their abundance is not readily appreciated.

Not all rhabdocoeles live in sandy substrates, some preferring pools and damp places on rocky shores. Thus *Microstomum* is a common genus living on the seaweeds *Ulva* and *Fucus*. *Monocelis* may be encountered in coralline rockpools where it is best observed at night. Under dark conditions it is luminescent and can be recognized as tiny specks of light swimming across the water surface or gliding along the air/water interface.

Some rhabdocoeles live in association with other organisms, possibly mirroring the adaptive changes shown by the parasitic classes of Platyhelminthes. Thus *Syndesmis franciscanus* is found in the digestive system of the sea urchin *Strongylocentrotus franciscanus*. It feeds on material taken in by the sea urchin rather than on its host itself and may therefore be classed as a commensal rather than a parasite. Other rhabdocoeles have been recorded as living in association with crustaceans, molluscs, and sipunculids.

Rhabdocoeles are so named because of the presence of strange rod-shaped bodies known as rhabdoids which are found in the epidermis. Their function is not understood, but they are discharged into the surrounding water upon irritation of the worm and may serve as a defense mechanism. Alternatively their discharge may be important in the production of slime, copious amounts of which are needed for the cilia to operate. Rhabdoids are found in all the orders of turbellarians except for the acoeles.

The primitive organization of turbellarians is probably an important factor in their ability to regenerate. The potential is enormous and is discussed more fully below. The rhabdocoeles have utilized their regenerative powers for reproductive purposes, new individuals being budded off posteriorly to produce a chain of zooids. When development is fairly well advanced the maturing zooid breaks free from the chain.

The **alloeocoeles** are a group of turbellarians showing slightly more advanced characteristics than the acoeles.

They are almost exclusively marine and are very common in sandy or muddy beaches but, because of their small size (generally less than ten millimeters) and difficulties in identification, they remain little known.

The next stage of turbellarian development is shown by the **triclads**, which get their name from the pattern of the gut, where three diverticula are visible through the body wall. They reach larger sizes than the lower turbellarians and have exploited a wide range of ecological niches. The freshwater triclads are probably the best known, mainly through the use of the laboratory planarian in teaching, but there are many marine representatives. These are littoral and are generally found beneath stones or inside empty shells. Although much time has been spent in studying the freshwater species, most aspects of marine triclads have been neglected and we have very little knowledge of the general biology of this group. Exceptional features have been noted however, and those species displaying unusual traits are better known. Thus *Procerodes ulvae* is unlike other triclads in its ability to tolerate fluctuating salinities. This is a small black worm about five millimeters in length which typically inhabits the tidal stretches of small streams. Physiological investigations have shown that in dilute seawater the worm takes up water and swells at first, but gradually equilibrates to the lowered salinity and actively pumps water out of its tissues in order to maintain its internal salt balance. *Procerodes* is absent from completely freshwater habitats and is also inhibited by soft water conditions, apparently because the calcium found in hard water is essential for the pumping process. Perhaps surprisingly, the worm is not found in truly marine conditions either, presumably because it cannot compete successfully with other better-adapted species. In fact, *Procerodes* is ideally suited to its restricted environment, its flattened shape offering minimal resistance to the flow of water.

Also of interest, but for very different reasons, is the triclad *Bdelloura*, several species of which are found in commensal association with the horseshoe crabs of the genus

Two common Caribbean corals: above, *Dichocoenia stellaris;* below, *Meandrina meandrites* (both Caribbean; P. Colin photos).

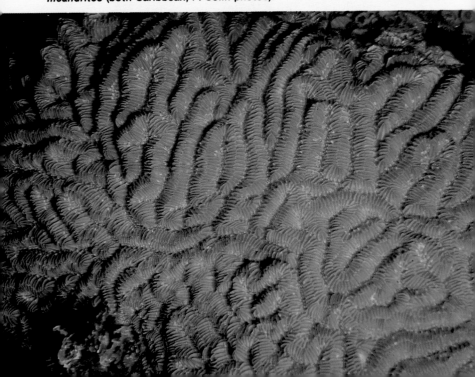

Scolymia, a solitary disc coral, is noted for its aggressiveness (Caribbean; P. Colin photo).

Pillar coral, *Dendrogyra cylindricus,* is common on shallow bottoms, and the colonies may reach a height of over three meters; the polyps are often expanded during the day (Caribbean; P. Colin photo).

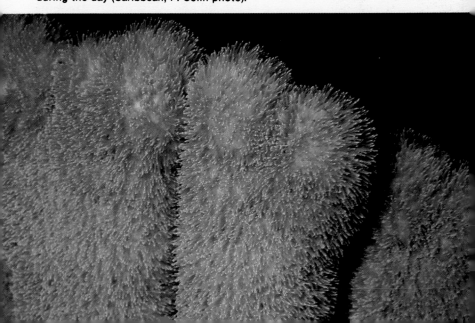

Limulus which are found on the eastern coast of North America. The worms live on the book-gills of these primitive arthropods, attaching themselves by suckers at the posterior end of the body. Nevertheless, they are not to be confused with the parasitic trematodes which also use suckers as a means of adhering to their hosts.

The last order of turbellarians, the **polyclads**, is exclusively marine. These worms, characterized by a highly branched gut, are the best known of all the marine turbellarians, their relatively large size making them more obvious to the casual observer.

They are more or less limited in their distribution to rocky shorelines although there are a few pelagic forms differing from their coastal relations in being transparent. Most polyclads are rather dull in color, blending well into their surroundings; in some cases their cryptic coloration makes them virtually impossible to see. The margins of the body may be uneven so that the shape of the worm becomes blurred and more difficult to distinguish. A few polyclads are brightly colored, however, and may be patterned.

Alloioplana is a large flatworm with blue, green, black, and white markings. *Thysanozoon* is covered with tiny papillae, known as cerata, all over its dorsal surface. These each contain extensions of the gut and may possibly function in the transport of respiratory gases across the body wall. Those on the middle of the back are cream in color turning to red along the sides and mauve around the edges. Mediterranean species of these beautifully colored turbellarians are known locally as skirt dancers because of their swimming motions.

Like the other flatworms, polyclads move across the rock surfaces with a gliding motion achieved by cilia, but their large size makes them relatively inefficient at this. Some are more commonly seen swimming instead and may be observed in rock pools, their bodies rippling with a wave-like action.

Polyclads are carnivorous and, although they appear innocuous enough, they are capable of attacking animals much larger than themselves. Polychaete worms appear to be most favored for food. The flatworm attaches itself to the annelid by an extensible pharynx. In these larger turbellarians digestion is extracellular and digestive enzymes are released which attack the prey and gradually dissolve it away. A polyclad has been observed to attack a clamworm many times its own length by engulfing the tail end in its mouth and slowly moving forward as the worm was dissolved away. Although armed with chitinous jaws, the polychaete was unable to defend itself.

Other polyclads may be quite restricted in their food requirements. *Cryptophallus magnus*, for example, is a very large polyclad living on mussel beds and feeding on these molluscs; *Alloioplana* has a similar diet. *Stylochus*, the oyster leech, is a common pest of commercial oyster beds.

Polyclads often take on the color of their prey. The skin is transparent and, as the gut ramifies throughout the body, brightly colored food material will show up clearly. Thus *Eurylepta* is generally a dull brownish color, but after a diet of eel grass it turns green and if fed on molluscan egg masses it appears pink.

Polyclads continue to show the regenerative abilities of lower turbellarians, but they do not use them for reproductive purposes. Experiments have shown that if a small piece is cut from the side of a polyclad the cells will reorganize themselves so that a miniature worm is produced. This will grow in size so that it comes to resemble the original animal.

Polyclads are also able to control their size according to the availability of food. If food is in short supply the worm gradually regresses but continues to function normally. With acute starvation certain non-essential organs degenerate, the reproductive organs being lost first. In severe cases only the nervous system remains intact and yet, if feeding is resumed, the worm is capable of regenerating all the missing parts and eventually attains its original size.

Reproduction is a sexual process and, although polyclads are hermaphroditic, cross-fertilization occurs. The egg masses form encrusting layers on the undersides of rocks. Small larvae hatch from these after a few days and eventually metamorphose into the adult stage.

In *Isophyllastrea rigida* the typically single polyp in each valley is isolated by a system of closed ridges and is contrastingly colored (Caribbean; P. Colin photo).

The plate-like species of *Mycetophyllia* have polyps that bud in chains with ridges developed only between adjacent chains, so they appear to radiate from the center (Caribbean; P. Colin photo).

Few solitary corals are found on reefs, but the small *Desmophyllum riisei* may occur in groups in reef caves and under ledges (Caribbean; P. Colin photo).

Plerogyra sinuosa, one of the Indo-Pacific bubble corals, is recognized by the large "bubbles" expanded during the day, possibly to protect the polyp tentacles (B. Carlson photo).

Phylum Nemertea

Nemertean worms are little-known to aquarists although they should be familiar to anyone who turns over rocks and rubble at the seashore. Many of the species in this phylum (variously called Nemertea, Nemertinea, or Rhynchocoela) are large and brightly colored, but they are virtually impossible to keep alive in normal aquaria as they require living food and have the disturbing habit of fragmenting into dozens of pieces when disturbed. This minor phylum of about 500 species (most coastal) has gained limited fame in biological circles for being the first group with a complete digestive tract—it not only has a mouth, but it also has an anus. The unique proboscis is the phylum's most obvious feature, with occasional genera such as Gorgonorhynchus having it extremely long and divided. A few species are found on land, while an even smaller number are freshwater.

Classification of this phylum is again based on internal structures visible only in serial sections, so it is meaningless to the aquarist. Some of the brighter and larger species of northern coasts are well illustrated in various local manuals —identifications based on color pattern are probably not accurate, but they will suffice for most aquarium studies.

The phylum is usually divided into two classes with four orders:

Class Anopla: Proboscis unarmed, without stylets; central nervous system external to the muscles. The two orders Palaeonemertini and Heteronemertini contain most of the familiar nemerteans.

Class Enopla: Proboscis usually armed with stylets; nervous system below the muscle layers. The order Hoplonemertini contains both rather common worms plus the unusually *flattened swimming pelagic and bathypelagic genera that often have anterior arm-like processes. The order Bdellomorpha contains a few species with leech-like form and habits.* Malacobdella, *the most commonly seen genus, lives within the mantle cavity of clams, including the edible quahogs, and feeds off the larger particles brought in by the filtering system of the clam. They are not harmful.*

The nemerteans, also known as ribbon worms or proboscis worms, are a small group of unsegmented worms which are very successful, in terms of abundance, in the marine environment. Although there are a very few freshwater species, nearly all are marine, being found in shallow waters and intertidally. They have a wide distribution but are more numerous in cooler temperate waters and are particularly common off the coastlines of North America.

Because they are a small group and because they are apparently a side branch off the main invertebrate lines of evolution, nemerteans are largely neglected. They are rather nondescript worms, often very similar to each other externally, and they are not immediately obvious to the casual observer of intertidal life. As a result they tend to be ignored by all except the most ardent collectors, and there is only a little information concerning their ecology and general way of life.

Nevertheless, their strange morphology and their adaptations to the marine environment make them a very interesting object of study for those prepared to devote sufficient time to them.

Ribbon worms are believed to have evolved along similar lines to the flatworms, but they show a more complex level of organization with clearly defined nerve cords and a circulatory system. They are rather unremarkable in appearance, looking—as their name suggests—rather like long strips of ribbon. The flattened outer surface is ciliated. There are no obvious external landmarks or appendages

Close-up of a single retracted polyp of *Eusmilia fastigata,* a species which forms small colonies of large individual polyps (Caribbean; P. Colin photo).

The striking orangish polyps of *Tubastrea* species are a familiar sight on coral reefs around the world, even though *Tubastrea* itself is not a reef-building coral.

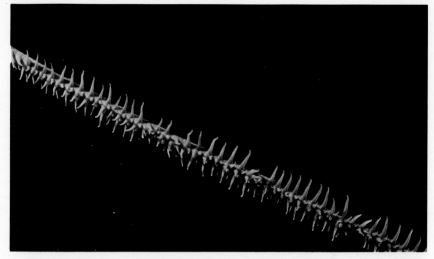

Black corals occur in several rather diverse forms. Above is the whip-like *Stichopathes lutkeni,* which forms forests of single stems that sometimes intertwine; below is the highly branched *Antipathes pennacea,* perhaps more typical of the group (both Caribbean; P. Colin photos).

apart from the presence of eyespots (often in numbers) in some species. Many nemerteans are remarkable for their great length. *Cerebratulus* and *Lineus*, for example, two common North American genera, may reach two meters, and a specimen of *Cerebratulus herculeus* from California has been recorded as more than three meters long.

In addition to attaining such considerable lengths, nemerteans are capable of great expansion so that a worm only a few centimeters long at rest may extend itself to over half a meter. *Cerebratulus*, in particular, shows this elastic property. When contracted, *Cerebratulus lacteus*, a species inhabiting muddy sand around Florida, is rounded in cross-section and a few centimeters long, but it can extend to over ten meters in length, at which stage it is completely flattened in appearance. *Lineus longissimus*, commonly known as the bootlace worm, is also capable of great extension and must be one of the longest living animals. On the other hand, some nemerteans are very small, species of *Tetrastemma* scarcely reaching half a centimeter in length.

Unfortunately for the collector of these animals, they tend to stretch themselves if captured. Furthermore, many show extreme tendencies to break up, and a large specimen one moment may be reduced to a cluster of rounded fragments the next.

Most nemerteans are dull in color, being buff, brown, or occasionally black, but bright red or green specimens can be found, and some species are strikingly patterned. The patterns may be quite distinctive, taking the form of stripes or perhaps of speckles of a contrasting color. The degree of pigmentation varies from one worm to another, especially according to sexual condition, and coloration alone cannot be used as an identification feature.

Many marine nemerteans are burrowers inhabiting clean or muddy sand, but a wide variety of species are found on rocky shores concealed deep in crevices, under stones, or among seaweed holdfasts. A few live in tubes abandoned by tubicolous polychaetes, and one or two produce mucous-lined tubes of their own. Their elastic anatomy, as described above, makes them particularly well adapted to living in tight situations so that they can exploit areas unsuitable for other worms.

Some species are more or less stationary, moving around only in search of food. *Lineus socialis*, a species often found gregariously under rocks and stones, behaves in this way. *Paranemertes peregrina*, on the other hand, moves about freely among the seaweed-covered rocks where it makes its home, searching for nereid worms and other small invertebrates for food.

Members of several genera can swim, and there are a very few pelagic species. Swimming is achieved by rapid undulations of the body brought about by muscular contraction. The pelagic species can swim in this way but generally float passively and have specially adapted broadly flattened bodies for this purpose. The pelagic species are all oceanic; most occur at great depths, sometimes below two thousand meters. They are rare animals, seldom caught in survey trawls, and there is very little information on their distribution, but the relatively constant conditions at abyssal depths suggest that they may be cosmopolitan.

The usual mode of locomotion shown by benthic nemerteans, however, is an instant pointer for identification purposes. The worm glides along in a characteristic manner in a slime produced by the epidermis, and waves of contraction can be seen passing down the length of the animal.

Perhaps the most characteristic feature of this phylum is the feeding mechanism. This involves the use of a proboscis structure unique in the animal kingdom. Usually when a proboscis is present it forms part of the gut, as in annelids, and opens into it, but in nemerteans the proboscis is entirely independent of the digestive tract at all stages of life. The proboscis is made up of a long blind tube which may be coiled and which is freely suspended in a fluid-filled cavity. The tube is attached to the cavity wall by means of a retractor muscle. In some species the tube may be armed with a barbed structure known as a stylet.

The proboscis is used for capturing prey and may also

Zoanthids are often confused with anemones but are usually colonial. Both symbiotic types (such as *Parazoanthus swifti,* above) and species forming large free-living colonies (like *Palythoa caribbea,* below) are common (Caribbean; P. Colin photos).

Corynactis californica, a beautiful corallimorpharian from the East Pacific (K. Lucas photo at Steinhart Aquarium).

serve as a defense mechanism. When a suitable invertebrate, usually an annelid but occasionally a mollusc or crustacean, is in range the proboscis is shot out from the cavity and wrapped around the unfortunate target. Sticky secretions from gland cells on the wall of the proboscis help to hold the animal fast, and it is possible that they have some paralyzing effect. Where a stylet is present this is used to spear the prey, which is then swallowed whole by rapidly sucking it into the mouth. The proboscis is drawn in by the retractor muscle.

The prey is thought to be located by chemoreceptors situated on the anterior end of the body, but these only work at close range so that initially hunting must be very much a random process.

The fragmentation of many nemerteans is associated with a varying ability to regenerate missing parts. Thus the proboscis will readily grow again if lost, but only the anterior end of a fragment of body tissue will regenerate. Like many other invertebrates which show a potential for regeneration, some nemerteans, such as *Lineus*, reproduce asexually by fragmentation, and in this instance new growth will occur at both ends of the fragment.

In general, however, sexual reproduction is the rule and the majority of nemerteans are dioecious. The reproductive system is very simple, and ripe gametes are exuded by squeezing them through short ducts developing after the maturation of the eggs or sperm. Pairing behavior occurs in some species whereby the male enters the female's burrow to spawn and occasionally crawls over her. In a few instances the male and female occupy a communal burrow for breeding purposes.

Some nemerteans develop directly to the adult, but in many there is a free-swimming larval stage known as the pilidium (so named because of its resemblance to a Roman helmet or pilidium) which undergoes a fairly complicated metamorphosis into a young adult.

Phylum Aschelminthes

This large phylum of generally small (0.5-10.0 mm in most species) animals is held together by a single character: the presence of what is called a pseudocoelom. This means that there is a true body cavity which is not lined with special membranes and does not contain membranes extending from the body wall to support the intestines. In aschelminths the stomach and intestine are well developed and usually curved, but the anus is situated either at the posterior end of the body or at least well behind the head. There are no muscular layers in the stomach.

The typical animals of this phylum are the nematode worms, elongate, unsegmented worms covered with a thin, hard cuticle which may bear spines or other processes. The rotifers and gastrotrichs are usually broader, may appear segmented at least posteriorly, and have secondary modifications for feeding. Many scientists believe that the five classes of Aschelminthes are sufficiently distinct to be recognized as separate phyla; since none of the animals in the phylum are really important to aquarists, they are lumped into a single phylum here.

Class Nematoda: Roundworms; body cylindrical, unsegmented or with superficial segments, the anus terminal or nearly terminal; feeding modifications various, the species free-living, parasitic, or parasitic on plants; cilia absent although bristles may be present. Over 10,000 species, perhaps as many as 500,000 species. There are as many as 17 orders recognized in this class, all very difficult to distinguish without magnification and specialized literature. None are of interest to the aquarist, except for the numerous parasitic forms which are not covered here.

Caribbean corallimorpharians: above, *Rhodactis sanctithomae;* below, *Ricordea florida* (both P. Colin photos).

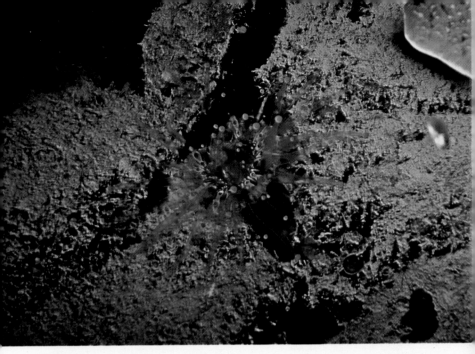

Tentacles ending in balls are typical of many corallimorpharians, such as this unidentified Caribbean genus (P. Colin photo).

Paradiscosoma neglecta, a corallimorpharian shaped like an inverted cone (Caribbean; P. Colin photo).

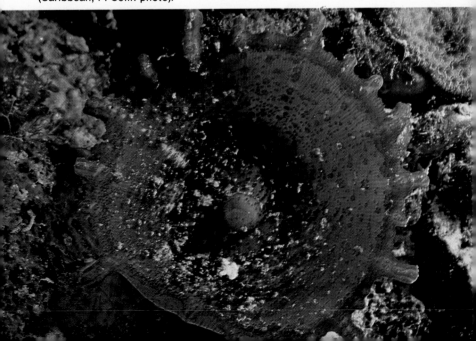

Class Nematomorpha: *Horsehair worms; very much like nematodes, but the digestive tract incomplete in adults, which do not feed; juvenile stages parasitic in arthropods, especially insects. About 200 species.*

Order Gordioidea: Smooth; terrestrial or freshwater.

Order Nectonematoidea: Body with a double row of bristles; marine.

Class Gastrotricha: *Gastrotrichs; microscopic, 0.5-1.5 mm long on average; digestive tract complete, without anterior jaws or ciliated disc; cuticle often formed into scales, plates, spines, or bristles, the cilia in patches. About 150 species known, but the group is poorly studied and many more species will probably be discovered.*

Order Macrodasyoidea: Marine; often with a long posterior tail; many adhesive tubes over body.

Order Chaetontoidea: Mostly freshwater; with long or short paired posterior projections; adhesive tubes few, 2-4.

Class Rotifera: *Rotifers; microscopic; digestive tract complete, with a ciliated disc (corona) anteriorly and jaws in the pharynx; cuticle usually smooth or with a few long processes. Over 1500 species, many very common in fresh water.*

Order Seisonacea: A single genus parasitic on gills and body of nebaliacean crustaceans; corona small, a long 'neck' before body; marine.

Order Bdelloidea: Typical rotifers with the corona double, the body superficially segmented posteriorly; males absent; typically non-swimming.

Order Monogonata: More advanced rotifers with a tendency to reduce the corona and develop a specialized lorica; often sessile or planktonic.

Class Kinorhyncha: *Aberrant microscopic worm-like animals with the body divided into 13 or 14 rings of cuticle; rings often bearing spines but no cilia; anterior segments with rows of slender spines, the segments capable of being retracted into the second or third segment. Strictly marine. About 100 or fewer species.*

166

NEMATODA

The Nematoda are a class of worm-like animals placed in the phylum Aschelminthes, characterized by the presence of a pseudocoelom. They are generally fusiform in shape, but a few are filiform or rounded. A wide range of ecological niches are occupied, including the parasitizing of fungi, higher plants, vertebrates, and invertebrates; they are found free-living in terrestrial, freshwater, and marine habitats. Nematodes have a great constancy of form, both internal and external, and it is differences in behavior and physiology superimposed on this basic plan which have led to their great success as a group. It has been estimated that there are at least half a million species of nematodes, of which half are free-living, but it is not merely numbers of species for which these worms are renowned, but numbers of individuals. 527,000 individuals per acre were found in the top three inches of sand on a beach in Massachusetts.

External features are usually few, but around the mouth are a number of sensory papillae arranged in definite patterns. These include a pair of special structures known as amphids which differ from the rest in structure and innervation. They are found in all nematodes although they are most highly developed in marine species. It is thought that they have a chemosensory function, possibly the equivalent of taste. Also found are structures called phasmids which occur in the tail region; their function is not known, but the taxonomy of nematodes is based on the presence or absence of these organs. Locomotion is brought about by sinusoidal (S-shaped) waves along the body which, according to the prevalent conditions, produce swimming, gliding, or wriggling movements.

Nematodes are mostly dioecious, but there are some hermaphrodites and some produce young by parthenogenesis (reproduction without fertilization by the male). In the marine environment the sexes are usually equal. The basic life history consists of an egg which hatches and then goes through five stages of growth and development. During the first four stages they are larvae; at the fifth stage the

Anemones and zoanthids are often difficult to identify from photographs. These beautiful animals are probably anemones, but the greenish ones above may be zoanthids (Pacific; R. Steene photo above, B. Carlson photo below).

Cerianthids look like anemones at first glance but are not directly related. At the right is an unidentified Caribbean species (P. Colin photo), and below is *Pachycerianthus torreyi* (East Pacific; K. Lucas photo at Steinhart Aquarium).

reproductive organs develop and function, leading to a mature animal. There is a molt between each stage. In parasitic forms this life history may vary to allow synchronization with the host species.

Obviously, since they are of great human importance as parasites (the roundworm *Ascaris* among many others), most work has been carried out on the parasitic species. In consequence the free-living ones, including the marine species, have been neglected. There are nine families which are predominantly marine, mostly without phasmids, many of which have well developed ornamentation in the form of spines. There are no known planktonic forms; most species are found in sands and gravels with a high organic content, and heavy concentrations may be found on algae and epiphytes. As one would expect from such adaptable animals, they occur almost everywhere they have been sought, from deep marine trenches to the intertidal zone and from the poles to the equator. The supply of food appears to determine the number and distribution of species. Some feed on the organic matter in sands, gravels, and mud, others are saprozoic or saprophytic, while others feed on living algae. Some are even predatory, feeding on other species of nematode.

There is a little variety in superficial form of the marine species, this mainly in the development of spines and bristles. The greatest modifications are found in the Epsilonematidae and the Draconematidae. In these families some of the bristles are hollow and adhesive, and the worms move in a manner similar to looping caterpillars. Some have light-sensitive pigment spots which in some species are covered by a sort of lens modified from the overlying cuticle.

There are of course numerous parasitic types found in the marine environment, among them *Contracaecum spiculigerum*, which invades invertebrates and fish before arriving at its final host, a fish-eating bird such as the cormorant. A similar cycle is followed in *Porrocaecum decipiens* before entering its final host, the seal. Not all are parasitic in vertebrates; there is at least one species that is parasitic in echinoderms.

170

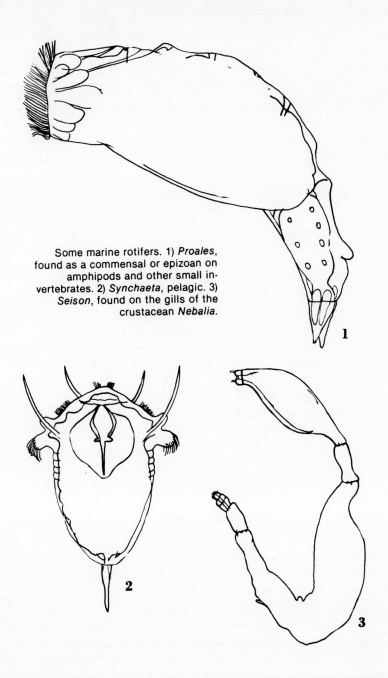

Some marine rotifers. 1) *Proales*, found as a commensal or epizoan on amphipods and other small invertebrates. 2) *Synchaeta*, pelagic. 3) *Seison*, found on the gills of the crustacean *Nebalia*.

1

2

3

A *Cerianthus* removed from its tube; note the pointed base, one of the characters that distinguish cerianthids from most anemones (R. Larson photo).

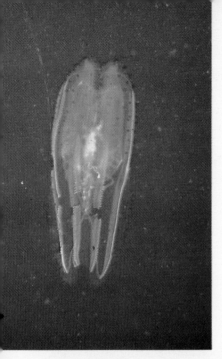

Two East Pacific ctenophores of typical shape; note the tentacles and comb plates (left, D. Gotshall photo; right, A. Kerstitch photo).

Beroe ovata, an iridescent sac-like ctenophore without trailing tentacles (Atlantic; D. L. Ballantine photo).

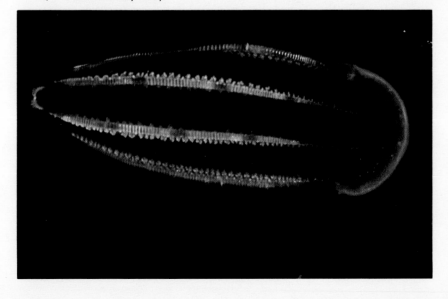

NEMATOMORPHA

The Nematomorpha are pseudocoelomous worms of filiform shape sometimes called hairworms or horsehair
worms. They greatly resemble the Nematoda, to which they
are closely related. All are free-living as adults and parasitic
in arthropods as juveniles.

There is only one genus which is marine, *Nectonema*, in
which there are only three species. Unlike the other
nematomorphs, the males of *Nectonema* are larger than the
females, the former being approximately 200 mm long and
the later about 60 mm. The adults have a pale translucent
appearance and have been found swimming close to the surface of the ocean at night. The adults not having been found
in daylight hours at the surface, it is assumed that they shun
the light. The young stages are incompletely known but
have been found in several species of hermit crabs and true
crabs.

The nematomorphs take no food into the digestive tract
at any time in the life cycle and therefore must obtain nutrition by absorption through the body surface. A small mouth
is present which leads into an intestine; however, this tapers
away without reaching the cloaca, which serves a purely
sexual function. There are no circulatory, respiratory, or excretory systems.

ROTIFERS, GASTROTRICHS, AND KINORHYNCHS

The three remaining classes of aschelminths are unlikely
to be noticed by the aquarist or casual collector as they are
microscopic or nearly microscopic in size. **Rotifers** are most
familiar because they are the well known ' wheel animacules '
of freshwater literature and may occasionally be cultured
for food along with protozoans. These multicellular animals
are probably the smallest known animals having complete
organs and organ systems.

Rotifers are microscopic multicellular animals recognized
by the presence in typical forms of an anterior disc bearing
cilia which are kept in constant motion; this is the corona,
an organ which serves to capture food either by drawing in

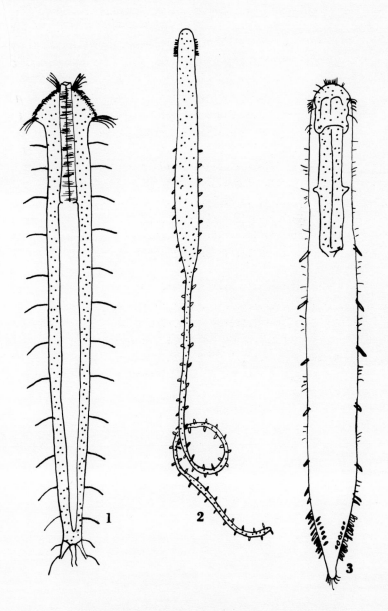

Representative marine gastrotrichs. 1) *Neodasys*.
2) *Urodasys*. 3) *Macrodasys*.

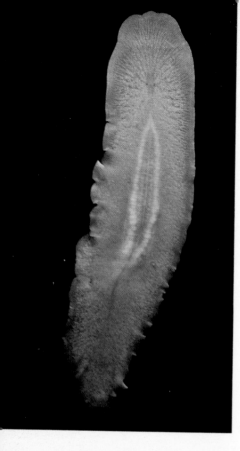

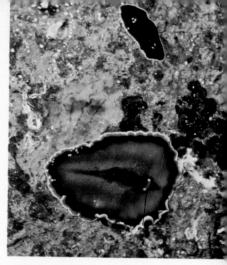

Many smaller flatworms occur on all shores but are of little interest to aquarists. Above is *Callioplana marginata* (Pacific; T.E. Thompson photo); at left is *Prosthiostomum* sp. (East Pacific; A. Kerstitch photo).

Prosthecereaus vittatus; note the strongly developed tentacles of this flatworm (East Atlantic; T.E. Thompson photo).

Some of the larger marine flatworms are quite active, such as this gracefully swimming *Pseudoceros* sp. (East Pacific; A. Kerstitch photo).

The Caribbean flatworm *Pseudoceros crozeri* feeding on sea squirts (C. Arneson photo).

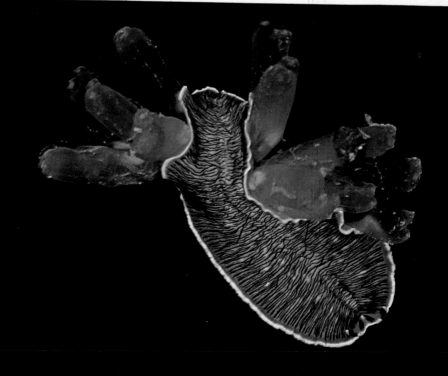

smaller organisms with water currents or by literally 'pouncing' on individual prey animals. Rotifers have a well developed set of jaws, the mastax, in the pharynx below the corona; these jaws consist of seven pieces, one unpaired and median, the others in three lateral pairs. The jaws are very important in the classification of rotifers and must be isolated and studied in detail before identifications can be certain. Since rotifers are usually less than 1.5 mm long, their study is very complicated and casual identifications are not to be trusted.

In most rotifers the body is a cylindrical or oval vase-shaped structure with a distinct cuticle; many species have thickened cuticles (more properly called lorica) bearing spines or various processes in planktonic species. There is a complete digestive system including a stomach and intestine with anus and an excretory system; the nervous system is well developed and includes a distinct brain. Posteriorly the body ends in a segmented foot portion usually bearing two toes. The toe or foot usually contains an adhesive gland and serves for temporary attachment. Some species are sessile, having lost the foot entirely, and remain attached to plants, other animals, or debris. A few species are parasitic in larger animals.

Although about 1500 species of rotifers have been described and the group is quite well known, there are few known marine species. Most rotifers are freshwater or semiterrestrial (found in wet mosses and liverworts). The marine forms may be planktonic or found in debris in shallow water. A few are found between sand grains like the gastrotrichs about to be discussed. Males are rare or absent in most species of rotifer.

Very similar at first glance to the rotifers, and probably closely related, are the less familiar but just as small **gastrotrichs**. These are cylindrical little multicellular animals which lack the corona and mastax of rotifers and have a scalloped head with long cilia. The cuticle commonly is developed as spines or scales with occasional patches of long or short cilia. At the end of the body are usually two long or short projections.

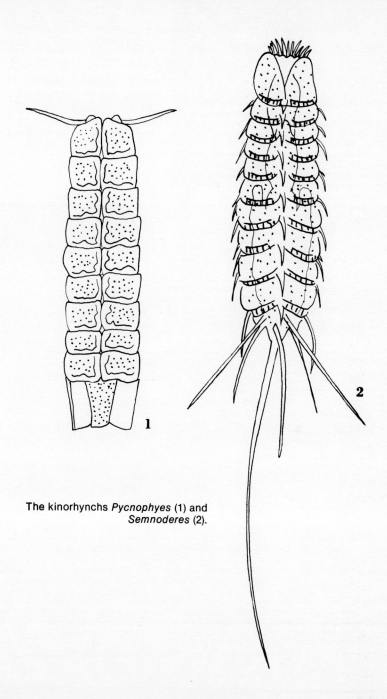

The kinorhynchs *Pycnophyes* (1) and
Semnoderes (2).

Three beautiful marine flat-worms (top: Pacific, S. Johnson photo; middle: Red Sea, W. Deas photo of *Thysanozoon flavomaculatum;* bottom: Pacific, R. Steene photo).

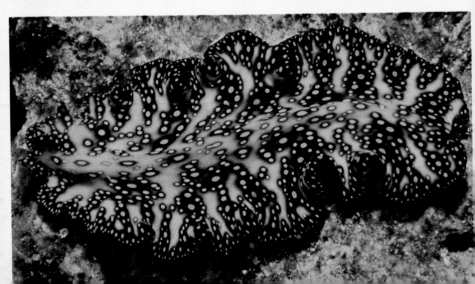

Two unusually patterned nemerteans: above, *Baseodiscus mexicanus* (East Pacific; A. Kerstitch photo); below, *Baseodiscus univittatus* (Pacific; S. Johnson photo).

Gastrotrichs are commonly seen in freshwater protozoan cultures but are less familiar in marine environments. Of the 150 species, only a few are marine and these are often greatly modified from the typical gastrotrich described above. Marine species usually have a large number of adhesive tubes which seem to be used for attachment to sand grains, may have very long tails, and commonly have rounded heads. Marine gastrotrichs are poorly known, although some live in what is called the interstitial fauna. This comprises the sometimes numerous and often weirdly modified animals that live in groundwater, either fresh or salt, trapped between sand grains in beaches above high tide. These can be found by digging down to water level a few feet above the tide line and roiling the water to dislodge the small or microscopic animals. Cultures from this habitat often reveal gastrotrichs, rotifers, small annelids, many copepods and isopods, and protozoans.

Like rotifers, gastrotrichs are predaceous on smaller animals, such as protozoans, as well as sucking up bacteria, diatoms, and debris. Many can be easily cultured for short periods by providing debris with food and changing the water often to prevent bacterial accumulations.

The last microscopic class of aschelminths is strictly marine (so far as known) and contains fewer than 100 species. These are the **kinorhynchs**, inhabitants of muddy and sandy bottoms in relatively shallow water. The body lacks cilia and is composed of about 14 broad rings often with lateral and posterior spines. The first two can be called the head or proboscis and have several rings of slender spines used for grasping prey; the first one or two segments can be withdrawn into the second or third segment. This class is very poorly known and seldom encountered.

Phylum Annelida

This is the worm phylum. Included here are such familiar animals as earthworms, whiteworms, and leeches, as well as the clamworms and fan worms more familiar to marine aquarists. Except for some of the very small or greatly modified species, the phylum is easily recognized by its generally 'worm-like' shape and the presence of distinct segments or annulations. Most marine polychaetes can be visualized by thinking of an earthworm with various bristles, parapodia, tentacles, etc., attached. Although some classifications recognize as many as five classes within the phylum (excluding the other minor groups here treated as separate phyla), only three are recognized here. The archiannelids, sometimes considered a class, are thought to be largely juvenile or otherwise modified oligochaetes. Myzostomes are highly modified parasites of crinoids but are obviously derived from polychaetes. There are almost 9,000 species in the phylum.

Class Polychaeta: The polychaetes or marine bristle worms; over 5,500 species, generally recognized by the presence of parapodia and distinct setae often of several different types; usually with obvious prostomial appendages such as tentacles, palps, etc.; only a few species are found in fresh water.

Class Oligochaeta: The earthworms and freshwater bristle worms; parapodia absent, the setae usually small and in several groups on each segment; prostomial appendages small or absent as a rule; generally freshwater (tubificids) and terrestrial (earthworms), but some groups marine.

Class Hirudinea: Leeches; ectoparasitic worms similar to oligochaetes but with a posterior sucker and often an

Typical nemerteans are rather dull reddish or brownish and not at all conspicuous against their background. Above is *Tubulanus polymorphus* (East
Pacific; K. Lucas photo at Steinhart Aquarium); below is an unidentified Caribbean species (P. Colin photo).

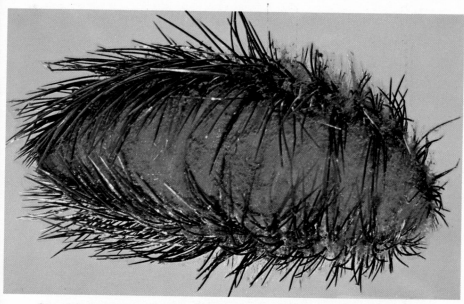

Sea mice, *Aphrodita*, are a type of large, burrowing scale worms with brittle hairs covering the elytra. Above is a dorsal view, below a ventral view showing the parapodia (East Pacific; A. Kerstitch photos).

anterior sucker; setae usually absent; found in all environ-
ments; about 300 species.

It is unfortunate that aquarists do not keep more worms in the aquarium. Usually only the fan worms are to be found on sale. These live well in the tank although they may be hard to feed because of the need for fine particulate foods; however, their beauty well deserves the extra care necessary. Aquarists generally do not distinguish between sabellid and serpulid fan worms as both groups have similar feeding habits and equally attractive crowns. Many of the nereid worms would make fine additions to the aquarium as they are large, obvious, active animals that feed on a variety of living and dead invertebrates including small crustaceans easily cultured as food. The major disadvantages of worms are that most are sensitive to light and tend to be active only in subdued light or at night, and that some of the larger clamworms can be aggressive with other tank inhabitants. These are also disadvantages found in crustaceans and some echinoderms and have not prevented these groups from becoming popular. The range of colors available in even common worms is large, including green, blue, red, black, and yellow in various combinations.

The other possible use of polychaetes in the marine aquarium, food, has also been ignored. Many of the smaller shore-dwelling polychaetes are easily cultured in a small tank and are excellent food for many crustaceans, other worms, and a large variety of snails and fish. No small worm added to a tank is likely to cause harm to larger inhabitants even in large numbers, and they are not so delicate as to die soon after addition to the aquarium. Certainly the polychaetes should receive more attention from the marine aquarist.

The phylum Annelida is made up of worms which are characterized by the division of the body into similar units known as segments. The process of segmentation begins early in the larval life, with each new segment forming in front of the tail, a process known as *metamerism*. Examination of

186

the external morphology shows that the divisions are reflected by the presence of annuli, rings which demark the segments. In some instances secondary annulation occurs, masking the pattern so that segments may be bi- or polyannulated.

Only the trunk region is segmented, the head being developed from a different embryological tissue and not a segment as such, although fusion with the first segments may occur as part of the cephalization process. The degree to which this takes place depends to some extent on the importance of the front region of the animal in its way of life. Similarly the tail of the worm, or *pygidium*, is a nonsegmented structure.

The Annelida comprises three main classes: the Oligochaeta, the Polychaeta, and the Hirudinea. Of these only the Polychaeta, or marine bristle worms, have expanded greatly in the marine environment. There are, in fact, over forty families of polychaetes, some containing a large number of species, although their relatively small size and cryptic habits make them less conspicuous than some other marine invertebrates. Nevertheless, a short time spent digging on a sandy beach will often yield large numbers of a variety of different species of these worms, many of them highly attractive in color and form.

Structurally, polychaetes are characterized by a cylindrical body made up of a few to many segments, each bearing fleshy appendages known as *parapodia* in varying extents of development. Each parapodium is basically biramous; that is, it can be divided into upper and lower regions known as the notopodium and neuropodium respectively. Each region is supplied with a bundle of setae or chaetae (hence the name Polychaeta) which play an important part in locomotion. Many types of setae are known, ranging from simple needle-shaped ones called capillaries to elaborate structures with protective hoods and numerous teeth. The degree of parapodial development and the precise setal pattern are closely related to the mode of life of the worm, active species such as the clamworms having

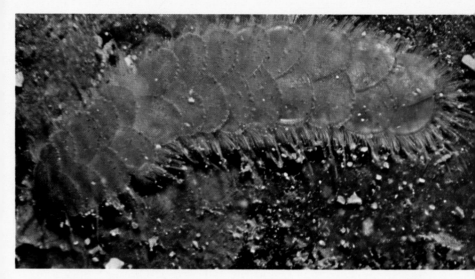

An unidentified Pacific scale worm (family Polynoidae) with obvious elytra or "scales" (S. Johnson photo).

Ventral view of a scale worm, *Scalisetosus longicirra,* that steals food from the grooves of crinoids. Note the large pharyngeal opening and the head appendages as well as the parapodia (Pacific; L.P. Zann photo).

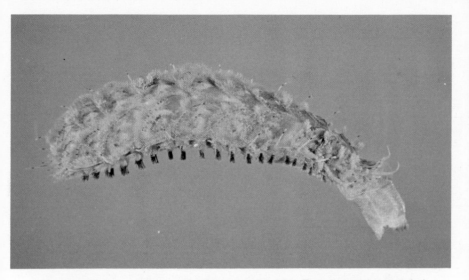

Chaetacanthus branchiatus, a Caribbean scale worm. The pharynx is extended and some teeth can be faintly made out at the margin (C. Arneson photo).

Hermodice carunculata, the fireworm, is a large (250mm) coral-eating worm with brittle setae that can cause intense pain when they break off in the skin of a diver (Caribbean; P. Colin photo).

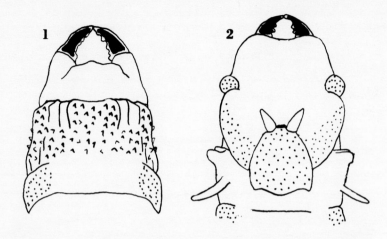

Above: Ventral (1) and dorsal (2) views of the head of a clamworm with the proboscis everted. Notice the strong jaws. **Below**: Various types of setae found in polychaetes. 3) Hooks, also called crotchets. 4) Various modifications of simple setae. 5) Jointed setae.

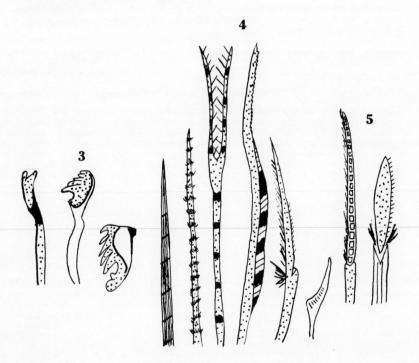

large fleshy parapodia with jointed or compound setae while some sedentary forms such as the lugworms have tiny parapodial ridges bearing hook-like setae for gripping the burrow wall.

The segmented trunk region of the polychaete body is capped by an anterior head region known as the *prostomium* and a posterior pygidium which carries the terminal anus. The prostomium may have eyes, chemical sense organs, and various antennae and palps in active worms requiring a high sensory input, or it may be an insignificant lobe in some burrowing forms such as the lugworms. The mouth is situated between the first segment, or peristomium, and the prostomium.

Polychaetes are one of the commonest groups of marine animals, occurring in most situations ranging from estuaries and inshore waters to the open sea and even hypersaline lagoons. Many species are found intertidally, especially on sandy or muddy beaches, but they have also been dredged from abyssal depths in marine canyons. The range of habitats is also enormous. Burrowing forms of many diverse types are to be found in sand flats and muddy areas in profusion. Rocky shorelines are less favorable, due largely to problems of desiccation, but several forms are found in crevices. Others occur in association with certain plants, kelp holdfasts providing a particularly good habitat for small active worms. Similarly the cracks and crevices so characteristic of coral reefs provide niches for many tropical species.

About six families of polychaetes contain pelagic or planktonic representatives. These are typically transparent and generally have well developed parapodial paddles for swimming.

While there are few parasitic polychaetes, many are commensals often living in association with other worms. Some scale worms, in particular, may often be found sheltering in the tubes of larger burrowing animals.

The extensive colonization of so many areas of the marine environment reflects the ability for adaptive radiation shown by this group. The basic structure as outlined above

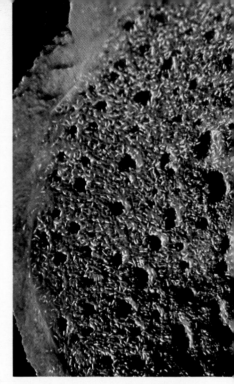

Two nereid worms: left, *Nereis* sp., a common clamworm of the East Pacific (K. Lucas photo at Steinhart Aquarium); right, large numbers of *Syllis spongicola,* a parasite or commensal of large sponges (Caribbean; P. Colin photo).

Chaetopterus variopedatus, the parchment tubeworm, is found in most seas in shallow water. Here the tube has been split to reveal the worm (L.P. Zann photo).

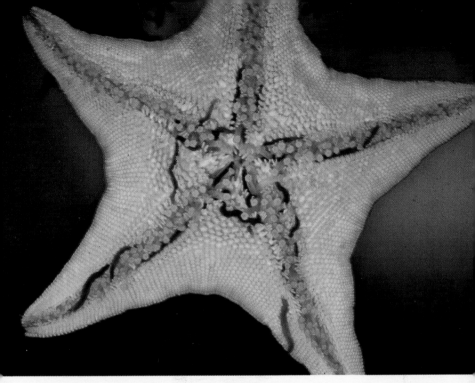

Commensal worms are of common occurrence on echinoderms, often feeding from the food grooves. Here many nereids search for food on a *Patiria* starfish from the East Pacific (T.E. Thompson photo).

The shell- and debris-encrusted tubes of *Diopatra* and related worms are visible at low tide on almost any warm muddy beach (Y. Takemura and K. Suzuki photo).

is sufficiently flexible to allow quite drastic modifications enabling successful penetration of many ecological niches. The main restriction on the group is, in fact, its reliance on the marine environment. The high surface-to-volume ratio, coupled with the permeability of the polychaete cuticle, results in severe desiccation if the worms are exposed for any length of time. Furthermore, the osmoregulatory system, which controls the balance of salts in the body fluids, is generally too primitive to withstand freshwater conditions, although there are many examples of estuarine species which can cope with fluctuating salinities.

The classification of polychaetes is rather arbitrary in that the class is broadly divided into two groups according to habit. Thus the errant polychaetes include those forms which are actively locomotive, including the clamworms or nereids. Characteristic features include well developed parapodia with compound setae, a large head with sensory appendages, and a jaw apparatus for feeding. The sedentary polychaetes are principally burrowers or tube-dwellers more or less stationary in habit, although some forms, such as the capitellids, do not form permanent burrows and can move freely through the substrate. Typically the body can be divided into regions showing different modifications. The parapodia are reduced and setal types are modified for locomotion within a burrow or tube. Biting structures are absent, feeding often being achieved by palps or tentacles situated on the head. On the other hand, prostomial sensory extensions are absent.

These are very much generalizations, however, there being almost as many modifications as species. For example, the adaptive radiation is reflected in feeding mechanisms and behavior, which are closely related to the general mode of life. A wide range of feeding methods have been adopted and there are representatives of each of the main types. Thus many errant worms are active carnivores devouring other polychaetes, molluscs, and other small invertebrates. The prostomium is well developed in association with the need to hunt for food so that eyes and sensory appendages are commonly encountered in these forms. Food is captured by an eversible proboscis which may be armed with jaws or teeth formed as a result of sclerotinization of the cuticle. Thus in clamworms (nereids) and in the catworm (*Nephtys*) large chitinous jaws are present, while syllids commonly have a ring of teeth. Once taken in the jaws, food is drawn into the gut by retraction of the proboscis, a process involving specially developed muscles. The arrangement of these is very precise and has been used as a basis for classification in the group. The carnivorous habit has been adopted by most families of polychaetes which crawl around, together with some errant burrowing and tube-dwelling forms such as the lumbrinereids and a few eunicids.

Similar adaptations are found in scavenging worms, which are usually closely related to carnivorous forms. Indeed, some carnivores will take dead and moribund material if it is available. A jaw apparatus may also be used by herbivores, which use the hard chitinous material as a rasp similar to the molluscan radula. Others may use the jaws as a cutting device for detaching pieces of seaweed; despite their carnivorous reputation, some nereids have been known to feed in this way. Indeed, they possess the enzymes necessary for the breakdown of cellulose, enabling them to gain maximum benefit from this food source.

Many sedentary species are direct deposit-feeders ingesting vast quantities of sediment and extracting nourishment from the microorganisms themselves feeding on the detritus attached to the mineral particles. Again the proboscis plays a part in digestion, but in this instance it is naked and very glandular so that mud particles stick to it. While this method of feeding appears very inefficient in that most of the material taken in is egested (for example as the familiar castings of the lugworm), there is some evidence that only certain particle sizes are selected for ingestion. This class of feeder includes worms which eat their way through the substrate, forming semi-permanent burrows (e.g., capitellids, opheliids), and some more sedentary burrowers such as the lugworms. However, some authorities

Terebellid worms are tube-dwellers with long tentacles that trap small food particles. The tentacles are often conspicuous, but the worms themselves must be removed from the tubes to be seen. Above, *Audouinia comosa* (Pacific; Y. Takemura and K. Suzuki photo); below, *Eupolymnia nebulosa* (Caribbean; P. Colin photo).

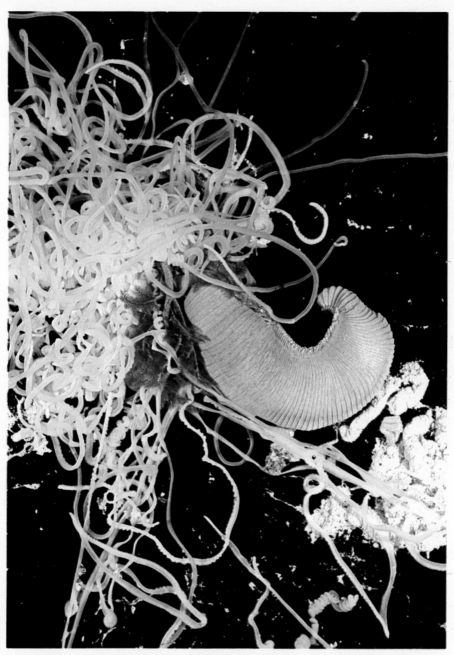

A large Australian terebellid, *Reteterebella queenslandica*, removed from its mucous tube; note the red gills just behind the tentacles (K. Gillett photo).

class the latter as a kind of filter-feeder since the sand they ingest has acted as a filter for water percolating through it. The bamboo worms of the family Maldanidae are examples of direct deposit-feeding tube-dwelling worms feeding in this way. The ariciids are also deposit-feeders but they utilize the proboscis, which is feathery, almost as a tongue to scoop up particles.

Another group of burrowers and tube-dwellers are indirect deposit-feeders. A proboscis is absent, food being collected by tentacles or palps carried on the prostomium or first few body segments. These are spread across the surface of the substrate so that only the top few millimeters of deposit are taken. Detritus adheres to mucus produced on the feeding tentacles and is drawn toward the mouth by ciliary currents acting along their length. Occasionally, as in the ampharetids, the tentacles may be retracted into the mouth so that food material may be removed directly. The tentacles are non-selective, but unsuitably sized particles may be rejected at the mouth. The cirratulids and terebellids are two large families of burrowing polychaetes showing these modifications, and spionids and oweniids have tubicolous representatives of this type.

These latter families also contain species able to feed by filtering suspended material from the seawater, although more spectacular representatives are given by the sabellids and the serpulids, collectively known as the fan worms. Again the proboscis is absent and specialized feeding devices have evolved as part of the head. In the fan worms these take the form of a crown which is composed of two opposed semicircles of rays or pinnules. Detritus and small plankton are drawn along these by ciliary action. Quite complex sorting mechanisms occur both in the pinnules themselves and at the collar-like structure around the mouth. In some instances material is retained for tube-building.

Recent research has shown that some polychaetes, at least, may obtain nutrients by the direct absorption of amino acids across the body wall. The importance of this process in the overall energy budget is not yet fully understood. In any event many species are not completely exclusive in their feeding behavior and may incorporate several of the above methods into their repertoire according to availability of food.

The general shape of the polychaete body makes it comparatively easy to obtain oxygen for respiratory purposes by direct diffusion across the body wall. The importance of this process in the overall energy budget is not yet fully understood. In any event many species are not completely exclusive in their feeding behavior and may incorporate several of the above methods into their repertoire according to availability of food.

However, the various specializations shown in response to environmental conditions have necessitated the development of respiratory aids. Most polychaetes, excluding the capitellids, glycerids, and some syllids, have a vascular system which transports oxygen around the body. In addition many possess a respiratory pigment, usually hemoglobin, which enables the animal to take up more oxygen. The red coloration of many species is due to the presence of hemoglobin, which may be present either in solution or in corpuscles carried in the coelomic fluid. Other polychaetes, notably the sabellids, use chlorocruorin as a respiratory pigment, while *Magelona* is very unusual in having hemerythrin, a pigment more commonly associated with sipunculids.

External aids to respiration also occur, and these often contribute to the beauty of form exhibited by many species. In some instances gills are found; these serve to present a larger surface area to the environment for diffusion purposes. They are generally expansions of the parapodia and may take the shape of flattened leaves as in the notopodial gills of many nereids or may be arborescent in form as in the lugworm. In other species there may be anterior respiratory tentacles formed from the first few segments of the body. These may be found, for example, in the cirratulids, where they may be distinguished from the feeding tentacles by the presence of blood vessels, and in terebellids. In fan worms the prostomially developed crown also serves a respiratory function.

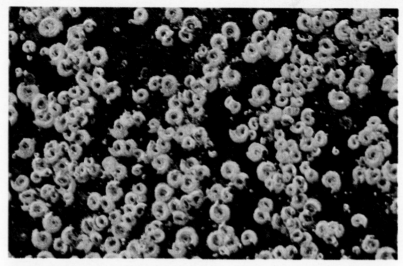

The snail-like tubes of *Spirorbis* worms are common on rocks at or just above low tide (Y. Takemura and K. Suzuki photo).

Salmacina, a small tube-building serpulid worm that lacks an operculum (Atlantic; T.E. Thompson photo).

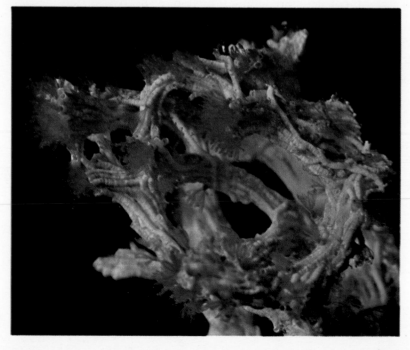

Left, *Pomatostegus stellatus,* a Caribbean species with the tentacles or radioles arranged in a "U" rather than the typical circle or spiral (P. Colin photo). Right, notice the white operculum, formed from a modified tentacle, of this Pacific serpulid (L.P. Zann photo).

Sabellid worms, such as this colorful *Sabellastarte indica,* have a flexible, rather parchmentous tube and lack an operculum (Dr. D. Terver photo courtesy Nancy Aquarium).

Many burrowing worms increase diffusion potential by irrigation, which draws a continuous flow of water across the respiratory surfaces. This is achieved by peristaltic waves induced by muscular activity.

Most polychaetes are dioecious. In primitive forms reproduction may involve a relatively simple process whereby eggs and sperm are released simultaneously into the surrounding water. This occurs, for example, in the lugworm and in the famous palolo worms and necessitates a precise synchronization of spawning behavior, often related to phases of the moon, to ensure a high rate of fertilization.

Polychaetes are able to regenerate missing parts with comparative ease, and in some groups this ability has been adapted for reproductive purposes. Thus syllids commonly reproduce by asexual fragmentation; each part regenerates the necessary segments to complete the whole. Stolons with over ten developing individuals may sometimes be encountered. Fragmentation also occurs in sexually reproducing forms such as the palolo worms (*Eunice*), in which the ripe tail portion breaks free and swarms at the surface with like individuals.

In other species sexually mature individuals metamorphose into specialized reproductive types such as the epitokes of nereids which are able to swim to the surface for spawning.

A few forms pair and exhibit brood protection. This occurs in some scale worms where the eggs are held on the back of the parent and in serpulids where they are housed in the tube or under the operculum. In some capitellids, where fertilization is believed to be internal, the eggs are extruded into the maternal tube in which they are irrigated by movements of the parent. Protection is commonly restricted to coating the eggs in a jelly-like capsule or cocoon (e.g., *Scoloplos*, phyllodocids) which affords protection against desiccation.

Development is via a trochophore larva which may have a free-swimming stage in the plankton (for example, *Spirorbis*, *Sabellaria*) or may be purely benthonic or demersal in

Late larva of a polychaete showing the development of body segmentation.

development as in the nereid *Neanthes diversicolor*. Its function is mainly one of dispersion, and feeding may or may not occur. The adult form is achieved by metamorphosis, which is often very gradual in nature.

With this brief review of the principal features of polychaetes we may now turn to the specialized adaptions shown by individual families and genera.

Perhaps the most familiar worms of sandy beaches, the lugworms are conspicuous because of their relatively large size and the presence of casts on the surface of the mud marking the opening of the burrow. They are of some commercial importance, being widely used as bait for fishing and also as a typical worm specimen in college zoology courses. Lugworms have a wide distribution, the most common genus being *Arenicola*. While some species are crevice-dwellers, the burrowing forms are best known and have been studied minutely.

Close-up of the tentacle crown (actually gills) of *Serpula vermicularis* (Caribbean; P. Colin photo).

Spirobranchus giganteus is the most colorful worm commonly kept in the marine aquarium; the radioles are in two spiralled groups (tropical; L.P. Zann photo).

"Living rock" is the aquarium name for groups of featherduster worms (like these *Spirobranchus*) still embedded in their coral clumps.

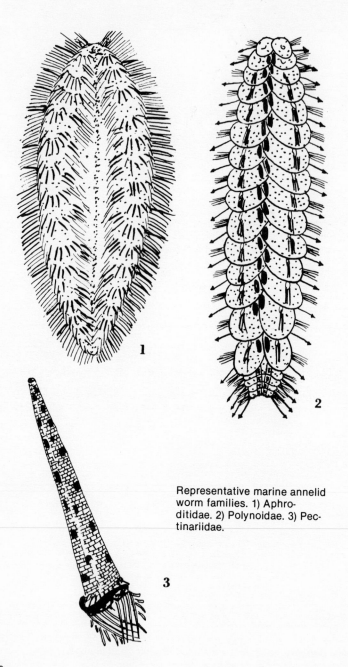

Representative marine annelid worm families. 1) Aphroditidae. 2) Polynoidae. 3) Pectinariidae.

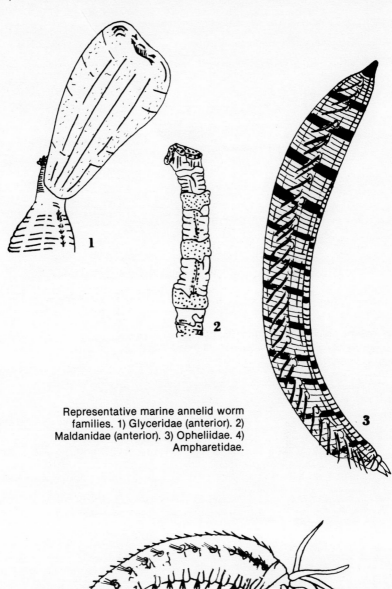

Representative marine annelid worm
families. 1) Glyceridae (anterior). 2)
Maldanidae (anterior). 3) Opheliidae. 4)
Ampharetidae.

Close-up of the end of the tube of *Spirobranchus giganteus* with the tentacles withdrawn (Caribbean; P. Colin photo).

A marine leech removed from the buccal cavity of a tiger shark. Shark leeches often have heavily sculptured bodies (Pacific; L.P. Zann photo).

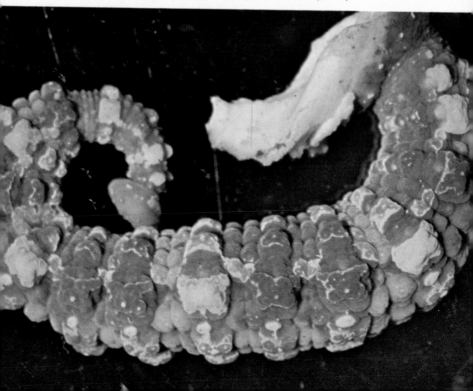

Siphonosoma cumanense, a Caribbean sipunculid worm (C. Arneson photo).

Contracted and extended stages of *Urechis caupo*, a common peanut worm or echiurid of the East Pacific (K. Lucas photos at Steinhart Aquarium).

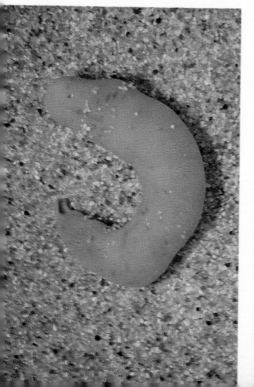

1) The small serpulid *Mercierella* removed from its tube. This genus is found in both marine waters and coastal brackish lakes and streams.
2) A typical terebellid worm.

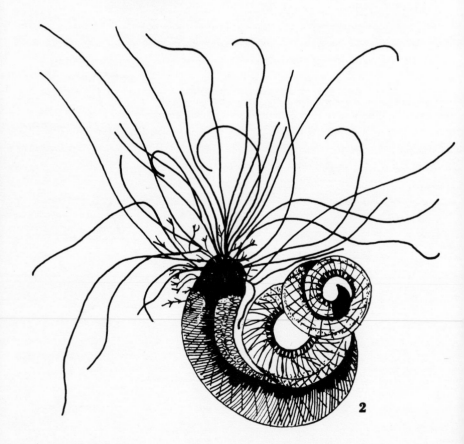

At first appearance the lugworm would seem to be the archetypal direct deposit-feeder. It has a large, very globular eversible proboscis with which it engulfs sand particles. As in all sand-eaters, the gut is long in order to remove the maximum amount of nourishment from the sediment. The percentage of digestible material may be very small, sometimes less than 1%, so that the animal must spend much of its time eating. Waste material is egested from the burrow as casts. Observations on the shape of the burrow and the behavior of the worm have led to the suggestion that the lugworm is really a filter-feeder. The burrow is a U-shaped structure composed of a head shaft, gallery, and tail shaft. The worm spends most of its time in the horizontal gallery, moving backward up the tail shaft only to defecate. The head shaft is filled with sand which is continuously moving downward as the worm ingests material from the bottom of the shaft. Irrigation by peristalsis draws water in through the tail shaft, but if the direction is reversed water flows down through the head shaft and any small particulate matter, such as plankton or detritus, is filtered off by the mineral particles, thus enriching the organic content of the sediment. In any event the tidal cycle results in a fresh layer of detrital material being deposited on the surface about twice every day so that a continuous supply of food is available for the worm.

The appearance of lugworms is rather unremarkable in that they lack any large appendages and give the impression of a fattish earthworm with a thin, drawn-out tail. Thus the prostomium is very small, being completely dwarfed by the everted proboscis, and lacks any sensory palps or cirri. The parapodia are similarly reduced, with setae designed for locomotion within a burrow, although the notopodia of the middle region of the body are modified into branchial tufts. The worm irrigates its burrow by producing peristaltic waves which move along the body and act against the burrow wall like a piston in a cylinder, drawing water over the gills.

While the lugworm is essentially a sedentary animal, it

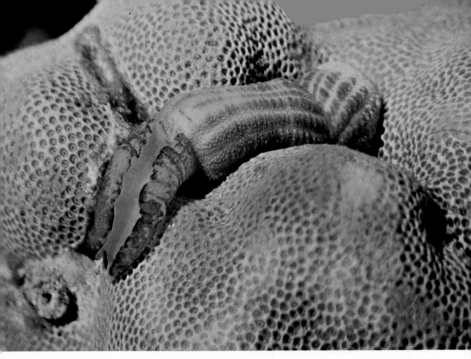

Two large echiurid worms showing the gutter-like, non-retractile proboscis characteristic of the group. Above, possibly *Echiurus* sp. (Pacific; J.H. O'Neill photo); below, possibly *Thalassema* sp. (Caribbean; C. Arneson photo).

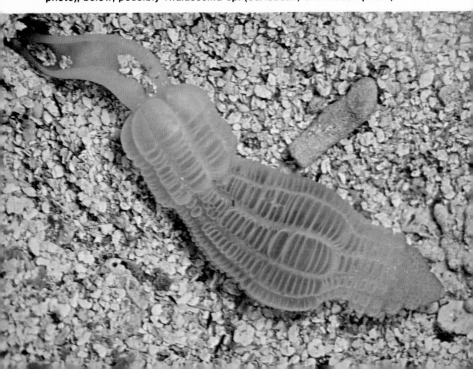

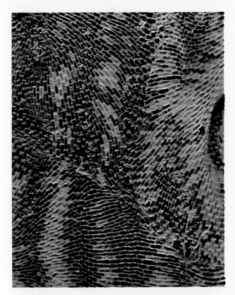

Two East Pacific encrusting bryozoans, genus *Membranipora:* at left, *M. serrata;* at right, *M. tehuelcha* (K. Lucas photos at Steinhart Aquarium).

Phoronids are seldom seen by collectors. *Phoronis vancouverensis,* shown here, is locally abundant in the East Pacific and of relatively large size (T.E. Thompson photo).

has been observed to swim short distances both under laboratory conditions and in the field. This is achieved by throwing the body into lateral waves. The function of swimming is obscure in this species, but it may play a part in dispersal of young worms, which are known to migrate down the shore. Spawning is synchronized roughly to phases of the moon, and eggs and sperm are liberated freely into the seawater. Fertilization results in simple planktonic trochophores.

The capitellids, as typified by the cosmopolitan *Capitella capitata*, are very similar to lugworms in some respects and are probably closely related. *Capitella* is easily recognizable by its blood red color and is very common, although its association with organically rich deposits, such as in harbors and polluted bays, makes it less accessible to the average layman. It is one of the few polychaetes in which fertilization is thought to occur internally, and it has several modifications toward this end. Most obvious in this respect are the genital spines which are enlarged setae believed to function as clasping tools. Because fertilization is internal, synchronized spawning is not necessary, so *Capitella* may reproduce throughout the year providing conditions are suitable. Furthermore, the eggs once spawned can be protected from predation by keeping them in the mucous-lined burrow of the parent. The larvae are lecithotrophic (non-feeders) and can develop in one of two ways. In some instances the young trochophores are released into the plankton after about one week, but alternatively development to young adults is carried out entirely within the parental tube. The latter method may be an adaptation enabling the species to rapidly colonize suitable areas where competition from other animals is minimal. This is an example of opportunism, a feature of many species described as indicators of polluted conditions. In fact *Capitella capitata* has aroused great interest among workers in this field, as knowledge of its distribution can be of considerable importance.

Other sand- and mud-eaters which probably arose from

the same stock as the arenicolids include the burrowing opheliids and the tubicolous maldanids, commonly known as bamboo worms due to their segmentation patterns. Less closely related but very interesting in its own right is *Sternaspis*, a very specialized form which bears a superficial resemblance to echiuroid worms. It is a very small dully colored animal most un-worm-like in its rounded shape. At the posterior end there is a sclerotized structure which covers numerous gills. The animal is found intertidally just under the surface of the mud with the gills projecting onto the surface. Little is known of its ecology, but the presence of specialized appendages suggests that copulation occurs.

The spionids are a family which exhibit intermediate stages between mud-eaters and filter-feeders. In burrowing forms such as *Scololepis* the proboscis is used for digging and feeding, but in tubicolous genera such as *Nerine* feeding is achieved by means of two ciliated tentacles. These are wafted around in the overlying water and drawn across the mud surface, any small particles being caught up in their ciliary current. Periodically the tentacles are wiped across the base of the prostomium and food is ingested.

A particularly interesting adaptation is shown by species of *Polydora* which bore into limestone rock and, in some instances, shells. The damage caused may reach high levels, so *Polydora* can be a pest in commercial shell fishery beds.

Closely related to the spionids are worms of the family Chaetopteridae, although the most familiar species, *Chaetopterus variopedatus*, has undergone such extensive modifications that the similarities are obscure. In fact this species must surely show the greatest range of specialized segments of any polychaete. The changes are all associated with the feeding behavior. *Chaetopterus* does not use the prostomial tentacles but instead filters particulate matter in a mucous bag strung across the U-shaped tube. The parapodia of segments fourteen to sixteen are greatly expanded to produce fan-shaped wings fitting tightly the tube diameter. These beat rhythmically and produce a current of water which flows across the worm from anterior to

Bryozoans occur in many different colony forms, from the ruffled *Hippodiplosia insculpta* (above) to the branched *Bugula neritina* (below) (both East Pacific; above, D. Gotshall photo; below, K. Lucas photo at Steinhart Aquarium).

Left, the bryozoan *Dendrobeania lichenoides* (East Pacific; K. Lucas photo at Steinhart Aquarium). Right, a delicate coralline bryozoan from the Caribbean (P. Colin photo).

Colonies of bryozoans serve as habitats for many animals, forming part of the large encrusting fauna on debris and dead coral boulders in any warm sea (Pacific; T.E. Thompson photo of *Tubucellaria*).

posterior. The notopodia of the twelfth segment, known as the aliform notopodia, show an even greater expansion and have a ciliated epithelium which is richly supplied with mucous glands. A sheet of mucus is stretched between the tips of these notopodia and formed into a bag. The end of this is held by a cup-like structure on the dorsal surface of the worm. The bag forms a very efficient filtering device, and most of the water drawn into the tube has to pass through it. Large particles are deflected by the peristomium and allowed to bypass the bag by an upward movement of the aliform notopodia. The cup rolls up the bag as it extends backward as more mucus is produced; when the resulting ball is sufficiently large, the bag is broken free and drawn into the roll. The ball is then transferred to a ciliated mid-dorsal furrow where it is conveyed to the mouth. The tentacles may assist in inserting the food inside the mouth. Other members of the chaetopterid family utilize similar feeding mechanisms, but in more primitive forms such as *Phyllochaetopterus* the current is produced by cilia.

Chaetopterus lives in a tough opaque tube sometimes buried in sand and sometimes attached to a rocky surface. While it is essentially a sublittoral form, occasional specimens may be found at extreme low water on a good spring tide.

Despite its tube-dwelling existence, *Chaetopterus* is one of the few worms producing a luminescent slime. This is exuded by specialized glands situated mainly on the aliform notopodia. Although the physiology of this has been intensively studied, no real conclusions can be drawn concerning the function of the slime.

Although it is so complex in design, *Chaetopterus* shows remarkable powers of regeneration. Providing no more than the first fourteen segments (the most specialized ones) are lost, the worm can regenerate completely. Furthermore, a whole new individual can be formed from a single isolated segment from this region. Beyond the fourteenth segment regeneration is no longer possible. The regenerative faculties of this worm are particularly surprising, not only because of its specialized nature, but also because, being protected within its tube, it is unlikely to need these talents very often. Possibly the integrated use of several segments in feeding makes it even more vital than normal that any missing part be replaced without delay.

Unlike the chaetopterids, the sabellariids, as illustrated by *Sabellaria*, have expanded the use of tentacles for feeding. This worm lives in a sand tube held together with mucus and uses muscular tentacles to filter food. The two larval tentacles so typical of spionids are surpassed by an array of tentacles borne on a stalk-like structure produced as a result of the fusion of the first two segments. This stalk also bears a hard setal crown which can be drawn down as an operculum to plug the tube. *Sabellaria* is notable for its colonial behavior. Under suitable conditions large reefs can develop as the tubes build up, the regular spacing of individuals giving rise to a spectacular honeycomb pattern.

Other groups of polychaetes using tentacular feeding mechanisms have evolved along the indirect deposit-feeding line rather than filtering. The cirratulids *Cirratulus* and *Cirriformia* are good examples. Typical of mud-eating worms, these are rather dull in appearance apart, of course, from the tentacles. These form a mass of coiling fingers around the anterior end and are spread out across the mud surface to collect particles. Although widespread in occurrence they are not immediately obvious and, when they are found intertidally at low tide, they do not show their full beauty. They inhabit mud flats or muddy areas between rocks so that only the anterior end is visible. Without a covering of water the tentacles tend to adhere to each other, presenting a sticky, rather unhealthy looking picture.

The terebellids are similarly afflicted, but here the bright colors and plumules of feathery gills provide a more interesting picture. They and the related pectinariids and ampharetids are short plump worms with little parapodial development, features typical of their sedentary habits. *Terebella* lives in a permanent mucous-lined burrow. It is a very colorful worm with a yellow body and bright red gills and tentacles. The gills are short tufts situated dorsally on

Two East Pacific brachiopods: left, *Laqueus californicus;* right, *Helithiris psittacea* (both K. Lucas photos at Steinhart Aquarium).

Opened specimen of the brachiopod *Terebratalia transversa* showing the lophophore occupying most of the shell (East Pacific; T.E. Thompson photo).

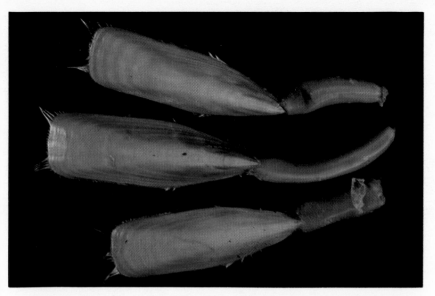

A lingulid brachiopod, *Glottidia palmeri*. These brachiopods are exceptional in that they burrow in the soft bottom and can withdraw the shell into the substrate (East Pacific; K. Lucas photo at Steinhart Aquarium).

An unidentified small, fragile brachiopod from deeper waters of the Caribbean; note the relatively small size of the lophophore (C. Arneson photo).

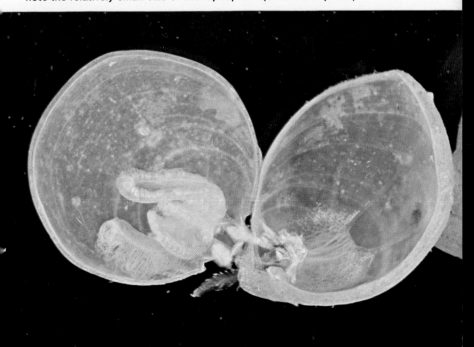

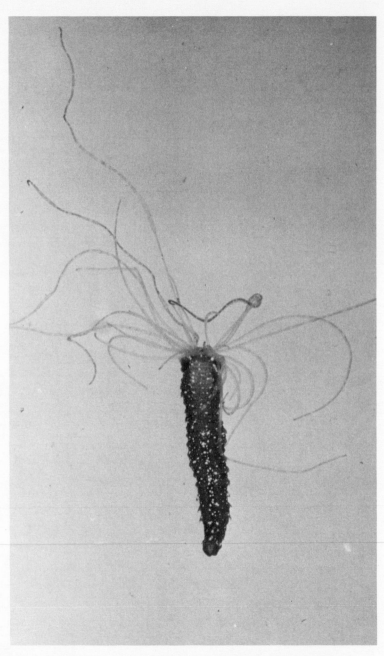

Thelepus setosus, a Caribbean terebellid (C. Arneson photo).

the first few segments, while the tentacles are lithe extendable structures reminiscent of those in cirratulids. Because of the large size of some terebellids, such as *Amphitrite*, the tentacles must be capable of gleaning a large surface area of mud. Feeding behavior in this worm has been studied in detail and is known to involve a quite complex coordination of parts. Although there are longitudinal muscles for the retraction of the tentacles, extension results from ciliary creeping, about half the tentacular surface being ciliated. The whole surface is provided with mucous cells and, once extended, attachment is secured by their secretion. The tentacle is then turned over by muscular activity and the ciliated portion becomes a gutter along which small food particles are wafted toward the mouth. Any large material is squeezed down the gutter, muscularly, and if the particles are very large the whole tentacle may be drawn toward the mouth area once the food is secured properly. At the base of the tentacles the cilia come to an end and food is conveyed to the mouth by wiping the tentacles across it. While the tentacles are non-selective in so far as they will transport all material within their means, some sorting may occur at the lips. These are formed from the first segment and their precise nature varies according to the feeding requirements of the species in question, but in each case their shape and musculature allow rejection of grossly unsuitable material.

While most terebellids are sedentary burrowers, some build a permanent tube. For example, the sandy tube of *Lanice conchilega*, the sand mason, has a feathery crown which acts as a filtering device. Small particles are trapped among the filaments and the entangled material is wiped off by the tentacles.

The related pectinariids, or ice cream cone worms, also build interesting tubes on sandy beaches, but whereas those of the sand mason are constructed from any available material, the pectinariids are much more selective. Sand grains and shell fragments are used, each piece being carefully aligned so that the tube has an overall smooth surface. Generally it is conical in shape. Both ends are open, that

Some common East Pacific chitons. Above: left, *Tonicella lineata;* right, *Placiphorella velata.* Below: *Stenoplax heathiana* (all K. Lucas photos at Steinhart Aquarium).

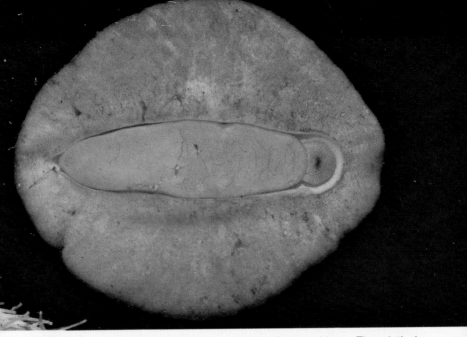

Ventral view of *Cryptochiton stelleri,* one of the largest chitons. The relatively narrow foot and the small "head" are clearly visible (East Pacific; T.E. Thompson photo).

The black chiton, *Katharina truncata,* an East Pacific species in which the dark girdle nearly covers the valves (T.E. Thompson photo).

with the wider diameter being projected down into the sand. The head can be protruded through this during feeding. This is achieved by means of short feeding tentacles which are protected from abrasion by the comb-like rows of setae projecting from each side of the anterior end. These are also used to dig the burrow, as the worm has no proboscis. Water is drawn down through the mud and through the feeding cage, where particulate matter is taken up by the tentacles. Egested sand is passed out of the burrow via the smaller opening of the tube which projects slightly above the sand surface. Because *Pectinaria* does not feed off surface detritus it has to move its burrow fairly frequently as the deeper layers of mud are not so rich in organic matter.

The tubes formed by these worms are among the most beautiful in the animal kingdom, ranging as they do from almost straight cylinders to broadly curved cones. Unfortunately their delicate construction makes them very fragile and, while the occasional intact specimen may be found by an ardent beachcomber, the chances of keeping it without damage are remote.

The ultimate in prostomial feeding structures is shown by the fan worms or feather duster worms. These are two separate families, the sabellids and serpulids, which show convergent filter-feeding mechanisms. That of the peacock worm, *Sabella pavonina*, has been studied in detail. The feeding crown consists of skeletally supported radioles, presumably derived from prostomial feeding tentacles, which are branched to give pinnules, thus increasing the surface area. The whole oral, or upper, surface is covered with short cilia beating toward the mouth and gland cells that produce mucus. The sides of the pinnules bear longer cilia which draw water toward the center of the crown where it flows up and out. Any particulate matter in the water stream becomes stuck in the mucus and under the action of the cilia is directed along a series of gutters to the mouth. Sorting occurs at the base of the tentacles and involves complex ciliary currents. The particles are directed into a gutter which acts as a feeler gauge. Very small par-

226

ticles fall to the bottom and are transported to the mouth, while large ones are pushed up over the side of the gutter and are removed along specialized filaments. When they reach the center of the crown they are swept away in the outgoing current. Intermediate sized particles may be retained for tube-building.

Similar feeding processes are believed to occur in other sabellids although there are variations in crown structure. Thus *Sabella spallanzanii* (also known as *Spirographis*) has a spiralled crown. This is a warm water species often encountered as a special attraction in tropical marine aquariums. In *Myxicola* the radioles are interconnected by a fine membrane so that only the tips are available for filtering. This may have evolved to prevent clogging, as *Myxicola* lies with its crown just above the surface of the fine mud which it typically inhabits.

The process of tube-building illustrates the complex operations which may be performed by such comparatively simple organisms. Sabellid tubes are open-ended affairs made up from sand and shell particles collected during feeding and bound together with mucus. The peristomium, which is folded back as a collar fitting over the tube, is instrumental in the building process. Suitably sized particles are stored in sacs located below the mouth where they are mixed with mucus secreted from the sac epithelium. Strings of mucus and sand are exuded from these sacs to the collar directly beneath them. This structure has two folds which ' handle' the string and mold it into position on the top of the tube as the worm slowly rotates. The tube is strengthened by an internal mucous covering produced from ventrally placed gland shields on each segment.

The precise orientation of the tube is of importance in some sabellids, and the worm is provided with the necessary sense organs to determine this accurately. Thus there are anteriorly placed statocysts which enable the worm to maintain posture. *Branchiomma* is a diurnal worm which orients its tube toward the light and shows a marked shadow reflex whereby it curls up the radioles and with-

In typical chitons the plates or valves are large and just touched by the girdle, which often bears many bristles or cilia. Above is the common Caribbean *Acanthopleura granulata* (K. Lucas photo at Steinhart Aquarium). Below is a ventral view of *Mopalia ciliata* showing the relatively broad foot (East Pacific; T.E. Thompson photo).

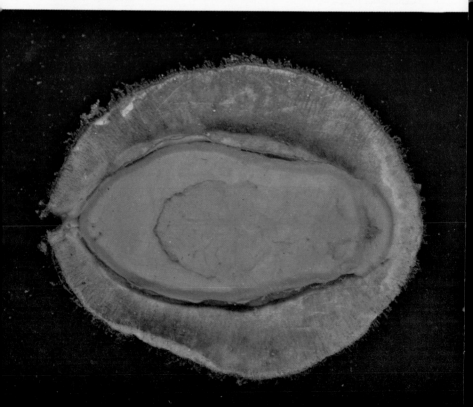

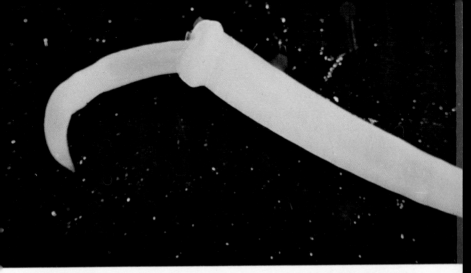

The tusk shells are common as dead shells but are seldom seen alive by the casual collector, probably because of their burrowing habits (Caribbean; T.E. Thompson photo).

Chitons seem to be larger and more abundant in cooler waters, but many, such as this *Ornithochiton quercinus,* are also found in the tropics (Pacific; T.E. Thompson photo).

draws the crown into the tube if the light intensity decreases. This is achieved by a sudden contraction of the whole body brought about by impulses from a system of ‘giant fibers’ which form a fast pathway in the nervous system. This is triggered as a response to stimulation of the large compound eyes on the crown.

Like *Chaetopterus*, the sabellids show remarkable powers of regeneration. Laboratory experiments have shown that the crown is important in respiration as well as feeding and it must be replaced as soon as possible if damaged. While a worm can survive without the crown if it can irrigate, the oxygen supply is insufficient if respiratory currents are prevented.

Associated with these powers of regeneration is the ability to reproduce by budding off new individuals, and many species will alternate a sexual generation with several asexual ones.

Feeding in the serpulid fan worms is very similar to that of sabellids, but the collar structure and ciliary currents are simpler because they are not involved in tube-building. Instead serpulids are characterized by a calcareous tube. Calcium carbonate is secreted by two large glands situated beneath the collar. This is exuded between the body wall and the collar, which acts as a mold. As in sabellids the internal surface of the tube is coated with mucus from the ventral gland shields. The tube can be closed by an operculum which is formed from one of the radioles. This may be quite exquisitely sculptured and is of importance in identification in some members of the family.

One important difference between the two families of fan worms lies in the fact that in serpulids the tube is blind. This means that a through-flowing irrigation current cannot be achieved and instead water is drawn in and out via the anterior opening. The relatively small size of these worms allows a sufficient current from ciliary activity. A further consequence of this type of tube construction is the problem of fouling. As in sabellids, egested material is expelled with the outgoing crown current, but quite complex modifications have been evolved for excretion. The segmental paired nephridia of the "basic polychaete" are restricted to a single pair placed anteriorly. The opening, or nephridiopore, is situated on the head so that excretory products may be readily removed into the water.

The reproductive biology of serpulids is also of interest, several species being hermaphroditic. *Spirorbis* and *Filograna* develop mature male and female gametes at the same time and are self-fertilizing, a process which may be an adaptation to chromosomal abnormalities. *Hydroides* and *Pomatoceros*, on the other hand, are protandrous hermaphrodites. This means that each individual starts life as a male and subsequently changes into a female. Obviously cross-fertilization is the rule in these genera.

Associated with hermaphroditism and the presence of a tube, many serpulids show some degree of larval protection. Thus in *Spirorbis* the eggs may form a string within the parental tube or may be housed in a special cavity made by the operculum. In *Microserpula* calcareous egg-containing capsules are attached to the side of the tube. Development does not proceed very far under parental protection however, most serpulids producing very beautiful free-swimming trochophore larvae which feed in the plankton.

The small size of serpulids, and the fact that they rapidly withdraw into their tubes when disturbed, makes it easy to miss their quite spectacular appearance, and the beautifully colored gametes are seldom observed at all. Nevertheless, the tube itself can be quite intricate in design, ranging from the ridged structure of *Pomatoceros* to the coiled *Spirorbis* tube. Because of the protection from predation offered by the tube, serpulids have been able to exploit apparently hostile environments. Thus many species are found intertidally on rocky shores where the white twisting tubes may be seen all over rocks on the lower shore. Lime is not impermeable, however, and resistance to desiccation is incomplete. Some species of *Spirorbis* have partially overcome this problem by attaching themselves to fucoid seaweeds which retain their moisture for some time when uncovered

A slit shell, *Pleurotomaria teramachi*, the dream of many collectors. Only a few species of these primitive but often large gastropods still survive (Pacific; K. Lucas photo at Steinhart Aquarium).

Haliotis asinina, a common small abalone of the Pacific area with a large and attractive animal (A. Power photo).

Ventral view of an abalone,
Haliotis sp., from the East
Pacific (K. Lucas photo at
Steinhart Aquarium).

Patella cochlear, the South African pear limpet, is extremely abundant at the lowest intertidal (T.E. Thompson photo).

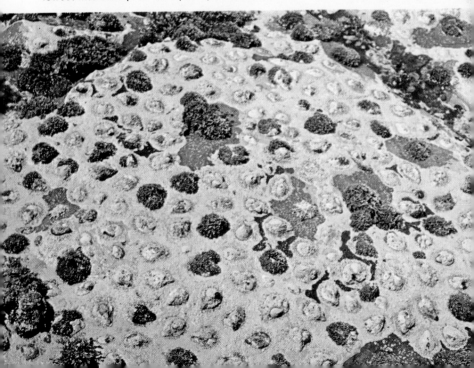

at low tide. Other serpulids have become colonial in habit, forming small reefs of limy tubes. *Serpula* and *Filograna* fit into this category. In some instances the colonial behavior can be of economic importance. A growth of limy tubes can cause a significant amount of fouling on ship hulls and populations can build up very quickly. *Mercierella enigmatica* has extended its range from Australia to Europe by hitching free rides on ship bottoms in this way.

The precise habitat preferences shown by these worms necessitate some method of ensuring that young worms arrive on the correct substrate. This is complicated, of course, by the period of planktonic development. The older larvae are supplied with chemical sense organs allowing them to recognize a suitable area, possibly by the presence of adults, and the length of the larval life is not definite so that settlement may be delayed until a suitable site is located.

Such problems are much less important for the errant polychaetes which are free to move around. Perhaps the best known of this group are the clamworms, members of the family Nereidae which are used for bait. Most nereids live in semi-permanent burrows which they may leave on feeding expeditions. There are exceptions, however. Thus *Neanthes fucata* is a commensal living inside the adopted shell of the hermit crab *Pagurus bernhardus*. The advantage is presumably one of protection since the worm may leave its host to feed.

Clamworms show many adaptations to an active existence, not least of which is the modification of the parapodia for locomotion. Three different phases of movement can be seen. Slow crawling is achieved solely by parapodial action. Each parapodium pushes backward against the substrate with setae extended and then swings forward with the setae withdrawn so that there is no contact with the ground. The parapodia of each segment work alternately and a wave of movement can be seen passing backward down the worm as it moves forward. Faster locomotion is achieved primarily by contractions of the longitudinal muscles acting against the points of contact made by the parapodia. This brings about the familiar lateral undulations of nereids, a feature also typical of swimming ones.

The adaptations for feeding also reflect the errant nature of nereids. There is a muscular eversible proboscis bearing numerous teeth or paragnaths, the arrangement of which is used taxonomically. The everted proboscis is surmounted by jaws which can be manipulated by special muscles. In carnivorous species the proboscis can be rapidly withdrawn by powerful retractor muscles, but these are absent from algae-eating forms.

Several types of well developed sense organs are associated with the active form of life, and in general nereids show a better developed nervous system than more sedentary species. There are two pairs of prostomial eyes with lenses, and light receptive spiral organs are present on the antennae. Nuchal organs, which are believed to be chemo-receptors important in the location of food, are also present on the prostomium. Other less specialized sensory structures are cells scattered over the epidermis, especially in regions associated with tactile stimulation such as the antennae and palps on the head and cirri on the peristomium.

The aggressive hunting behavior of nereids is also reflected in their behavior among individuals of the same species. Thus *Neanthes caudata* is known to fight with members of its own sex, a phenomenon which is presumably related in some way to reproductive behavior, and *Nereis pelagica* will fight over territory under laboratory conditions.

While many species of nereids are restricted to the marine environment, being common both on sandy shores and within crevices, a few species are tolerant of lower salinities and are more typically estuarine in their habitat. *Neanthes diversicolor* is the best known in this respect and is known to regulate the salt balance of its coelomic fluid. This ability is stretched still further by *Nereis limnicola*, a very closely related species which even penetrates into fresh water.

Two giant limpets: above, *Megathura crenulata,* which may reach over 150mm in length (East Pacific; K. Lucas photo at Steinhart Aquarium); below, the black animal of *Scutus antipodes* covers the smaller white shell in life (Pacific; T.E. Thompson photo).

Two typical keyhole limpets: right, *Diodora inaequalis* (East Pacific; A. Kerstitch photo); below, *Lucapina suffusa* (Caribbean; T.E. Thompson photo).

The reproductive biology of nereids is also influenced by their motile life style. Swarming is the normal situation, with ripe individuals collecting at the surface of the water prior to spawning. In some species, such as *Platynereis dumerili*, ripe adults metamorphose into reproductive forms or epitokes which exhibit several secondary sexual characteristics such as enlarged parapodia. Indeed the swimming epitokous forms are so different from the bottom-living stages that they were once assigned to a separate genus, *Heteronereis*. The expanded parapodia act as paddles for swimming and greatly enlarged eyes increase visual responses. As this is a non-feeding, non-bottom-dwelling stage, tactile sense organs such as the palps and cirri of the head region become reduced. Usually only the rear end of the body undergoes metamorphosis, although in some instances changes are restricted to the mid-body region. In *Tylorrhynchus* for example, the unmetamorphosed tail is discarded prior to swarming. The external modifications are paralleled by internal ones. Some muscles degenerate while others, notably those required for swimming, are developed, and the gut is eroded away so that gametes can be released via the anus.

Swarming behavior in male and female epitokes is synchronized with the lunar cycle in order to ensure fertilization. In most species the males congregate at the surface first and await the coming of the females before shedding their sperm. It is believed that the females may produce a substance eliciting spawning behavior, the eggs themselves being released upon stimulation by the sperm. In some species, for instance *Neanthes succinea*, the worms perform a nuptial dance in which the male twirls around the female, thus increasing the chances of successful fertilization. Experiments suggest that sense organs situated dorsally on the parapodia of the male may enable him to orient with respect to the female.

Platynereis megalops, a species found at Woods Hole, Massachusetts, takes the pairing behavior a stage further. The eggs in this species are rendered infertile after only thirty

238

seconds of exposure to seawater; fertilization is, in fact, internal. The male performs the same nuptial dance as *Neanthes succinea* but comes much closer to the female so that it actually wraps itself around her. The anus is then inserted into the female's mouth and sperm passes directly into the coelom (the gut having been digested away), where fertilization occurs. The egge are then released into the water through the female's anus.

The estuarine species of clamworm have abandoned swarming behavior and epitoky, possibly as a response to conditions of reduced salinities. Thus *Neanthes diversicolor* spawns on the surface of the mud, or even in the burrows, and the eggs develop into demersal (bottom-swimming) larvae.

Epitoky is also a feature of the reproductive biology of eunicids and syllids, and the swarming behavior of the palolo worm is very famous. This name was originally given by natives to the Samoan species *Eunice viridis* but has subsequently been applied to several other swarming eunicids as well as nereids. The behavior of the Atlantic palolo (*Eunice scemacephala*) is well known. It is a West Indian species living in coral crevices and among rocks at subtidal levels. Like all swarming polychaetes the non-sexual, or atokous, form is negatively phototropic, leaving its crevice only to feed at night.

The timing of the swarming is very precise and can be accurately predicted. In this species it occurs in July during the first and third quarters of the moon. The time of day is also synchronized among individuals, mature worms backing out from their burrow at three or four in the morning. Only the tail region undergoes metamorphosis, and this portion breaks free and swims to the surface tail first. Unlike the atoke, the headless portions are very sensitive to light. A swarming dance ensues at the surface with many hundreds of worms spiralling rhythmically around in the water. At dawn the body wall ruptures and ripe gametes are shed into the water, where fertilization occurs. The developing trochophore larva remains in the plankton for three days and then sinks to the bottom.

A group of periwinkles, *Littorina scutulata* (East Pacific; K. Lucas photo at Steinhart Aquarium).

The majority of marine snails are small, inconspicuous animals like these *Neritina* (rounded shells) and *Cerithium* (elongated shells) (Australia; L.P. Zann photo).

The bright orange animal and delicate shell colors make *Calliostoma annulatum* one of the more collectable "primitive" gastropods (East Pacific; K. Lucas photo at Steinhart Aquarium).

Snails are common but inconspicuous parasites or commensals of other invertebrates, such as this *Hipponyx* species attached to the arm of the starfish *Linckia* (Pacific; S. Johnson photo).

The Samoan palolo behaves similarly except that swarming occurs at the beginning of the first lunar cycle of October or November. The natives take a heavy toll of the epitokes, eagerly awaiting the chance to net such a great delicacy.

Apart from the spectacular palolo worms, the eunicids, or rock worms as they are sometimes known, are remarkable also for their size, one species of *Eunice* attaining a length of three meters. The group as a whole has undergone extensive adaptive radiation so that some authorities prefer to divide it up into separate families. Many, like the palolo, are crevice-dwellers superficially similar to the nereids in their carnivorous adaptations, while others are algal eaters, and some are tubicolous carnivores unlike other tube-dwellers in that they can move around carrying their tube with them. *Diopatra*, for example, builds a heavy membranous parchment tube which it secretes from glands situated on the ventral surface and disguises the protruding tip of the tube with pieces of seaweed or shell which are held by the jaws and manipulated into position. The quill worm, *Hyalinoecia*, drags its tube along with it, leaving a distinctive trail along the sea bottom. It moves forward by extending its front end from the tube and pushing its head into the substrate to form an anchor. The tube is then drawn forward by looping the body and pulling against the point of contact. The process is then repeated, with the head end being extended from side to side so that the straight track of the tube is patterned with side chains caused by the head. Because of their active life the tubicolous eunicids do not show the usual sedentary attributes but remain similar to other errant species.

The syllids are superficially similar to the eunicids in many respects, notably their reproductive behavior, but this indicates a convergence of evolution rather than any close underlying relationship. The family as a whole shows remarkable powers of regeneration and, by linking this ability to a process of fragmentation, many species are able to reproduce asexually. Development of new individuals may occur prior to the disintegration of the worms so that chains of individuals can be found. *Myrianida*, for example, has been recorded with twenty-nine individuals attached to the parent stock.

The site of stolon proliferation varies according to the species, as is amply illustrated by the genus *Trypanosyllis*. *Trypanosyllis prolifera*, as its name suggests, can produce new heads from most segments and sometimes chains of new individuals may be found on consecutive segments. *Trypanosyllis asterobia*, on the other hand, forms clusters of stolons rather than chains, and *Trypanosyllis crosslandi* produces bunches of new worms in the pygidial region. Some species of *Syllis* also produce lateral buds; in *Syllis ramosa* they may remain attached until they bear stolons.

It is not far from the production of new individuals in this manner to the formation of specialized sexual individuals or epitokes, and a number of syllids reproduce in this manner. Although the method of production is different, the syllid epitoke is adapted similarly to that of the nereid and eunicid worms with its long swimming setae and well developed eyes. The epitoke of *Syllis hyalina* shows further similarities in that it lacks a head and is thus reminiscent of the palolos. Not all syllids produce epitokes by fragmentation; *Eusyllis* and *Odontosyllis*, for example, develop epitokes by metamorphosis.

However it is formed, the epitoke undergoes the unusual swarming procedure, a process often beautiful to watch. In *Odontosyllis*, for example, the highly luminescent females attract the males. Once spawning has occurred and the gametes are liberated into the seawater, the light fades away.

Spawning is not the dramatically fatal process of the nereids and, indeed, larval protection may occur in some species. In *Autolytus*, for example, the eggs are retained in a sac-like structure on the ventral surface of the female. In other types, such as *Grubea*, the eggs are attached individually to the dorsal surface of the parent.

Despite their interesting reproductive biology, syllids are

Wormshells, *Serpulorbis imbricatus;* note the opercula (Pacific; Y. Takemura and K. Suzuki photo).

Wentletraps, *Epitonium,* are usually heavily sculptured, elongated white shells that feed on anemones and other coelenterates (Pacific; L.P. Zann photo).

Janthina, the purple snails, are delicate pelagic snails that hang from the surface by a raft of air bubbles. In the photo below can be seen the snail's reddish egg cases attached to the bubble raft (tropical; C. Arneson photo above, T.E. Thompson photo below).

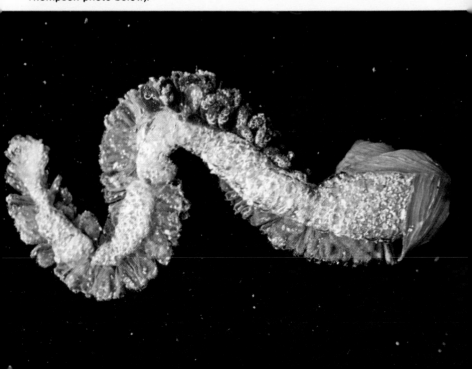

not among the most familiar of polychaetes, although they are widely distributed. They are small active worms commonly found in kelp holdfasts, inside sponges, or entangled among hydroids. The latter form an important item in the syllid diet, and in some species the feeding apparatus is specialized accordingly. Thus *Autolytus*, which lives in tubes attached to hydroids, carries out feeding expeditions in which it crawls over the hydroid and eats the polyps. These are cut off by a circle of chitinous teeth and sucked into the body by a muscular pumping process. In *Procerastea* the proboscis is inserted directly into the polyp via the mouth and the contents sucked out.

While the epitokous syllids show adaptations for swimming, some families have evolved a totally pelagic existence. Among the most specialized are the alciopids and the tomopterids, but all show some modifications associated with the oceanic way of life. Protection from predation can no longer be attained by hiding in tubes or burrows, and instead the pelagic worms are transparent and thus less conspicuous. They are all carnivorous, feeding on smaller planktonic material, and have well developed sense organs. The eyes in particular are very prominent and especially remarkable in that the lens can be moved and focused. This suggests that the eye is capable of image formation and does not simply function as a shadow detector. Many alciopids and tomopterids are capable of producing light. As some live at abyssal depths where sunlight cannot penetrate, it is possible that the luminescence functions to attract members of the same species.

The tomopterids differ from other polychaetes in the absence of setae. The parapodia are modified into greatly exaggerated paddles used for swimming.

Some species of polychaetes are venomous and can cause an unpleasant sting if handled carelessly. The amphinomids are notable in this respect as their common name, fireworms, suggests. These are warm water animals often found in corals and can be found in a variety of shapes and sizes. They feed on sedentary organisms, browsing with a rasp-like proboscis. Some species have become so specialized that they are virtually parasitic on sponges and, due to their flattened shape and cryptic coloration, are difficult to distinguish from their hosts.

The venomous properties of these worms are defense mechanisms. When disturbed the worm flexes its body so that the setae become erect. If touched they discharge a noxious substance which is believed to act on the nervous system. Further discomfort is caused by the setae themselves. These are of a calcareous nature somewhat reminiscent of sponge spicules and are very brittle so that they are readily dislodged and become embedded in the skin of the offender, where they cause intense irritation. *Hermodice carunculata* is a common amphinomid which frequently causes distress to skin divers. It is a large species, growing up to twenty-five centimeters in length, and is equipped with tufts of white setae. Upon contact with the skin these cause a burning feeling rapidly followed by a numbing sensation which may be spread far from the initial point of contact and persist for several hours.

Some species of sea mice (Aphroditidae) also cause discomfort if treated without due respect. Thus the strange but highly attractive *Aphrodita* can erect its hollow spine-like setae and inflict injury to an intruder. *Aphrodita*, the common sea mouse, is most unworm-like in its shape, being fairly short and swollen in the middle region. The body is covered with scales (called elytra) which are hidden by a dense mat of iridescent hair-like setae which arise from the notopodia. The poisonous spines are arranged laterally. It is a fairly common worm occurring at or just below extreme low water mark where it may be found hiding under stones or buried in the mud. Despite its dumpy appearance it is an active carnivore and its head, hidden from dorsal view by the felt mat, is equipped with the typical sense organs associated with this life style. Studies of behavior in *Aphrodita* and the closely related *Hermione* show a positive thigmotactic response helping to keep it under stones where it is sheltered.

Carrier shells, *Xenophora,* are very different from the conchs in shell shape, but the animal bears a strong superficial resemblance to *Strombus* (Caribbean; D.L. Ballantine photo).

Strombus gracilior, a common tropical East Pacific conch (A. Kerstitch photo).

Close-up of the partially retracted animal of *Strombus luhuanus*, a very common Indo-Pacific conch. Note the serrated operculum, the brown undersurface of the foot, and the large, colorful eyes (T.E. Thompson photo).

Closely related to the sea mice, and once included in the same family, are the scale worms. These are characterized by the elytra also shown by the sea mice but lack the setal mat. The scales afford some protection to the worm and function as points of attachment for eggs in those genera showing brood behavior (for instance, *Halosydna*).

The thigmotactic behavior of sea mice is also shown by scale worms and may form the basis of commensalism in this group. Many of these worms are found in the burrows or tubes of other polychaetes (e.g., *Gattyana* with *Neoamphitrite*), a relationship which may have its roots in a desire to find a dark secluded habitat. Naturally, during the course of evolution the association has become better established so that the scale worm may be actively attracted to its host, although the mechanism of this behavior, believed to be basically chemical in nature, is not fully understood.

Not all species are very specific about their host, however. *Halosydna brevisetosa*, for example, has been found in association with terebellids and nereids and is also known as a free-living species.

Other worms are not the only hosts. Some scale worms are commensals on sea cucumbers and have modified hook setae to ensure a secure hold; *Hesperonoe adventor* lives in the burrow of the echiuroid *Urechis*. Laboratory observations show that it feeds on particulate matter rejected by the host and positions itself in the burrow so that it is virtually touching the echiuroid's proboscis, ready to pounce on any material that is too large for its host.

While many polychaetes show commensal associations like this, only a few cases of parasitism are known. Many of these belong to the eunicid group of families. Thus *Oligognathus* is found internally in an echiuroid worm and *Haematacleptes* lives inside the vascular system of a terebellid.

Ectoparasites are also found, a famous example being *Ichthyotomus*, which feeds on the gills of an eel by attaching itself with a scissor-like jaw apparatus. *Calamyzas*, a syllid, lives on the ampharetid *Amphicteis*. *Histriobdella*, which is parasitic on lobsters, is highly specialized for its way of life. All the parapodia are lost except for a single pair at the posterior end of the body which are used to propel the worm over its host.

Although parasitism is rare among polychaetes, the potential is obviously there and has been developed by the myzostomids, a group of annelids totally parasitic on echinoderms. The adults are highly modified flattened discs with hooked setae for hanging onto their hosts, but their polychaete affinities are clearly shown in the larval development.

The other major group of annelids with marine representatives, the leeches, also contains some parasitic forms. The balance between predation and parasitism is very fine, but those leeches which feed only on a specific host and remain attached to one particular individual for some time may be said to be parasitic. Hosts vary from invertebrates such as the oyster, which is attacked by *Ostreobdella*, to vertebrates, sharks and rays being the major hosts of the marine piscicolids.

While the myzostomids and leeches are of much interest to the parasitologist, they are less exciting to the marine biologist who is more concerned with modifications associated with the marine environment. The role of the polychaete as *the* marine annelid is therefore largely unchallenged by these other groups.

Cowry shells are popular with collectors because of their beauty, but the animals may be equally attractive. Above is the California brown cowry, *Cypraea spadicea* (T.E. Thompson photo); below is the tropical East Pacific *Cypraea albuginosa* (A. Kerstitch photo).

The mantle of cowries may be nearly smooth, as in the *Cypraea isabella* above (Pacific; A. Kerstitch photo), or heavily covered with papillae, as in the *Cypraea chinensis* below (Pacific; S. Johnson photo).

Minor Worm Phyla

Under this convenient classification are included four phyla of marine worm-like animals that are probably related to the annelids or, in the case of the priapulid worms, nematodes.

Phylum Sipunculida or Sipuncula consists of rather cylindrical worms with a retractile proboscis and an anteriorly located anus. Many are tube-dwellers and commensals, some rather attractive. There are about 275 species, but only a few are at all common.

Echiuroids, phylum Echiuroida, are generally similar to sipunculids but the proboscis is not retractile and the anus is posterior as normal. There are about 150 species. Both these phyla are very closely related to Annelida and are often considered classes of that group. Both phyla contain a few species that would make interesting aquarium additions, but the problems of feeding them and, especially, finding them would be insurmountable to most aquarists.

The beard-worms, phylum Pogonophora, are one of the most amazing phyla and the last generally recognized phylum to be discovered. Most of the known species, about 75 and growing, are extremely slender worms of the deep seas, but lately an increasing number of species have been described from shallower waters along the Atlantic coast of the United States, and a species with tubes over an inch in diameter and three feet long has recently been found off the Galapagos. Once considered close relatives of the acorn worms and thus of the chordates, it is now generally suspected that beard-worms are highly modified annelids and perhaps should be placed within that phylum. They seem to be unique in having no digestive system at any stage *of development and no recognizable way of digesting food. Pogonophores are currently enjoying a great deal of attention from biologists, so anything said about them now is subject to modification on short notice.*

Nine species of priapulids, phylum Priapulida, are currently recognized, although none is really common. The retractile proboscis with small spines is distinctive, but the actual relationships of this phylum are very uncertain. They are often considered to be closely related to the aschelminths. Most are coldwater species, but the recently described Tubiluchus *is from the West Indies.*

MINOR WORM GROUPS

While the vast majority of marine invertebrates can be grouped together in phyla such as the Annelida or Mollusca, a few worm-like organisms remain which cannot easily be fitted into any of these categories. These were once grouped together for convenience sake under the name Gephyrea, but as more information became available it was apparent that three distinct types could be distinguished.

The first of these are the **sipunculids**, sometimes known by the common name of peanut worms. All are marine. They are benthic animals living at a wide range of depths from the intertidal zone to abyssal depths of over 4500 m, although most are shallow-water species.

The body is cylindrical and can be divided into a thinner anterior introvert, which includes the head, and a posterior trunk. The terminal mouth is ringed with tentacles or lobes used in feeding and possibly as "gills" for respiratory purposes. In genera such as *Phascolosoma*, specialized sensory cells, possibly functioning as chemoreceptors, are present on the introvert. The anus is not terminal as in the annelids, but is located anteriorly on the trunk. This means that the worms do not need to turn around in their burrows to defecate.

The whole of the introvert can be withdrawn into the body by the contraction of special bands of retractor

Jenneria pustulata, a false cowry more closely related to *Cyphoma* than to *Cypraea* (East Pacific; A. Kerstitch photo).

Although the shells of *Cyphoma* are white or pale salmon in color, the mantles are often attractively patterned, as in these *Cyphoma signatum* (Caribbean; D.L. Ballantine photo).

Ovula ovum, a large egg cowry with a starkly black animal; the large white shell is often used for ornamentation by Pacific island natives (A. Power photo).

Phenacovolva piragus, a rare Caribbean false cowry (C. Arneson photo).

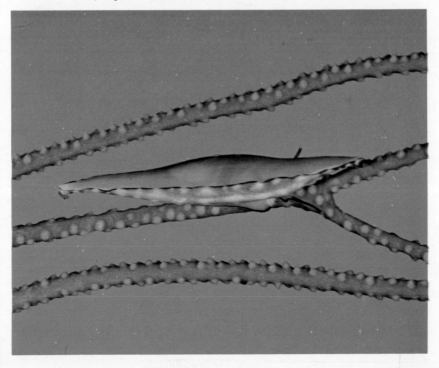

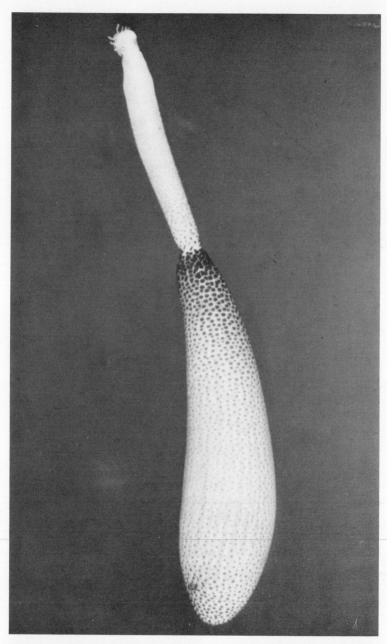

The sipunculid *Phascolosoma antillarum* with the introvert extended (C. Arneson photo).

muscles running from the trunk to the esophagus. These play an important part in the classification of sipunculids. Extrusion of the introvert is achieved by an increase in coelomic fluid pressure brought about by contraction of the body wall. Withdrawal of the introvert obviously has an important protective function but it may also be involved in the feeding behavior in some species. When in a contracted state, some of the crevice-dwelling intertidal forms such as *Dendrostoma petraeum* take on the shape of a peanut kernel, complete with the pointed tip. Hence the common name.

Although sipunculids are exclusively marine they exhibit a wide range of habitats within this environment. Some are burrowers in sand, either forming permanent burrows like *Dendrostoma zostericulum*, which lives in sand among the root system of eelgrass, or moving through the deposit as in the case of *Sipunculus nudus. Dendrostoma petraeum* lives in sand beneath stones or in rock crevices, and there are many sipunculids living in holes in rocks or in coral crevices. *Aspidosiphon*, for example, inhabits coral crevices and has a thickening of the trunk which presumably acts as a plug to block the crevice when the introvert is withdrawn. None of them build tubes as such but the burrowing forms produce a mucous lining which gives a firmer surface to the burrow. The body wall of sipunculids is tough and leathery in texture and may bear spines so that the protection offered by a tube may be unnecessary.

Sipunculids are generally deposit-feeders although they vary in their precise feeding methods. Those species living in a permanent burrow extend their tentacles over the surface of the sand. Particulate matter becomes entangled in the mucus secreted by the tentacles and the beating of cilia creates a current carrying the food to the mouth. Members of the genus *Sipunculus* do not build permanent burrows but eat their way through the sand. The feeding tentacles are replaced with a fluted fold which acts as a scoop. Sipunculids are largely nonselective feeders ingesting all material that is collected by the tentacles, but in practice the

Trivia pediculus, a heavily ridged cowry ally (Caribbean; T.E. Thomspon photo).

Moonsnails are large-footed burrowing predators that drill distinctive beveled holes in their bivalve prey. Shown in *Natica alderi,* an attractive species of the East Atlantic (T.E. Thompson photo).

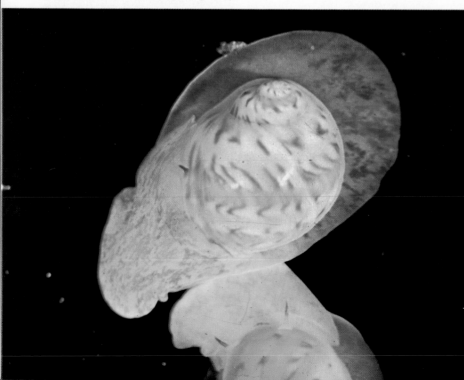

The spectacularly colored animal of *Cymatium intermedium* puts its dirty brown shell to shame (Pacific; S. Jazwinski photo).

Cassis tenius, a colorful helmet shell of the East Pacific (A. Kerstitch photo).

mechanics of feeding ensure that really large particles do not reach the mouth. One species, *Golfingia procera* from the North Sea, is thought to be carnivorous and is believed to attack the sea mouse *Aphrodita*. It penetrates the body wall in some manner and sucks out the contents of the polychaete.

Respiration in sipunculids is largely a matter of diffusion across the body surface, although the feeding tentacles may assist by providing an increased surface area. There is no vascular transport system, but an unusual respiratory pigment, hemerythrin, is present.

Sipunculids are dioecious and ripe gametes are shed into the seawater. Swarming behavior is known to occur in at least one species, *Dendrostoma petraeum*, the worms forming compact masses among the rocks just prior to spawning. Fertilization results in a trochophore larva which spends a variable length of time in the plankton before descending to the bottom and metamorphosing into the adult state.

Some species of sipunculid are commensals living in association with other animals. *Phascolosoma hesperum* is one such example which is quite catholic in its choice of host. It is often found in the tube of the polychaete *Chaetopterus variopedatus* and other chaetopterids, but it has also been recorded in association with the sea anemone *Edwardsiella* and the brachiopod *Glottidia*.

The affinities of the sipunculids are somewhat obscure but the embryology and certain anatomical features suggest an annelid stock. The complete absence of metamerism (segmentation) implies that the sipunculids must have departed from the annelids at a very early stage.

The **echiuroids** are perhaps closer to the annelids as they do show metamerism in their larval development. They are exclusively marine worms superficially similar to the sipunculids in their general appearance and habits. Most live in shallow waters, but deepsea forms have been recorded and they have been dredged from over 8000 m.

Commonly known as sausage worms, echiuroids have a cylindrical unsegmented body with an anterior spade-like proboscis which, unlike that of the sipunculids, cannot be drawn into the rest of the body. Study of the internal anatomy shows that the proboscis is, in fact, a cephalic structure housing the brain and is possibly homologous with the annelid prostomium. The proboscis is used as a feeding structure and is modified accordingly. The edges are rolled under to give a ventral furrow and the distal end may be bifurcated as in *Bonellia*. Size relationships vary considerably, and the proboscis may be very large. In the Japanese *Ikeda*, for example, the proboscis is 1½ m long and the trunk only 40 cm. There are no obvious sense organs, but the proboscis is presumed to have a tactile function. A pair of large chitinous setae are situated ventrally on the anterior end of the trunk, and in some species there is a ring of setae near the anus around the posterior end of the trunk.

By and large echiuroids are dull animals, but a few are brightly colored, *Listriolobus pelodes* being a translucent green.

Echiuroids commonly live in U-shaped burrows in sand and mud, *Echiurus* and *Urechis* being two common genera. The North Carolina *Thalassema mellita* shows an interesting adaptation and is often found in the tests of dead sand dollars. When very small the worm enters the test and grows too large to leave it.

Most echiuroids are detritus-feeders using the proboscis as a kind of shovel. *Echiurus*, for example, extends the proboscis from the burrow and spreads the ventral surface over the substrate. Small particles adhere to the mucus on the proboscis surface and are swept into the ciliated furrow and from there transferred to the mouth. The proboscis is then withdrawn into the burrow and extended again in a different direction.

Urechis, which lives in a permanent U-shaped burrow, feeds in a completely different manner more reminiscent of chaetopterid polychaetes. The proboscis is much shorter than that of *Echiurus* and there is a ring of mucous glands around the top of the trunk. The worm expands this region

The thin shell of *Ficus ventricosa* is almost covered by its large mantle (East Pacific; A. Kerstitch photo).

The predatory Triton's trumpet, *Charonia tritonis,* is one of the few animals known to feed on crown of thorns starfish (Pacific; S. Johnson photo).

Tonna perdix, a common large Indo-Pacific tun shell.

muscularly so that the glands are pressed close against the sides of the burrow. It then secretes a ring of mucus and slowly retreats down its burrow, spinning a mucous net as it goes. This is attached anteriorly to the burrow wall and posteriorly to the worm itself so that any incoming water must pass through this. Water is drawn into the burrow by peristalsis and small detrital material is collected in the bag. As this becomes choked up the filtering rate rapidly diminishes and the worm then moves forward in its burrow and devours the bag and its contents. A new bag is then built and the process repeated. The time cycle of this behavior is related to the richness of the substrate, and in very productive areas a new net may be constructed every few minutes. *Urechis* is a fussy eater and any large particles retained by the net are rejected. These supply a rich food supply for other animals living as commensals with the echiuroid. These include a scale worm, a small crab, and occasionally a fish which hides in the mouth of the burrow. Feeding activity is naturally restricted in intertidal worms to periods of high water, and at low tide the worms lie quiescent at the bottom of their burrows.

Ochetostoma shows yet another feeding mechanism. The proboscis wipes detritus from the burrow wall and collects material when a new burrow is formed. It is extremely flexible so that it can penetrate into small crevices in the search for food. Again unwanted material is rejected, coarse particles passing along a shallow groove around the base of the proboscis instead of into the mouth.

When the proboscis is extended out of the burrow, as in *Echiurus*, it is at great risk from predators such as fish. It has been shown that in some species the worm protects itself from such dangers by casting off the proboscis if it is damaged in any way and regenerating a new one.

The irrigation produced for feeding purposes also serves a respiratory function. There are no external respiratory organs, diffusion presumably occurring across the body wall. In *Urechis* and some other genera there is a siphon running alongside the intestine and attached at each end to a ciliated intestinal groove. The precise function of this is not clear, but as it is known to fill with water it may act as an internal respiratory surface.

A closed vascular system is present in most genera except *Urechis* and a respiratory pigment occurs in cells in the coelomic fluid, both of which improve the availability of oxygen to the tissues.

The sexes are separate and in most cases fertilization is external as in sipunculids. In *Bonellia*, however, there are extensive modifications resulting in internal fertilization. There is a very marked sexual dimorphism, the male being much smaller than the female, only a few millimeters in *Bonellia viridis* compared with 100 mm in the female. The male has no proboscis and the digestive system is greatly reduced at the expansion of a large reproductive system. There is no circulatory system. The whole body surface is ciliated. The tiny male enters the body of the female and lives in the uterus, a modified excretory organ, absorbing nourishment through its body wall. As in all echiuroids fertilization results in the development of a free-living trochophore larva. If such a larva comes into contact with a female *Bonellia* and enters her body it develops as a male under the influence of a hormone produced by the female. Female individuals develop from larvae not receiving this stimulation.

The third member of the Gephyrea, the phylum **Priapulida**, is rather puzzling and its origins remain completely obscure. It is a very small group containing only a handful of species. These are restricted in their range, being found mostly in cold waters around the American coast, in the Baltic, and in Antarctica. They are littoral organisms living buried in mud and sand.

The body is cylindrical and may be several centimeters long. It is covered with a cuticle which contains chitin and is molted at regular intervals. It can be divided into an anterior proboscis and a posterior trunk region. The mouth is placed anteriorly on the proboscis, which is a bulbous structure covered with longitudinal papillae. The trunk is

An East Pacific muricid, *Pterynotus pinniger* (A. Kerstitch photo).

Muricids are successful predators found in all the warm seas. Here *Murex erythrostoma* (foreground) and *Murex recurvirostris* feed on a large scallop (East Pacific; A. Kerstitch photo).

Murex trunculus, a Mediterranean shell from which the valuable dye Tyrian purple was once obtained (S. Frank photo).

A group of large mud whelks, *Nassarius papillosus,* common shallow-water scavengers (Indo-Pacific; S. Johnson photo).

A female *Bonellia viridis*, one of the largest echiuroids (G. Marcuse photo).

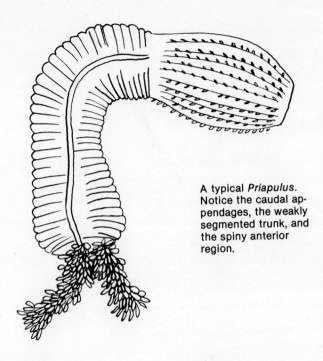

A typical *Priapulus*. Notice the caudal appendages, the weakly segmented trunk, and the spiny anterior region.

Anterior end of *Priapulus* with the proboscis everted to show the tooth-like spines.

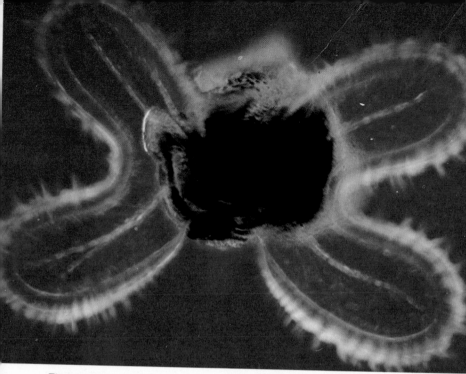

Thais species, the drills, are found in both tropical and cool waters. Above is the larva of *Thais clavigera* (Pacific; Y. Takemura and K. Suzuki photo); below is an adult *Thais intermedia* (Pacific; S. Johnson photo).

The color of a molluscan animal seldom has anything to do with the color of the shell. Thus the dull shell of the *Fasciolaria* whelk above (Atlantic) contrasts with the bright red animal, while the attractive shell of *Tritonoturris robillardi* below (Pacific; S. Johnson photo) shames the nearly colorless animal.

annulated but the resulting segmented structure does not reflect an internal metamerism. The tail end of the trunk bears one or two caudal appendages in the genus *Priapulus*. These consist of a stalk carrying a series of spherical vesicles. The function of these is not known, but they may have a respiratory role or function as chemoreceptors.

Priapulids live buried in the substrate with the mouth just beneath the surface. They are largely sedentary but can move through the substrate by contracting and extending the body. They are carnivorous animals feeding on polychaetes and other slow-moving invertebrates. The mouth, which lies back in the proboscis, is surrounded by several rings of curved spines. These are pushed forward during feeding and are used to seize prey.

There is very little information concerning the reproductive biology of priapulids. The sexes are separate and it is assumed, with no evidence, that fertilization occurs externally. The embryology is very different from that of the annelids and the other gephyreans in that the cells are laid down by radial cleavage instead of spirally. The developing larva is very similar to that of rotifers, suggesting relationships with the pseudocoelomate phyla. The molting of the cuticle is also suggestive of such an association. However, the presence of a coelom would seem to refute this concept. Their relationship to other coelomates remains completely obscure and one can only assume that they diverged from the main stock at a very early stage.

Recently a fourth group of minor worm phyla has been discovered. This is the **Pogonophora** or beard-bearers, so named because of the anteriorly placed tentacles. Pogonophores are very long and very thin so that they present a thread-like appearance; this, coupled with their restricted range, led to their being overlooked for many years. They were first discovered at the turn of the century in waters around the Malayan Archipelago, but their systematic importance was not discovered until the early 1950's. Since then there have been more and more records of pogonophores and there are now about a hundred described species.

As they are so little known it is not possible to outline their world distribution, but from available records it would appear that they favor cold waters. Because of this they are mainly deepwater species living on the continental slopes and extending down into deep trenches. A few species are found on continental shelves, and they occur at depths as shallow as twenty meters in the Norwegian fjords. Although sampling is by no means complete, it is known that pogonophores inhabit all the major oceans of the world.

Pogonophores are tube-dwelling worms usually found with the tube buried vertically in mud, although one species, *Lamellibrachia barhami*, has been found anchored in silt on a rocky surface with the tube free in the water; *Sclerolinum brattstromi* is found in rotting wood. The gigantic *Riftia* has been found in large colonies in the Galapagos Rift and is strongly suspected of using bacteria to metabolize sulfur compounds as food; its 3 cm tubes over a meter long make it the largest pogonophore as yet discovered.

The tubes are usually stiffened structures longer than the worms themselves and are of a brownish color, although a few species have limp colorless tubes. Close examination of the tube shows that it is made up of several layers of a secretion laid down by the worm and spread evenly by structures on the forepart of the animal. The tube sometimes has an annulated appearance due to changes in coloration. The significance of this is not known.

The external morphology of pogonophores has only recently been studied in detail due to difficulties in obtaining complete specimens, but all seem to be based on the same pattern except for the somewhat divergent genera related to *Lamellibrachia* and *Riftia*.

The forepart of the worm consists of a cephalic lobe which bears numerous long tentacles which may be soft and flexible or rigid and tightly held together. Sometimes the tentacles are arranged in a definite pattern.

The cephalic lobe is followed by a thickened region of cuticle, known as the bridle, which is believed to function in

Kelletia, a common East Pacific whelk that can be kept in the aquarium (K. Lucas photo).

Ancillista velesiana, an olive with a thin, high-spired shell and a very large animal (Australia; W. Deas photo).

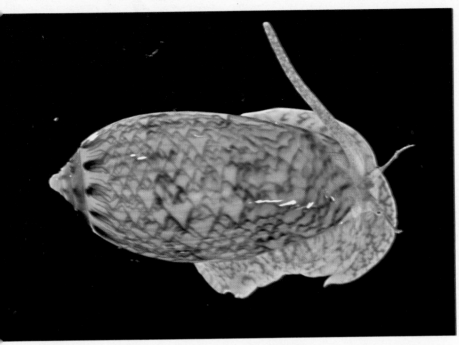

Two attractive olives: above, *Oliva reticularis* (Caribbean; C. Arneson photo); below, *Oliva porphyria* (East Pacific; A. Kerstitch photo).

Large colony of the pogonophoran *Riftia pachyptila*, the largest known member of the phylum Pogonophora. This species is known from deep-sea vents off the western coast of North and Central America. The white crabs are *Bythograea thermydron*. (Photo by Dr. J. Edmond, MIT, courtesy of Woods Hole Oceanographic Institute.)

Harps, genus *Harpa*, are recognized by the strongly ridged shell and gigantic animal. The posterior part of the foot can be shed when danger threatens (Pacific; A. Power photo).

Cymbiola vespertilio, the Philippine bat volute, one of the few common volutes (A. Norman photo).

The spectacular animal of this Australian *Amoria* volute is typical of the brightly patterned animals of many volutes.

The large Pacific volute *Melo amphora* laying its eggs. Called the bailer shell, this species was once used to bail water from native canoes (R. Steene photo).

tube-building. It may be associated with a glandular area which may be supposed to produce the tube. The trunk is a very long cylinder extending from the forepart. Anteriorly it is covered with two rows of metamerically placed papillae which demark a shallow groove. Further down the trunk the groove disappears and the papillae become random in arrangement. They appear regularly again just anterior to the girdle region. The papillae are raised epidermal structures bearing cuticular plaques and function in locomotion within the tube. A ciliated band is found in association with the papillae but on the opposite side of the body.

The girdle region is characterized by epidermal ridges, reminiscent of the uncinal ridges of polychaetes, which bear two or more bands of short toothed setae which also aid in movements within the tube.

The trunk terminates in a short segmented region known as the opisthosoma. There are segmentally placed setae occurring, usually singly, at four distinct points so that longitudinal rows can be seen passing down the opisthosoma. The precise function of this region can only be guessed at. Originally it was thought to act as an anchor, but observations of living pogonophores suggest a burrowing function as it can be distended by an increased blood supply.

Pogonophores show many strange and interesting features, but the most unusual is the complete absence of a gut at any stage of development. It has been suggested that the tentacles might group together to form an external stomach, but the absence of enzyme-producing cells in them disproves this theory. The most likely answer is that the worm absorbs dissolved organic matter directly across the body wall; the attenuation of the body would certainly assist in this.

Respiration is also of great interest in these worms. Diffusion is presumably directly across the body wall and oxygen is transported around the body by a well developed blood system. This contains an enormous quantity of hemoglobin in solution, and recent experiments on this pigment show that is has a remarkably high affinity for oxygen and could function at very low oxygen levels.

The reproductive biology has been little studied, but it is known that the sexes are separate and that the male packages the sperm in spermatophores. It is thought that the worms do not leave their tubes to spawn but that the spermatophore is released into the water, where it floats around until it becomes entangled by a large filament to another tube. The spermatophore walls gradually break down in seawater and release the sperm into the water. With luck these will locate the eggs which are held by the female's tentacles. The eggs are usually large and yolky so that they do not need to feed and the larvae develop within the tentacular mass. When they are sufficiently large the larvae are expelled by the adult and presumably settle to the bottom. In some species the eggs are much smaller and there is no evidence of brood protection, so a free-swimming larva may be assumed.

The phylogenetic relationships of pogonophores have been hotly debated for some time. An affinity with the chordates has been suggested because of certain morphological features, but recently, with the discovery of the opisthosoma, a closer relationship with the annelids seems more likely.

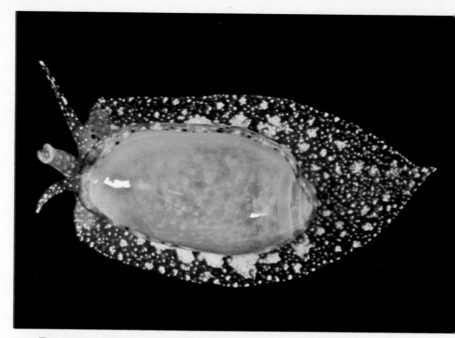

The marginellas are usually small, rather plain shells, though some are large or attractively patterned. Above is *Marginella capensis* (South Africa; T.E. Thompson photo); below is the more attractive *Prunum guttatum* (Caribbean; T.E. Thompson photo).

Head-on view of the venomous *Conus striatus* showing the siphon at the top, the proboscis with eye stalks at the middle, and the foot below (Indo-Pacific; W. Deas photo).

Conus marmoreus, a very common Indo-Pacific cone with a distinctive shell pattern (T.E. Thompson photo).

Phylum Chaetognatha

Chaetognaths, commonly known as arrowworms, are among the most common of planktonic animals although there is one genus, *Spadella*, which is benthic. They are a small group consisting of about fifty species.

In shape a chaetognath is rather like an arrow, hence the common name, and the body can be divided into a head, a trunk, and a tail region. The head is surrounded by a group of recurved spines which are used for capturing prey and several rows of spines like teeth which are arranged around the front of the head. The entire head region can be covered with a hood which is a fold of the body wall. It is believed that this protects the spines when they are not in use and aids in streamlining the body to reduce water resistance.

The trunk and the tail are characterized by the presence of horizontal fins. In most species there is a single pair projecting from the trunk and overlapping the tail region, but in the common genus *Sagitta* there are two pairs of lateral fins. The end of the tail is splayed out to produce a caudal fin. The fins play no role in the locomotion of chaetognaths and merely act as flotation devices. Chaetognaths can swim, but their locomotory powers are small and they spend much of their time drifting with the current. For this reason they are very useful as indicators of the origins of water bodies. Every chaetognath lives in a more-or-less definite region and at a certain depth; temperature is believed to be an important factor limiting their distribution. There are thus cold and warm water species which can be separated on a horizontal as well as a depth basis. In addition, some species are very sensitive to changes in salinity and are limited by this factor.

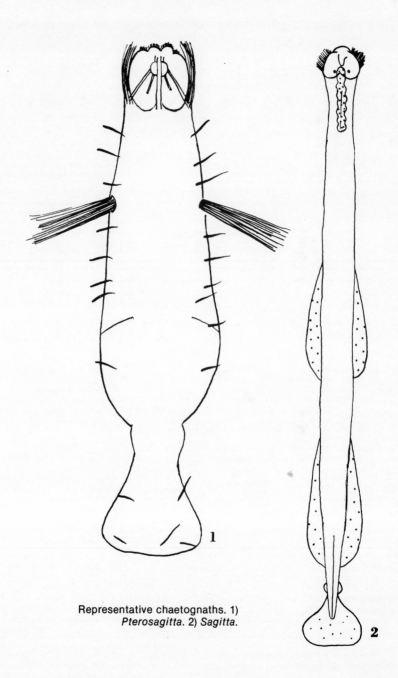

Representative chaetognaths. 1)
Pterosagitta. 2) *Sagitta*.

Two East Pacific cones: above, the uncommon *Conus xanthicus* (A. Kerstitch photo); below, the common California cone, *Conus californicus* (W. Farmer photo).

Some "advanced" gastropods. Turrids (1-5) and terebras or auger shells (6-10) and close relatives of the cones. Pyramidellids and ringiculids are strongly shell-ed forerunners of the opisthobranchs (11-14). Bubbleshells (15-17) are the reduc-ed shells of primitive opisthobranchs.

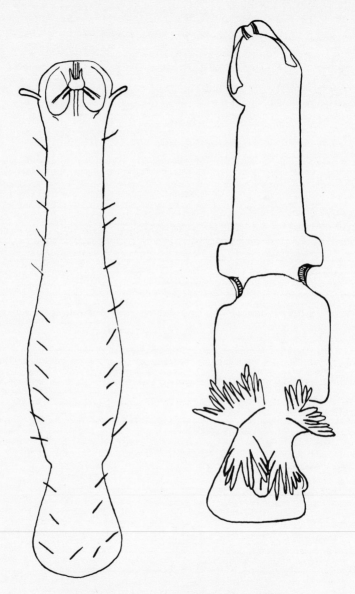

Two species of the chaetognath genus *Spadella*.

Seasonal changes in water masses can thus be monitored by noting the species of chaetognath present. Thus *Sagitta serratodentata* is principally a warm water species living in all tropical and subtropical waters, but it is tolerant of cooler temperate waters and may be carried northward across the Atlantic with the Gulf Stream. Thus in summer it is frequently found off Newfoundland but disappears under the colder winter conditions.

Chaetognaths are all carnivorous, feeding on other planktonic organisms, especially copepods, although there are reports of *Sagitta* catching fish almost as large as itself. Prey is caught by the chaetognath suddently darting forward, withdrawing its hood, and rapidly gripping the animal in its teeth. These hold the prey while the larger spines are brought down, crushing it. After capture the prey is pushed down into the mouth.

Chaetognaths are hermaphroditic, with ovaries in the trunk region and testes in the tail. When mature the sperm are stored in special spermatophores which are released from the body. Self-fertilization is believed to occur in at least some species. Once released the spermatophore adheres to the body and the sperm are released. They then migrate to the female opening and move inside the animal to fertilize the eggs. In those species where cross-fertilization occurs, the chaetognaths presumably pair adjacent to one another.

The eggs may be planktonic as in *Sagitta*, which produces eggs covered in jelly, or they may be carried around with the parent for some time. The larvae are in fact really miniature adults and no metamorphosis occurs.

The affinities of chaetognaths are somewhat obscure. They are superficially similar to aschelminths but the nature of the coelom and their embryology show that they are nearer to the primitive vertebrate stock or deuterostome line, although the relationship is probably remote.

NOTE: The following photos of opisthobranchs and sea slugs are all by T.E. Thompson.

A parasitic white pyramidellid, *Odostomia columbianus* (attched just above the host's aperture), attacking a larger snail. Pyramidellids are offshoots of the sea slug line that retained the shell (East Pacific).

Acteon tornatilis, a burrowing opisthobranch similar to the remote ancestors of the sea slugs. The stout shell and spade-like head are good adaptations for pursuing burrowing worms (East Atlantic).

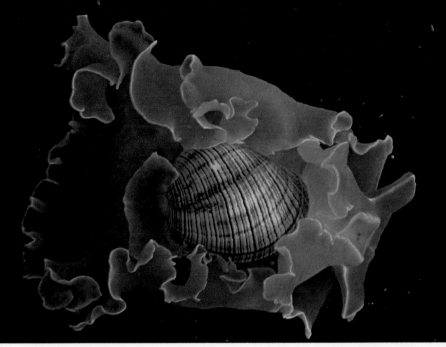

The small, fragile shell of *Hydatina physis* can no longer accommodate all the foot and body. Such opisthobranchs hunt their prey on the surface of the substrate (Pacific).

Some opisthobranchs, such as *Micromelo undata,* have developed virulent defensive glands and flamboyant markings to advertise their distasteful qualities (Caribbean).

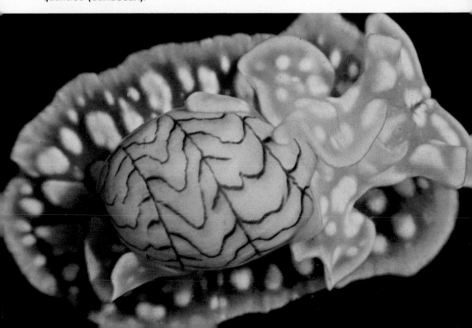

Lophophorate Phyla

Four small phyla are rather unique in possessing what is known as a lophophore, a semicircle of ciliated tentacles used to create water currents for filter-feeding. Whether these phyla are actually related or not is unimportant to us, and most aquarists will only see members of the phylum Bryozoa unless they actively collect in the few areas where the other phyla are exceptionally common.

PHYLUM BRYOZOA: The moss animals or bryozoans, also called ectoprocts, are mostly encrusting colonial animals of very small, almost microscopic size, although colonies can often be quite large. There are many hundred species, but the taxonomy is so complicated as to be beyond the reach of any but the specialist. Virtually all species are marine, belonging to the class Gymnolaemata, with the dozen or so freshwater species placed in the class Phylactolaemata. In true bryozoans the mouth opening is within the circle of tentacles but the anus is outside, a good adaption to prevent fouling of the tentacles.

PHYLUM ENTOPROCTA: This very small phylum (about 60 species) is very similar in appearance to some true bryozoans, although it forms somewhat different types of colonies. The major obvious difference between the two phyla is that in the entoprocts the anus is still within the tentacular ring. A more technical difference concerns the embryological origin of the tissues surrounding the digestive tract, but the similarity of form would seem to rather overwhelm the embryological differences.

PHYLUM PHORONIDA: Phoronid worms. These mostly small worms are often colonial in shallow marine habitats; they produce thin leathery tubes from which the *feeding crowns project until the animals are disturbed. The anus is outside the lophophore. Uncommon in most areas, one or two of the dozen or so species are large enough to be of theoretical interest to aquarists.*

PHYLUM BRACHIOPODA: Lamp shells. The almost 300 species of lamp shells or brachiopods look very much like small clams at first glance but they are not related to molluscs, instead feeding by means of a lophophore. Few living lamp shells will be seen by aquarists, although several rather large species occur in shallow water. Some are burrowers, but most occur attached to rocks or algae in small clumps or as single specimens. Since they are filter-feeders, they usually do not do well in aquaria.

PHYLUM PHORONIDA

The phoronids are a small group of animals all contained within a couple of genera. They belong to a group called the lophophorate coelomates. This rather complicated name refers in part to the feeding mechanism of the group, which involves the use of a lophophore. This is a horseshoe-shaped outgrowth of the body wall which surrounds the mouth. It is provided with hollow tentacles and is ciliated. By the action of the cilia water is drawn through the lophophore and any suspended material, particularly plankton, is caught up in the mucus on the tentacles. The food is then conveyed to a groove by cilia, from where it is passed to the mouth.

The second part of the group name, ' coelomate, ' refers to the presence of a lined body cavity within the animal. Such a cavity is present in many invertebrates and in vertebrates, and its embryological development and final state are important features in the discussion of invertebrate evolution. The coelom in phoronids shows certain similarities to the vertebrate condition in that there are two separate coelomic compartments which can be compared with the mesocoel and metacoel of primitive vertebrate stock. For this reason phoronids, and the lophophorate

In *Umbraculum sinicum* the shell has become reduced to an umbrella-like disc that does not even protect the delicate flattened gill visible beneath the edge of the shell (Pacific).

In aglajids such as *Philinopsis pilsbryi* the shell is so reduced that the animal has to rely on chemical glands for defense. The vivid epidermal markings ensure that inquisitive predators quickly learn to leave it alone (Australia).

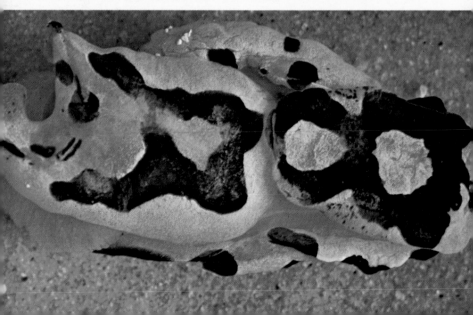

Onchidium damelii, an air-breathing slug related to the common garden slugs. *Onchidium* species are common inhabitants of mangrove swamps and are technically marine animals (Australia).

Several unrelated lines of "sea slugs" have evolved. For instance, the prosobranch sea slug *Lamellaria perspicua* is more closely related to the cowries than to true opisthobranchs (East Atlantic).

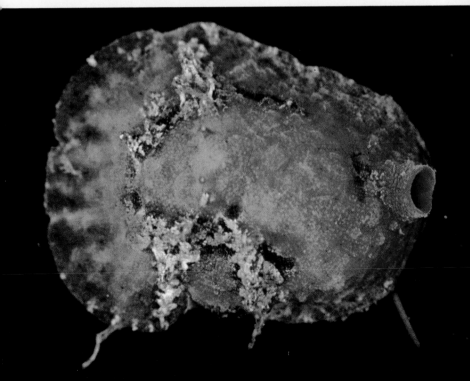

coelomates in general, have aroused the interest of biologists and their embryology and larval development have been studied in detail.

All phoronids are marine, living in shallow water to about 200 m. Most live in a blind-ended leathery tube which is either buried in sand or attached to rocks, but some bore into mollusc shells. They are worm-like in shape, with a slight swelling at the posterior end and no appendages apart from the lophophore, and show many features associated with their sessile way of life. In their feeding behavior, for example, they rely on the food coming to them rather than actively hunting. The gut is U-shaped so that the anus opens near the mouth. This means that the tube does not get fouled. The anus is situated outside of the lophophore ring so that there is no danger of the animal ingesting fecal material.

Most phoronids are fairly small, seldom exceeding 10 cm in length, but *Phoronopsis californica* is up to 30 cm long. This is a bright orange species living singly in estuaries. A small Australian species, *Phoronis australis*, shows an interesting adaptation. This species lives in delicate tubes which are embedded in the wall of the tube of *Cerianthus*, a tube-dwelling anemone.

The majority of phoronids are hermaphroditic. There are no clearly developed gonads, but instead the cells aggregated around the ventral blood vessel develop into testis or ovary according to their position. Eggs and sperm are released into the coelom and leave the body via the kidney opening on the head. In most cases fertilization is external and is followed by the development of a planktonic trochophore larva. In a few cases the eggs are brooded within the lophophore ring where they are held in place by the inner tentacles. The larvae remain in the plankton for a few weeks, after which time they undergo a rapid metamorphosis and sink to the bottom where they secrete tubes and take up an adult existence. *Phoronis ovalis*, a species found in large aggregations, is capable of asexual reproduction. New individuals are produced by budding or in extreme cases simply by splitting in half.

PHYLA BRYOZOA
AND ENTOPROCTA

The phylum Bryozoa constitutes one of the major animal groups although it has aroused little interest except among specialists. At one time the bryozoans, or the Polyzoa as they were also known, included the Entoprocta as well, but as we shall see below these are not really closely related. The use of the term Ectoprocta has been suggested to cover the remaining bryozoans but is little used.

Bryozoans are colonial animals consisting of many united individuals or zooids. These are very small, seldom more than a millimeter in length, although the colony may extend over several feet in diameter. In most zooids the body is encapsulated in an envelope which contains an opening for the extrusion of the lophophore. In marine bryozoans of the class Gymnolaemata the box-like envelope is largely chitinous. The shape of the case varies, ranging from vase-like to tubular so that the colony as a whole may be very beautiful. The chitinous layer overlies a layer of calcium carbonate giving the whole a very rigid structure. In many species the opening for the lophophore is provided with a lid or operculum which closes down when the zooid is not feeding.

The exoskeleton is secreted by the underlying epidermis. The rigid nature of the casement prevents any locomotion and the only muscles of any importance are those which draw in the lophophore.

Internally, the zooid consists of a large coelom which surrounds the gut. The lophophore is circular and is made up of a ridge bearing a few to many ciliated tentacles which fan out when the lophophore is extended. The presence of the lophophore links bryozoans with other groups of animals such as the phoronids, which also use this feeding device. The mouth lies at the center of the lophophore, and food particles filtered by the tentacles are channeled down into the mouth.

Perhaps because of their very small size bryozoans possess no respiratory, circulatory, or excretory organs.

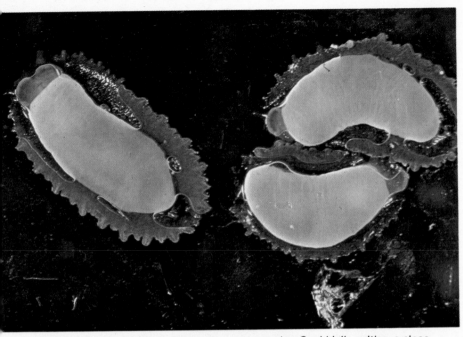

Ventral and dorsal views of the pulmonate sea slug *Onchidella celtica*, a close relative of *Onchidium* (East Atlantic).

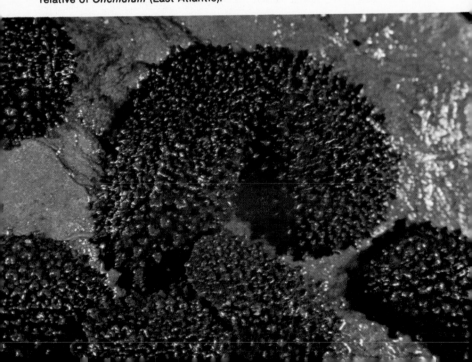

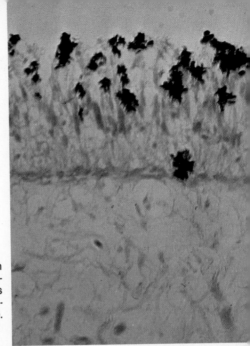

Section through the skin of an Australian pleurobranch, *Pleurobranchus peroni*, stained to show sites of sulphuric acid storage (black blotches) near the surface.

A pleurobranch sea slug, *Berthella plumula*, near its natural food, the sea squirt *Botryllus schlosseri*. *Berthella* depends on emission of strong sulphuric acid if attacked; most fish hate acidified food (East Atlantic).

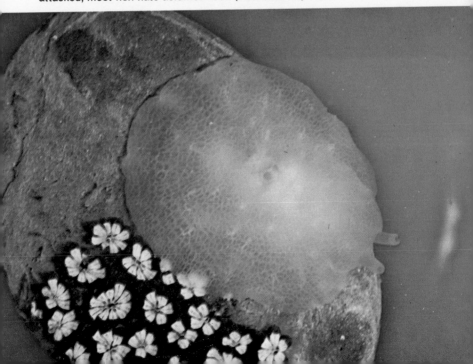

There are very many species of colonial Gymnolaemata, making them very abundant marine animals. Some have been recorded from great depths but most are found encrusting on rocks in the littoral or attached to seaweed or even to other animals. Some types such as *Bugula* are economically important as ship foulers.

A few species are not encrusting but have a stoloniferous life form. The stolons are produced from modified zooids which attach to the substrate by means of a normal zooid. The vast majority of bryozoans are encrusting, however. The orientation of the body is such that the dorsal surface makes contact with the substrate, the lateral surfaces are attached to other zooids, and the ventral surface contains the opening for feeding.

The actual pattern of growth varies considerably. The calcareous forms such as the very common *Membranipora* form encrusting colonies, while other calcified forms produce erect coral-like growths. Among the non-calcified varieties *Bugula* takes on a plant-like appearance, looking like a branching seaweed.

Each zooid can communicate with its neighbor through pores in the side walls of the case. These contain epidermal cells so that only a slow rate of diffusion is possible through the colony.

In most gymnolaemates the colonies are polymorphic. This means that certain zooids are modified to perform a particular function. Such zooids are known as heterozooids. The most common type is the one used in stolon formation or for attachment discs. Avicularia are another common type of heterozooids. These may be sessile or stalked and in the latter case they are flexible. They are very common in *Bugula* and are believed to function as cleansing agents removing small particles, such as larvae from other organisms, from the surface of the colony.

Some bryozoans have a vibraculum in which the operculum is modified to form a long bristle which presumably functions like the avicularia to fend off unwanted material.

There are also modified zooids with a reproductive function. Most marine bryozoans are hermaphroditic. Although eggs and sperm may be released together, there is some evidence that male gametes develop first. Only a small number of eggs are produced. Some species are known to self-fertilize, but it has been proved that cross-fertilization occurs in at least some species.

Very many bryozoans brood their young. In some cases the eggs are incubated within the coelom or they may be held in place by spines. *Bugula* has a special ovicell which is an external chamber like a hood on the outside of the skelton. The coelomic method is the most common, however, and some individuals degenerate so that they are merely egg nurseries.

Wherever fertilization occurs and whether brooding takes place or not, the egg gives rise to a modified trochophore larva. Most are non-feeding stages lasting for only a few hours prior to settling. The larva attaches to the substrate by the eversion of an adhesive sac which produces a sticky secretion. The larva then develops into the adult. The first zooid is called an ancestrula; by budding it gives rise to new individuals. The colony is then build up by asexual reproduction.

Thus a bryozoan is a sedentary colonial coelomate with each zooid feeding by means of a typical lophophore.

Entoprocts, on the other hand, do not possess a true coelom and belong to a large group of pseudocoelomate animals. Nearly all entoprocts are marine, living attached to rocks, pilings, etc. or epizoically on other animals. Most are colonial. The body consists of a structure known as the calyx which is attached to the substrate by a stalk. There is a ring of tentacles around the calyx which encloses both the mouth and the anus, hence the term Entoprocta or inner anus, as opposed to the Ectoprocta or outer anus.

PHYLUM BRACHIOPODA

Brachiopods belong to a group of animals known as lophophorate coelomates, which also include the bryozoans

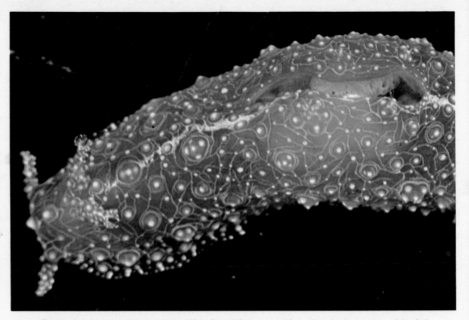

Green herbivorous sea slugs like *Petalifera petalifera* live in lagoons with rich growths of green plants. There is no evidence that the green pigment actually comes from the food (Caribbean).

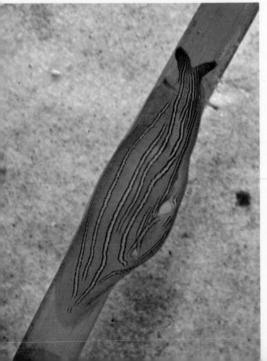

The aplysiomorph sea slug (sea hare) *Phyllaplysia taylori* bears markings that render it well concealed when feeding on eel grass (East Pacific).

Many sea hares, such as this *Aplysia punctata,* secrete a purple dye when distrubed; the dye tastes foul to human tongues and may be unpleasant to fish (East Atlantic).

Not all sea slugs have lost the shell during their evolution. At left is *Oxynoe antillarum* (Caribbean), whose stiletto-like radular teeth puncture algal cells so the sap can be sucked out; at right is the bivalved *Berthelinia,* whose shells were once thought to be clams rather than snails (Caribbean).

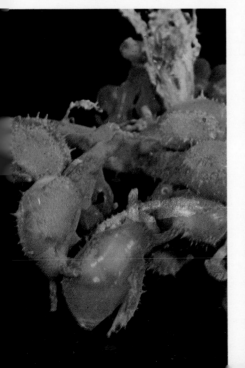

Views of the shells of representative brachiopods. Above is *Terebratalia transversa* from tidal waters of California. Below is *Discinisca cumingii*, an unusually rounded species with a coarse periostracum like that of many molluscs. (K. Lucas photo at Steinhart Aquarium.)

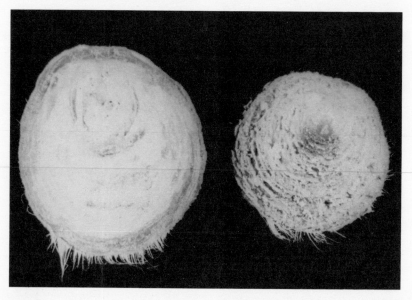

and the phoronids. These are related by the presence of a lophophore, which is a specialized feeding device.

Externally brachiopods bear a resemblance to bivalve molluscs in that they have a calcareous shell consisting of two valves and a device equivalent to the molluscan mantle. Indeed, at one time brachiopods were classified as molluscs. Once one examines the internal anatomy of a brachiopod, however, it becomes obvious that the relationship is only superficial. The body lies in a dorsoventral plane within its shell and is not compressed laterally like a bivalve. The ventral valve, which is usually larger, is nearly always attached to the substrate either directly or by means of a stalk known as the pedicle. The general shape and appearance of brachiopods have led to their common name of lamp shells, which relates to their resemblance to a Roman lamp.

Brachiopods are divided into two classes according to the articulation of the shell. In inarticulate brachiopods such as *Lingula* the valves are only held together by muscles and the shell can gape open quite widely. Articulate brachiopods have processes on the valves which cause them to lock together so that the valves can only open a little bit. There are also differences in the shell itself. This is secreted from the body wall and in inarticulate brachiopods consists largely of calcium phosphate, whereas in articulate forms it is mainly calcium carbonate. In both instances the mineral layer is covered with a chitinous periostracum.

Brachiopods feed by means of the lophophore. This is akin to a circle of tentacles around the mouth which filter suspended material which is drawn through the lophophore by ciliary currents. To increase the surface area for catching food the lophophore has anterior tentacle-bearing arms which may be coiled. Once trapped in the lophophore, food is passed to a groove and thence to the mouth.

Brachiopods are an entirely marine group living mainly in shallow waters on the continental shelf. Most species are attached to rocky surfaces. The pedicle is usually a short stumpy piece of connective tissue which is positioned in the valves such that the brachiopod comes to lie upside down in

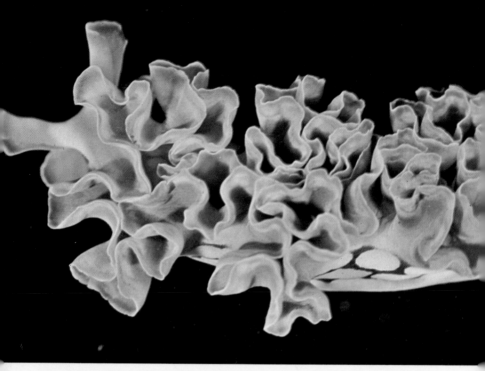

Tridachia crispata, a Caribbean sea slug that cultivates green algae in skin folds. It is often seen basking in the sunlight on top of corals.

The rather undistinguished *Adalaria proxima* and its prey, the bryozoan *Electra pilosa.* The development and life cycle of this sea slug have been extensively studied (East Atlantic).

The red warning colors of *Chromodoris splendida* appear black under the sea (Australia).

Hexabranchus sanguineus, a large coral reef nudibranch that swims with the motion of "wings" from the mantle (Pacific).

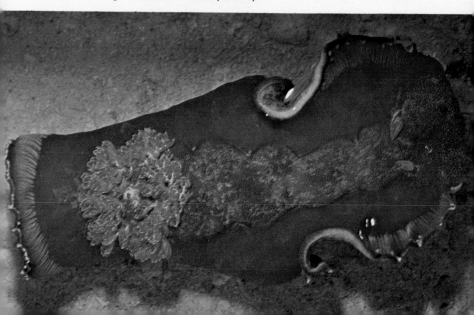

a horizontal plane. Movement is very restricted in these brachiopods and completely so in those which attach directly without a pedicle. The situation is somewhat different in *Lingula*. Unlike other brachiopods this animal lives in vertical burrows in sand or mud. The pedicle extends to the bottom of the burrow and is buried in sand. It is muscular so that the animal can extend from the burrow to feed.

Most brachiopods are dioecious, producing gametes which are released into the seawater via the kidney opening. Several animals spawn at the same time, and fertilization occurs in the water, although there are a few species which brood their young. The larva is a free-swimming feeding stage which differs in type according to the class. Inarticulates produce larvae resembling miniature adults which eventually settle and grow into full sized brachiopods, but the articulate larva undergoes a metamorphosis before reaching the adult stage.

Brachiopods, and *Lingula* in particular, are of considerable interest to paleontologists since they are living fossils. This means that species very like the modern ones have been found in rocks of great age. Fossil brachiopods have in fact been recorded back to pre-Cambrian times, and *Lingula* itself is known from the Ordovician. Because of the wealth of fossil material but relative scarcity of living species, it is thought that brachiopods are on the wane and that they reached their evolutionary peak in the Paleozoic and Mesozoic eras.

Phylum Mollusca

The familiar snails and clams are typical representatives of the Mollusca, the second largest phylum. With well over 120,000 species (and a taxonomy so confused it is really difficult to determine exactly how many species there really are), it ranks second only to the arthropods, unless the belief by some that the nematodes contain over a million species is correct. Regardless, most of its members are large and easily placed in the phylum, with some being of great economic importance.

Although the phylum name is derived from the Latin ' mollis ' or soft, this is in reference to the soft body only, as most species are covered with a heavy calcium carbonate shell secreted by the specialized mantle that covers the body. The foot is usually large, broad, and smooth, but sometimes it is small or divided and almost unrecognizable. In most species that do not feed by filtering water the teeth are very small and in multiple rows forming an organ called a radula that is able to scrape or cut off small pieces of food. In the filter-feeding bivalves the radula is absent and the organization of the body differs greatly from that of a snail or even a chiton.

The number of classes recognized varies among authorities from five to seven. We will take the larger number as more likely correct.

Class Polyplacophora: Chitons. Primitive molluscs with a broad foot and a shell usually composed of eight rather similar overlapping valves bordered by a thickened layer of tissue called the girdle. The thousand species are often small and inconspicuous, but a few are large and brightly colored. They are rather sedentary and scrape microscopic algae for food. All are marine.

311

The brightly marked papillae on the head and around the gills of *Polycera tricolor* (above) contain defensive glands that take a blue stain (below). Such defensive glands are common in nudibranchs (East Pacific).

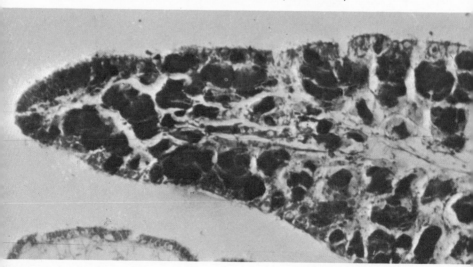

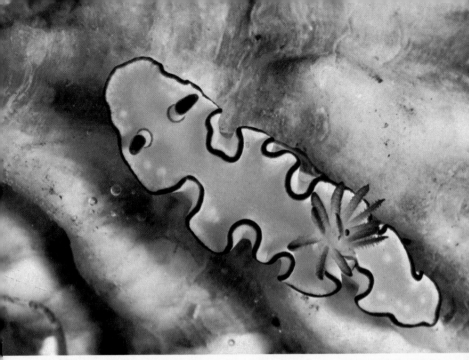

Although most dorid nudibranchs feel prickly or slimy to the touch, *Casella atromarginata* feels like tough wet leather. Little is known about its mode of life (Indo-Pacific).

Warning colors in the form of bright yellow-tipped glandular papillae are common in many European dorid nudibranchs, such as this *Limacia clavigera*. Are such color patterns merely coincidences?

*Class **Aplacophora**: Solenogastres. Worm-like molluscs of small size and usually rather deep-water marine habitat, with a reduced foot and no shell, only spicules of calcium embedded in the mantle. Often included with the chitons because of internal anatomy, they are perhaps best considered as a separate class. The few (less than 200) species are poorly known. When Polyplacophora and Aplacophora are combined they are known as Amphineura.*

*Class **Monoplacophora**: Neopilinas. Limpet-like molluscs with a cap-like single-valved shell but with repetition of some internal structures. Usually hailed as a 'living fossil' since the discovery of a living species (several more are now known) a quarter of a century ago, it has sometimes been suggested that neopilinas are not really very primitive molluscs as usually asserted, just very primitive gastropods. No species have been found in shallow water as yet, and all are small.*

*Class **Gastropoda**: Snails, slugs, nudibranchs. Over 100,000 species of snails and allies are known, with many more being described every year. These are the dominant molluscs, being found in marine, freshwater, and terrestrial habitats, often in large numbers. Size varies from very small to quite large, and a coiled single-valved shell may be present or absent. In most groups there is a distinctly twisted nervous system, the result of a complicated process called torsion. Radular teeth are usually well developed, although they are occasionally absent, the feeding habits varying from meek scrapers of algae to active carnivores with venom glands. Three major subclasses are usually recognized, each very variable in shell and general appearance.*

SUBCLASS PROSOBRANCHIA: Most marine snails with distinct shells belong here. The sexes are usually separate, and an operculum is usually present (absent in quite a few families, however). It is to this group that most of the shells popular with shell collectors belong, such as the cones (Conidae, with about 300 species), cowries (Cypraeidae, about 180 species), and volutes (Volutidae, about 200 species). With few exceptions, marine limpet-like shells also belong in this subclass. Over 60,000 species.

SUBCLASS OPISTHOBRANCHIA: Sea slugs and nudibranchs. Hermaphroditic snails and slugs, the shell usually absent, reduced, or rather simple and thin. Marine with very few exceptions. This group includes some attractive snails with very large animals (compared to the shell size) and the popular nudibranchs, as well as several other groups including some very small planktonic forms.

SUBCLASS PULMONATA: The familiar freshwater and terrestrial snails and slugs belong to this subclass with very few exceptions. There is no operculum, the sexes are together, and the shell is usually thin, simple in structure, and coiled or absent, Garden slugs are of course just snails with the shell reduced, internal, or absent. There are about 35,000 species of pulmonate snails, with a few marine representatives.

Class Scaphopoda: The tusk shells or elephant tooth shells. Simple tubular shells open at both ends, the animal with a large conical foot and feeding tentacles. Most of the 300-400 species are small, but a few are over 100 mm long and brightly colored. All are marine, many in deep water.

Class Bivalvia (Lamellibranchia): Bivalves, the clams and oysters. About 20,000 species comprise this large class easily recognized by the presence of two valves joined by a hinge and often by complex arrangements of hinge teeth. Although the head and radula are absent, the typical molluscan mantle is present, forming two siphons which function in filter-feeding. The foot is usually quite large and compressed, enabling the animal to rapidly burrow through sand and silt. Although there are many freshwater species, most are marine.

The economic importance of this class is large, as it includes the clams, oysters, scallops, mussels, and quahogs on the plus side and the destructive shipworms on the negative side. Because they are filter-feeders, few are of

Similar yellow-tipped papillae can be seen in this *Trapania maculata* (also European). Perhaps such patterns serve to speedily "educate" potential predators to avoid such unpleasant morsels.

Fresh spawn mass of *Archidoris pseudoargus*. Such a mass may contain hundreds of thousands of eggs and be more visible than the nudibranch that laid it (East Atlantic).

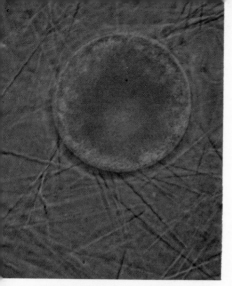
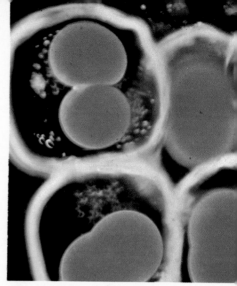

Left, fertilization of an egg of the sea hare *Bursatella leachi;* numerous sperm are also visible on the slide (East Atlantic). Right, fertilized eggs of *Dendrodoris miniata* undergoing their first cell division (Australia).

Shelled veligers of *Hancockia burni* are typical of the early swimming stage of most nudibranchs. The larvae will settle and metamorphose only where the bottom "smells" right (Australia).

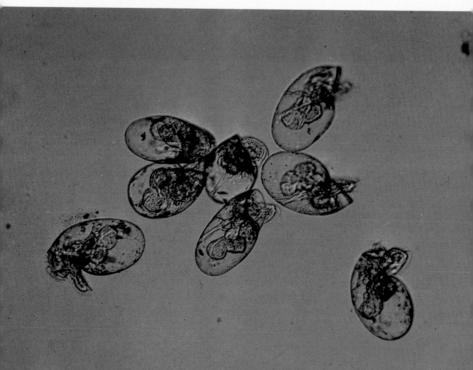

great interest in the aquarium and, because they are mostly brownish or grayish, there are also few of interest to shell collectors.

Class Cephalopoda: *Literally meaning ' head footed,' this class of course comprises the squids and octopuses and their allies. The arms are formed from a highly modified and divided foot, while the shell is usually lacking or represented by mere remnants. The 700-800 species are all marine, many being very active predators.*

Molluscs seem to be one of those groups that for some reason have never proved very popular in the aquarium. Although a few cowries and cones are sold in some shops and collectors enjoy great success with various other families of prosobranch snails, only a few scallops and allies are kept in tanks, while octopuses represent the cephalopods. There are a great many marine snails that would be suitable for the aquarium if they were stocked by dealers, especially the carnivores that will take dead food without any problems. Some of the herbivores, such as the smaller conchs, would make excellent algae-scrapers, while even the limpets and chitons (although often difficult to keep) have their attractive features. Bivalves in general are difficult to keep because they require fine food in suspension and because many or most spend their time burrowed below the surface with only the tips of the siphons visible. Surprisingly, the popular nudibranchs are not actually good tank inhabitants because many of them are very delicate and require very unusual diets, such as living soft corals or sponges; yet for some reason—probably the often bizarre colors—nudibranchs probably are the best known marine molluscs among aquarists.

The literature on identification of molluscs is immense and beyond the reach or interest of the average aquarist. There are many local manuals dealing with shells, but few that emphasize the living animal, which is often more interesting and colorful than the shell. There is an especially large literature concerned with shell collecting, many shells of popular groups being quite beautiful and expensive.

MINOR MOLLUSCAN GROUPS

Generally when one considers the phylum Mollusca three main types spring to mind: the Gastropoda or snails, the Bivalvia or clams, and the Cephalopoda or squids. These do not complete the assemblage, however, and there are other smaller groups to consider.

The first of these is the **Monoplacophora**. Prior to 1952 this class was known only from very early fossil records, but in that year ten living specimens were dredged from great depths in a Pacific Ocean trench off Costa Rica. Since that time a few more specimens of several species have been collected.

The name Monoplacophora refers to the single symmetrical shell borne by these molluscs. This varies in shape and has led to the class being grouped with both chitons and gastropods at one time or another. Examination of the live material, however, confirms the view that the monoplacophorans warrant a class of their own.

The specimens of the only living genus, *Neopilina*, are small molluscs ranging from a few millimeters to little over three centimeters long and anatomically resembling a cross between a chiton and a gastropod but like a limpet externally. The shell is large and symmetrical and in a single piece unlike that of a chiton. The body is remarkably chiton-like. The head is greatly reduced and the foot is very broad and surrounded by a pallial groove corresponding to the mantle cavity. The pallial groove houses five pairs of unipectinate gills which are much branched. The digestive system shows both chiton and gastropod characteristics, but the muscular system is unlike that of any other mollusc in that there are eight pairs of pedal retractor muscles. This multiplication of organs is shown in other systems as well; there are, for example, six pairs of nephridia or excretory organs.

The replication of organs has led to many authors speculating that *Neopilina* shows metamerism. This means that, like an annelid, it is basically made up of like segments which are duplicated to give a whole animal. This is a very important concept in our understanding of molluscan

319

Trinchesia caerulea, typical of many aeolid nudibranchs specialized for feeding on marine coelenterates. These slugs are not only immune to the nematocysts of the prey but can convert them to their own use (East Atlantic).

Aeolidia papillosa, one of the world's largest aeolid nudibranchs, is also one of the most cosmopolitan, occurring in the shallow seas of Europe and on both sides of the Americas.

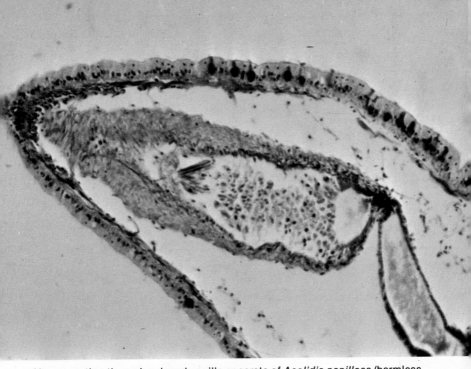

Above, section through a dorsal papilla or cerata of *Aeolidia papillosa* (harmless to human skin) showing the sac in which nematocysts are stored. Below, greatly enlarged view of some of the nematocysts ejected from one of the cerata after it was nipped with forceps.

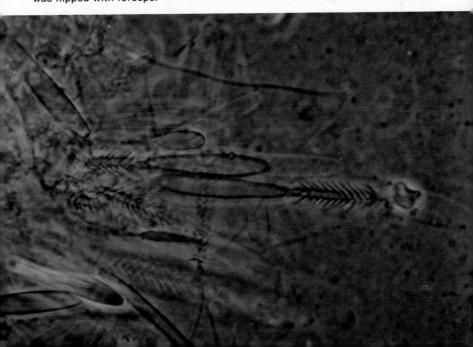

origins. There seems little doubt that there is a definite affinity between molluscs and annelids since their embryology and larval development are so similar. There has been much dispute, however, as to which came first and where they came from. If *Neopilina* does show metamerism, then this points to an annelid origin for molluscs; many eminent zoologists favor this view. There is some doubt, however, as to whether *Neopilina* is metameric. It is true that various organs are replicated, but they are not repeated in the same numbers. If the animal is metameric one should be able to pinpoint and describe a distinct *Neopilina* segment, which is not possible. Furthermore, while there are several pairs of gills they are not of the primitive bipectinate type and their replication could be a secondary characteristic. Also it is strange that there is no other evidence at all of metamerism in any other mollusc or in the embryology of *Neopilina*. So, while it remains of considerable interest as a living fossil, it seems likely that *Neopilina* does not make the impact on molluscan zoology that was once thought.

One of the groups that *Neopilina* resembles is the **chitons**. Chitons belong to the class Polyplacophora (meaning many plated), which is related to an aberrant worm-like group, the Aplacophora. Chitons are all marine and are characterized by their shell. This is divided into eight overlapping transverse plates which are wide and convex. These are fitted with projections such that each plate overlaps the one behind. The shell is covered to varying degrees by a fold of mantle and in some forms, such as *Amicula*, the whole plate is covered with mantle. Unlike some other molluscs the shell consists of only two layers, an upper tegmentum of an organic base impregnated with mineral and an underlying layer of calcium carbonate. In some species the heavy girdle extends well beyond the shell margins and may be very hairy or scaly. This is believed to help in the camouflage of chitons, which are generally rather dully colored creatures.

The foot is large and broad and functions in locomotion and in adhesion. Propulsion is achieved entirely by muscular contraction of the foot, but chitons are rather sluggish animals and do not move about much. Ordinary attachment to the rock is achieved by the foot, but if the animal is disturbed the mantle girdle is employed as well. It is clamped down very hard against the substrate and the inner margin raised so that a degree of suction is achieved. A further protective method is related to the division of the shell into plates. If a chiton is dislodged it is able to curl up into a ball, although this probably happens rarely under normal circumstances.

As a function of their sedentary existence, the mantle cavity has developed as two grooves down the side of the body, each containing a large number of gills, the precise number of which varies according to the species.

Most chitons are herbivorous, scraping algal material from the rock on which they live. They are mainly nocturnal and many appear to have a definite home which they return to after feeding expeditions, in a manner similar to limpets.

All chitons are dioecious and contain a single median gonad. Specialized tubes have been developed to transport the ripe gametes to the exterior. Copulation does not occur, and fertilization usually takes place in the mantle cavity, which is, of course, part of the external environment. In order to achieve a high success rate by this method chitons have to be gregarious. Eggs are laid singly in a protective envelope or in chains. Some species retain the fertilized eggs in the gills, where early development takes place; in *Nuttallina* the complete process of development takes place here. *Callistochiton viviparus* gives birth to young individuals which have undergone development in the ovary.

Development is typically molluscan with a free-swimming trochophore always present except where brooding takes place. There is, however, no veliger, and this is thought to be a primitive feature.

The third small molluscan class is the Scaphopoda. Scaphopods are commonly called **tooth shells** or elephant tusk shells because of their shape. The shells are tubular, tapering, and slightly curved with an opening at each end.

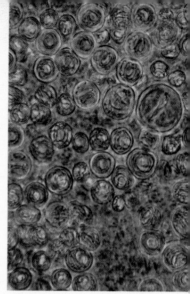

The venomous aeolid *Glaucilla marginata* feeds on *Physalia* and stores the nematocysts (right). This nudibranch can cause painful stings when handled by humans (Pacific).

Many aeolids, like the *Eubranchus farrani* shown here, possess vivid colors that warn inquisitive fishes of the presence of nematocysts in the tips of the dorsal papillae (East Atlantic).

The Caribbean *Learchis poica* feeds on the stinging hydroid *Pennaria distichia* and can utilize its nematocysts; it is potentially dangerous to handle.

Dirona albolineata, a "false" aeolid (actually an arminacean) that looks venomous at first glance but actually has a different diet and cannot utilize nematocysts; note the parasitic copepod *(Splanchnotropus)* on its back (East Pacific).

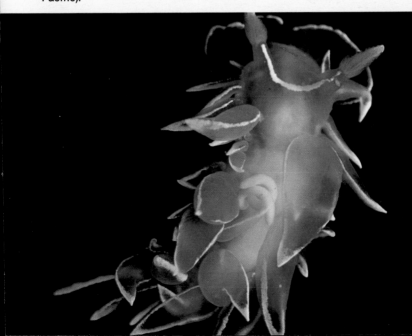

The radula of *Octopus hummelincki* is similar to the radula of other cephalopods and most gastropods. Notice that the shape of the central tooth is quite different from that of the lateral teeth in each row (W. E. Burgess photo).

Cleaned plates of the chiton *Cryptochiton stelleri* (K. Lucas photo at Steinhart Aquarium).

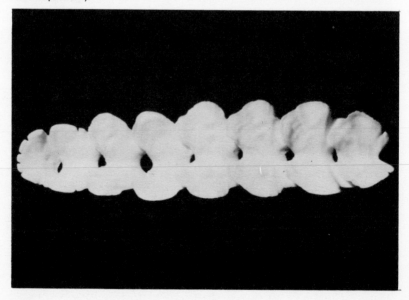

Most average about 5 cm in length, but some, such as the genus *Cadulus*, are much smaller. *Dentalium vernedi* from Japan is one of the largest species, achieving over 13 cm in length.

Scaphopods are bilaterally symmetrical animals with a large foot used for digging. The animal lies head downward in the sand with the tube placed at an angle to the surface. The smaller posterior opening is above the surface and provides the inlet and outlet for water.

Feeding in scaphopods is little understood. They possess a radula, and it is likely that they feed on small organisms and organic debris in the sand.

Most scaphopods live below tide mark so that they are seldom encountered. Some live at great depths and, judging by the numbers of shells washed up on beaches, they are not an uncommon group. The shells have in fact been used for jewelry, American Indians using them in particular.

CLASS GASTROPODA

The class Gastropoda is the largest group in the phylum Mollusca. While most people can recognize a mollusc such as the familiar garden snail or slug and can readily put a name on it, it is difficult to pinpoint the basic features which typify a mollusc since there has been so much variation from the basic stock. This is particularly true of the gastropods, which contain both the snails and the slugs and are typified in the marine environment by the limpets, periwinkles, cones, cowries, nudibranchs, and thousands of other types.

Because of this variety of form most authorities do not attempt a description of an idealized gastropod, but instead refer to a hypothetical ancestral mollusc from which all the present day classes can be derived. A complex description is obviously not desirable in this context, but the main features must be discussed briefly before the individual modifications shown by different gastropods can be understood.

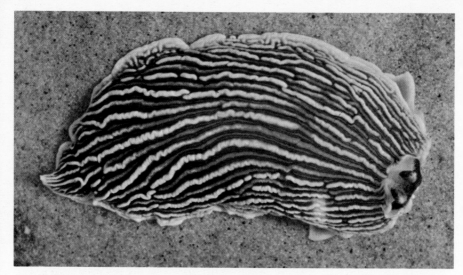

The flattened, burrowing arminacean nudibranch *Armina californica* feeds on the sea pansy *Renilla* in shallow water of the East Pacific.

The delicate-appearing yet voracious dendronotacean sea-slug *Tritonia festiva* feeds on either gorgonians or sea pens, digesting all the soft parts including the nematocysts (East Pacific).

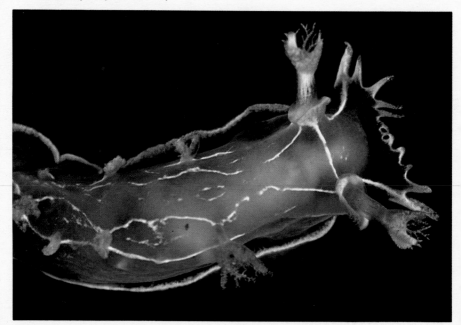

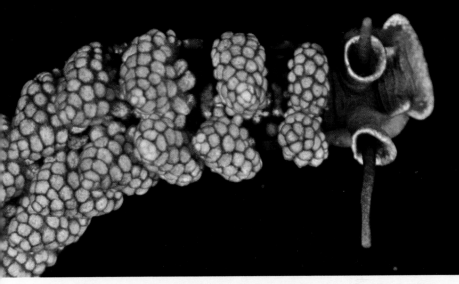

Doto fragilis is a member of the Dotoidae, which includes some of the smallest nudibranchs. Its dorsal papillae are sculptured and the tentacles can be retracted; they feed on delicate hydroids (East Atlantic).

The flamboyant Caribbean dendronotacean *Bornella calcarata* swims by sinuous lateral contorsions of the whole body. The tree-like dorsal processes aid both digestion and respiration.

The ancestral mollusc is thought to have been a slow-moving snail-like animal with a broad creeping foot and a low conical shell. The muscular foot is fused with an anterior head that bears sense organs, and the whole can be contracted inside the shell in times of danger. The remainder of the body is nonmuscular and never leaves the shell; it forms a visceral mass, attached to which is a fringe or skirt known as the mantle. This is very important in the life of the animal. Firstly it secretes the shell and secondly it envelops a space known as the mantle cavity in which several important organs are housed. Water is drawn into the cavity, which is placed ventrally in this creature, and ' tested ' by the osphradium, a chemical sense organ. Water then passes over the gills or ctenidia, which produce the water current. The exhalent current carries with it nitrogenous wastes and genital products. Two hypobranchial glands around the rectum coat the feces in mucus before it is expelled as pellets.

Feeding is achieved by means of the radula, a device unique to the molluscs. It is a broad tongue armed with rows of teeth which are drawn over the bottom to scrape off food particles. As the surface wears down, the radula moves forward and new teeth replace the old ones from behind.

The primitive mollusc, then, was a browsing shelled animal not dissimilar to the modern snails but with one major difference. An essential feature of gastropod phylogeny is torsion, a process whereby the visceral mass is twisted through 180° with respect to the head and foot. This occurs early in the larval development of the mollusc and possibly originated as a protective device for the young animals. With the mantle cavity at the back, the head could only be withdrawn after the foot, which might be too late if a hungry predator was near. After torsion the mantle cavity is at the front so that the head can be immediately protected. There are flaws in this argument, however, as most of the predators of molluscan larvae are fish such as herring, which engulf the animal whole, so the shell would afford little protection.

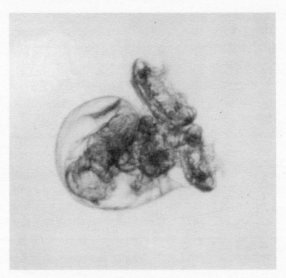

Veliger larva of *Conus purpurascens* (J. Nybakken photo).

Whatever its origin, torsion has resulted in profound changes in the morphology and behavior of the adult together with certain problems. The mollusc that has undergone torsion finds its mantle cavity around the head end, an ideal situation for protection but rather unfortunate in that the openings of the rectum and kidney now come to lie over the head. So, while the animal can now utilize its osphradium to the full in sampling water in front of it, this same water will be contaminated with wastes and genital products. Much of the evolution within the snails can be related to modifications to compensate for this problem.

Another typical gastropod feature absent from the ancestral mollusc is the spiral shell. The twisted growth pattern must not be confused with torsion as it is an entirely separate process related to the size of the visceral mass. As this grows larger the shell would become wider and heavier until it became unmanageable unless the increase in size could be kept compact and well balanced. The development of an asymmetrical spiral growth pattern enables this to take place.

With this knowledge of the basic molluscan features it is

An assortment of colorful Australian bivalves (K. Gillett photos). 1) *Anomia* (jingle shell); 2) *Bassina* (venus); 3) *Neosolen* (razor clam); 4) *Tellinota* (tellin); 5) *Hemicardium* (cockle); 6) *Austromactra* (surf clam); 7) *Chlamys* (scallop); 8) *Neotrigonia* (brooch shell); 9) *Veletuceta* (cockle); 10) *Fulvia* (cockle); 11) *Plebidonax* (coquina).

12) *Amusium* (scallop); 13) *Laciolina* (tellin); 14) *Trisidos* (mussel); 15) *Gloripallium* (scallop); 16) *Fragum* (cockle); 17) *Lioconcha* (venus); 18) *Regozara* (cockle); 19) *Tapes* (venus).

possible to examine the different subclasses of gastropods in more detail. The first subclass is the Prosobranchia, which contains most of the familiar intertidal marine gastropods from the limpets to the whelks. Solution of the sanitation problem controls the final shape and form of the animal.

In the more primitive prosobranchs the direction of water flow is altered so that the waste products are deflected away from the head and out of the animal via a secondary opening. This occurs in the keyhole limpets such as *Diodora*. This genus has a conical secondarily symmetrical shell reminiscent of that of the true limpets, but with a hole at the apex. Water is drawn into the mantle cavity in the usual way but is expelled through a siphon formed from the mantle which projects through this new opening. The abalone, *Haliotis*, has a similar adaptation but with a series of holes in the flattened shell.

The true limpets such as *Patella* have solved the problem more drastically. Their conical shells have no openings, but there are no ctenidia either. Instead a series of pallial gills have been developed around the foot, avoiding the issue altogether.

Most prosobranchs have undergone a more dramatic change in structure to solve the problem. They have lost the right gill and by differential growth the whole symmetry of the internal organs is changed. Water enters the mollusc to the left of the head and leaves by the right side. In many groups of snails the current is improved by the development of an inhalent siphon from the mantle. This type of prosobranch has undergone considerable modifications. Primitively it is, perhaps, most like the periwinkles of the littorinid family with their top-shaped spiral shells, but some have been secondarily changed and become conical like the limpets (slipper shells (*Crepidula*) and the Chinaman's hat (*Calyptraea*) for instance), and others have developed slender pointed spires. A few have evolved a pelagic way of life.

While the prosobranchs show such a variety in their mode of life, the next subclass, the Opisthobranchia, is more conservative. There are three main adaptations, all closely interrelated. Basically opisthobranchs are characterized by a reduction of the shell and visceral mass and, as a consequence, of the mantle cavity and its associated organs. Shelled forms, the most primitive, are somewhat similar to the prosobranchs except that the foot is extended into parapodia which are used to propel the mollusc across or into sand and the body is generally more streamlined. *Aplysia*, the sea hare, is a good example of a fairly primitive opisthobranch. At first glance the shell appears to have been lost, but closer examination reveals it as a small hard disc on the back of the animal. The next stage of evolution is to lose the shell altogether. This has occurred in the nudibranch gastropods, a group of very colorful, highly specialized molluscs which are dealt with elsewhere.

Finally, one order of opisthobranchs, the pteropods, is adapted for a pelagic life. Members of most opisthobranch orders can swim by using the flaps of the foot to propel themselves through the water, but in the pteropods or sea butterflies there are much more extensive swimming modifications. One group is shelled and the other is naked.

The last gastropod subclass is the Pulmonata. Here sanitation problems are solved by a complete loss of the gills and the modification of the mantle into a highly vascularized lung. Most pulmonates are terrestrial, but there are marine representatives such as the pulmonate limpets which are very similar in appearance to the true limpets.

This brief review has examined only one aspect of gastropod biology, namely the reactions shown to the problem of torsion. Nevertheless, it is immediately obvious that there is an enormous adaptive radiation of types and forms together with several examples of convergent evolution such as the limpet shape. A similar pattern of radiation can be obtained by studying any basic feature of the group; feeding behavior is a good example. The radula is the prime feeding device in this group, and it is the flexibility in its design and operation which enables gastropods to feed on such a variety of diets and to colonize such a range of habitats. The only real

335

Mytilus edulis, a large and nearly cosmopolitan mussel (D. Gotshall photo).

Pen shells are usually seen standing erect in the soft bottom, but occasionally they are found in crevices in coral heads. Shown here is the Atlantic *Pinna carnea* (P. Colin photo).

Large colonies of tree oysters, *Isognomon,* are familiar sights on shallow silty bottoms in the tropics (P. Colin photo).

Winged oysters, such as this *Austropteria lata,* tend to be associated with gorgonians and other sessile animals, wedging the shell into crevices and branches (Australia; L.P. Zann photo).

restriction is the inability of the feeding process to cope with large food particles, although there are exceptions even to this.

The more primitive prosobranchs tend to show the simplest feeding methods such as browsing and grazing on algae. Both *Haliotis* and *Diodora* feed in this way, using the radula to tear off small pieces of filamentous algae attached to the rocks on which they live. *Patella*, a true limpet, has a stronger, more robust radula adapted to rasping algal material directly off the rocks.

The more advanced prosobranchs show a very wide range of feeding habits. The littorinids are characteristically algae raspers, different species of *Littorina* showing preferences for different algae. Some genera, such as *Lambis*, which lives on coral reefs, use their radula to nibble at small pieces of algae growing on the reef flat. Deposit-feeders are common among the prosobranchs, *Hydrobia* being a good example. This snail shovels up the top layers of detritus on mudflats and gains nourishment from the microorganisms living in the decaying matter. Plankton-feeding can be achieved by utilizing the respiratory current for collecting purposes. *Crepidula*, the slipper shell, feeds in this way. Planktonic material is trapped on the gill filaments and conveyed on a mucous thread to the radula which tears off suitable sized particles for ingestion. *Vermetus* is another planktonic feeder. This genus traps food in mucous threads which are extended out into the surrounding water.

Other prosobranchs are carnivorous, feeding on a variety of prey. Some, such as *Cerithiopsis*, feed on sessile animals like sponges and ascidians, a process not dissimilar to eating pieces of algae or weed. But others, such as the moon snails of the genera *Natica* and *Polynices*, are hunters. These gastropods drill a hole in the shell of the bivalve prey and draw out the flesh from within. The method of hole boring is very well known, and it is probably a mixture of mechanical rasping by the radula and secretion of an acid. The planktonic prosobranchs such as *Janthina* are also carnivorous. *Janthina* feeds on the coelenterate *Velella*, possibly anesthetizing its prey with a burst of purple dye. *Pterotrachea*, a heteropod or ' different foot, ' uses its foot as a sucker to capture such prey as small fish or jellyfish. Once caught, the prey is grasped by the radula and swallowed whole.

The most advanced prosobranchs, the Neogastropoda, are largely carnivorous as well, although a few are scavengers (such as *Buccinum, Nassarius, Bullia*). Of the hunting forms, some feed in a manner similar to *Natica*, while the most advanced, members of the families Conidae, Turridae, and Terebridae, have reduced the radula to two rows of arrow-like hollow teeth used individually to pierce worms, fish, or other prey. The captured animal is killed with a highly toxic poison which has in a few instances proved fatal to man. Other prosobranchs (such as wentletraps) are so closely associated with their echinoderm and anemone prey that they could be called parasites with little argument.

The opisthobranchs show a similar range of modifications although these are perhaps less extensive. The more primitive forms are typically algal browsers, and the relatively primitive *Aplysia* feeds by cutting off pieces of algae. *Acteon* is a deposit-feeder burrowing in clean sand at the low water mark. The radula is reduced in size and the precise feeding mechanism is not known. A more advanced form feeds by sucking the contents from algal cells which are slit open with the radula. The pelagic pteropods are planktonic feeders collecting small particles in a ciliary current (*Limacina*) or actively hunting. *Clio*, for example, feeds mainly on filter-feeding pteropods such as *Limacina*. Some opisthobranchs are benthic hunters. Thus *Philine* swallows minute molluscs whole and crushes them up in the gizzard.

The pulmonate limpets such as *Siphonaria* rasp algae in a manner similar to that of primitive prosobranchs.

The reproductive system in gastropods ranges from the simple to the very complicated. Many are dioecious, but a large number are hermaphrodites. In the simplest cases the eggs and sperm are both released into the water, where fer-

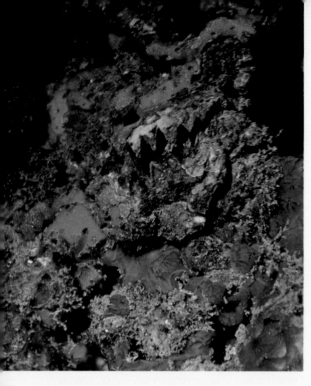

The jagged shell edges
of *Lopha frons* and
related oysters are
familiar sights to the
diver. These shells can
cause serious cuts
(Caribbean; P. Colin
photo).

Chlamys hericia, a common East Pacific scallop, displays an array of gorgeous
eyes (T.E. Thompson photo).

Scallops are among the few bivalves considered collectable by conchologists, and some are admittedly hard to surpass in beauty. These two relatively simple and plain species are typical of the *Chlamys* group. Above is the Hawaiian *Chlamys cookei (= C. irregularis)* (S. Johnson photo), below the Caribbean *Chlamys ornata* (T.E. Thompson photo).

A common tented cone shell, *Conus pennaceus*, with several of its egg cases (S. Johnson photo).

Pseudosimnia, an allied cowry, on its food animal, *Dendronephthya*. Notice the close resemblance of the cowry's mantle pattern and its host (R. A. Birtles photo).

tilization occurs. More commonly, however, the eggs are coated in a protective jelly or wrapped in a capsule and copulation is necessary to ensure fertilization before these protective layers are deposited. The resulting sexual behavior may be quite complex, especially in hermaphroditic forms, which include some prosobranchs and all opisthobranchs. Copulation with mutual fertilization occurs in most of these, but *Aplysia* is an exception. Here the molluscs form copulating chains where each individual receives sperm from its neighbor behind and gives it to the one in front, much as in earthworms.

Crepidula, the slipper shell, is a protandric hermaphrodite, each individual starting life as a male. The sex of an older individual is affected by that of its neighbor. These limpets live attached to one another in clusters of a few to many individuals. A male will retain its sex as long as it is attached to a female. If, however, there are a lot of males then a few may become female; if an old male is isolated it will develop into a female. Once this has occurred the change is irreversible.

Whatever the reproductive biology of the adults, the young stages of gastropods are characterized by the veliger larva. The usual trochophore larva, as found in the annelids, is repressed except in primitive prosobranchs. The veliger is derived from it by the development of a swimming organ, the velum, which consists of two semicircular ciliated folds. The velum also functions as a feeding organ, its current drawing fine particles into a food groove. As growth proceeds the larva reaches a stage where its foot is able to take on a creeping form of locomotion and the larva gradually develops into the adult.

Where the trochophore exists it is always followed by a veliger, but usually the eggs hatch directly into the latter. Indeed in some forms, particularly the more advanced prosobranchs such as the whelks, there is no free-swimming stage, both trochophore and veliger stages occurring in the egg. In these cases the young hatch out as tiny miniatures of the adults.

Flame scallops, *Lima scabra,* are common Caribbean scallops often seen in aquaria. The mantle and/or tentacles are usually bright red, contrasting with the plain white shell (above, C. Arneson photo; below, P. Colin photo).

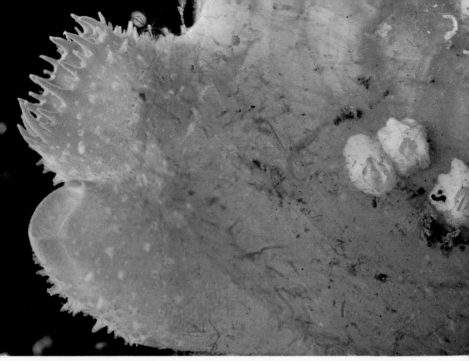

Views of the siphons of the "ugly clam," *Entodesma saxicola* (above) and an unidentified venus clam (below). The short siphons indicate these species are not deep burrowers. Notice also the barnacles and hydroids encrusting the shell of the *Entodesma* (East Pacific, T.E. Thompson photo above; Australia, L.P. Zann photo below).

Following this brief review of marine gastropods, we may now turn to individual groups and describe the ways in which they have adapted to their environment.

Starting with the primitive prosobranchs, we find those species with a cleft shell or a shell with a hole in it. These are sometimes grouped as the Archaeogastropoda, together with the true limpets. The keyhole limpet *Diodora* has already been discussed, but other members of the group exhibit a cleft shell instead of an exhalent hole. Thus *Emarginula* has a slit in front of the shell although the more advanced forms such as *Fissurella* bear a strong resemblance to *Diodora*. The abalones are another group with holes in the shell. In the European *Haliotis tuberculata*, commonly known as the ormer, there is a series of holes. As the ormer increases in size a new hole develops and an earlier one closes up. The ormer is commercially important both as a food and because its nacreous shell makes it one of the most attractive European shells. In America some of the West Coast abalones are a seafood delicacy. They are so popular in fact that there is a danger of over-collecting and protective measures such as minimum sizes and bag limits have been taken to prevent their sudden decrease.

The true limpets such as *Patella* and *Acmaea* are very common inhabitants of rocky shores. While these animals can and do move around at high tide, the foot is adapted as a sucker to maintain a firm anchorage at low tide. This is facilitated by each animal inhabiting a definite ' home. ' This means that the margin of the shell grows to the shape of the underlying rock so that a flush fit is achieved. Anyone who has clambered over a rocky shore will know how difficult it is to dislodge a limpet once it has pulled down hard onto its rock. At high tide, however, the limpets move around leaving a trail of cleaned rock as they rasp off the encrusting algae. Experiments have shown that limpets have a good homing behavior which enables them to find their own territory again after feeding. The shape of the limpet shell also tells us something of its way of life. The broad conical shell is well adapted to withstand the onslaught of stormy seas, and its generally smooth surface offers little resistance. It is in fact possible to relate the angle of the cone to the degree of exposure, the flatter limpets occurring under particularly rough conditions. Most limpets are dull in color, but the tiny *Patina pellucida* is one of several exceptions. This limpet lives on seaweed near the low water mark and is immediately recognizable by the shiny blue rays across its shell.

The trochids, commonly known as topshells, are another group of primitive prosobranchs so named from the shape of their shells. *Gibbula* and *Monodonta* are common forms recognized by the gray and mauve stripes across the shell, which when worn reveals a beautiful mother-of-pearl layer underneath. As a consequence these shells are often collected for the manufacture of seaside souvenirs. Large tropical species of trochids and the related turban shells are used commercially to supply mother-of-pearl for buttons and inlays.

The more advanced prosobranchs, or Mesogastropoda, are well represented by members of the genus *Littorina*, the periwinkles. There are many species, each well adapted to its position on the seashore. In Europe there are four common types: *Littorina littorea*, *L. littoralis*, *L. neritoides*, and the *L. saxatilis* group of species. The edible periwinkle (*L. littorea*) occurs at midshore levels on rocks and the flat periwinkle (*L. littoralis*) lives in damper conditions on seaweed at the same level. *L. saxatilis*, the rough periwinkle, occurs higher up the shore, and *L. neritoides*, the smallest one of the four, is to be found higher up still in the splash zone. A similar zonation occurs in America although the species may be different. Whatever the location, however, it seems clear that as a group the littorinids show a progressive adaptation to life on land. The lower shore species excrete ammonia as a waste product. This is very toxic unless it is rapidly diluted with water and washed away. Higher up the shore, however, there is a tendency to produce uric acid and to conserve water. *L. neritoides* lives so high up that it has to rely on aerial respiration most of the

Close-ups of the mantles of two species of giant clams, *Tridacna*. The brightly colored mantles vary in pattern and color somewhat with the species but are also quite variable individually and among populations (both Australia; G. Allen photo above, L.P. Zann photo below).

Two familiar but divergent Caribbean bivalves. Above, the common *Donax denticulata,* a coquina, is found by the thousands in loose sand at the tide's edge and is an extremely fast burrower. Below, *Spengleria rostrata* uses the ridges on the shell to abrade a burrow in coral heads (both C. Arneson photos).

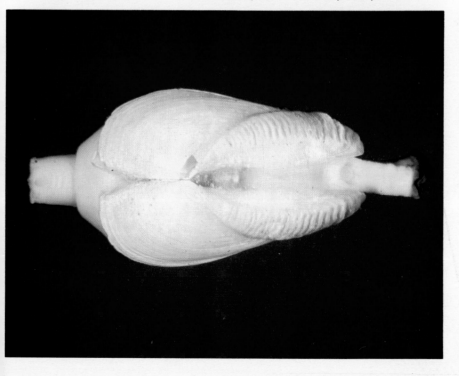

Above: Shell collecting is probably still the most popular hobby involving collecting natural objects. Advanced shell collectors often amass collections of literally thousands of species of molluscs (L. Easland photo).

Facing page: Some of the larger gastropod shells popular with collectors. 1) *Fasciolaria tulipa*, a Caribbean whelk. 2) *Murex ramosus*, a murex shell. 3) *Fasciolaria trapezium*, a Pacific Ocean whelk. (G.v.d. Bossche photos.)

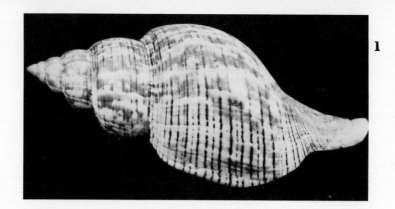

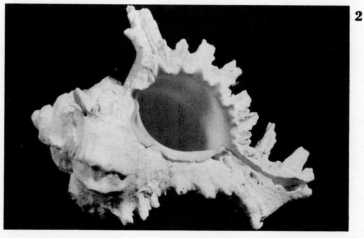

Preserved animal of *Nautilus pompilius* with its coiled external shell (Pacific; K. Lucas photo at Steinhart Aquarium).

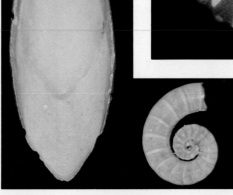

Internal shells of *Sepia*, a cuttlefish (left), and *Spirula*, a small pelagic squid (right) (Pacific, K. Gillett photos).

Living *Nautilus pompilius.* Note especially the "hood," the numerous tentacles, and the large eye (K. Lucas photos at Steinhart Aquarium).

time; to keep its mantle moist it shelters deep in rock crevices or between barnacles.

Hydrobia is another very small prosobranch as common on muddy shores as the littorinids are on rocky ones. Many hundreds of these tiny black snails can be seen in a square meter of mud at low tide, and their trails can be readily picked out. Like the limpets, they feed by eating the top layers of the substrate as they pass over it, but in this instance it is not algae but detritus that they ingest.

Several mesogastropods have evolved a pelagic existence. *Janthina* is a small snail with a thin purple shell. It cannot swim but floats upside down on a raft of bubbles which it secretes on the surface of the water. Like many pelagic gastropods, *Janthina* is carnivorous and feeds on other organisms floating in the surface currents such as *Velella*, the by-the-wind sailor. While feeding it abandons its raft and attaches onto the coelenterate. Not a lot is known about the biology of *Janthina*, but some species have been found towing rafts of egg capsules behind them.

Several other pelagic forms are grouped together as the Heteropoda. There are three common genera, *Atlanta*, *Carinaria*, and *Pterotrachea*, which present a good series of stages in the reduction of the shell, possibly indicative of opisthobranch origins. *Atlanta* is a small transparent snail with a highly compressed light shell which has a keel to keep it upright as the animal swims along. Swimming is achieved by undulations of the sole of the foot, which is greatly extended. *Carinaria* is very long, thin, and rather jelly-like in appearance with only a tiny cap-shaped shell. The body is held rigid by a dorsal crest. *Pterotrachea* has no shell and the foot is modified into a muscular fin which is held upward in the water. This heteropod can move very rapidly through the water, rhythmic undulations of the body assisting the sculling action of the fin. Because of their active way of life heteropods need a high sensory input and all are equipped with very well developed eyes. They are predacious animals feeding on jellyfish and small fish. *Pterotrachea* uses its foot as a sucker to capture such prey, and *Carinaria* pierces them with its sharp radula.

Some mesogastropods resemble the limpets in general shape, *Calyptraea* being a good example. The slipper shell *Crepidula* is closely related to this genus. The similarities to the true limpets are only superficial, however. While the latter browse for their food, the Calyptraeidae are more or less sessile forms relying on ciliary currents to waft plankton into their filtering systems for food. *Crepidula fornicata* was introduced into Britain with American oysters and rapidly spread although the oyster failed to successfully colonize. In fact it can be a pest in oyster fisheries, as it tends to smother the oysters and can compete very effectively for food.

From filter-feeders we turn to the predacious carnivores of the family Naticidae. These live buried in sand and show interesting modifications to this end. The shell is very smooth and a streamlined effect is achieved by extensions of the foot which wrap around the shell, giving it a wedged appearance ideal for burrowing. *Natica* feeds on burrowing bivalves which it grasps with the flaps of the foot. It then bores into them and sucks out the contents. *Natica* is liable to attack from starfish but it can protect itself by completely covering the shell with the foot so that the echinoderm cannot obtain a purchase on its slippery surface.

The cowries (*Cypraea*) are similar to *Natica* in that the foot wraps around the shell throughout life. This is somewhat of a mystery as cowries have a very colorfully marked shell which has no obvious function since it is always covered. Cowries may be omnivorous, carnivorous, or scavengers, but their feeding habits are poorly known.

Aporrhais, the pelican foot, and *Strombus*, the conchs, are capable of a rapid form of locomotion in addition to the normal crawling. The body is reared up off the ground as an extended 'wrist' so that the heavy shell falls forward at each step. The long operculum projects over the edge of the foot and is strengthened by a ridge for use as a digging tool.

In complete contrast are the worm shells comprising two families, Siliquariidae and Vermetidae. These are elongated forms whose shells have become so loosely spiralled that they look like corkscrews. They live on rocky surfaces, and

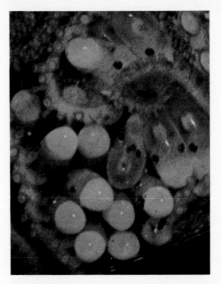 

Left, eggs of *Octopus ocellatus* (Pacific; Y. Takemura and K. Suzuki photo). Right, the brightly patterned dwarf *Octopus chierchiae* (East Pacific; A. Kerstitch photo).

Octopus ornatus (Pacific; S. Johnson photo).

Active specimen of *Octopus cyaneus* showing the large web between the arms and the ocellated spot (below and in front of eye) (Pacific; M. Goto photo).

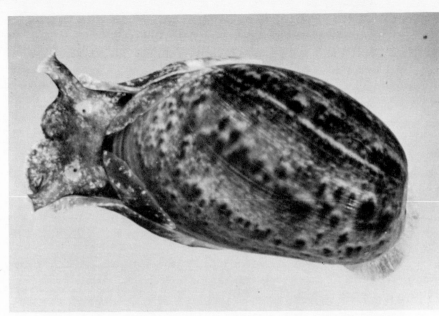

Bulla striata, a bubble shell, one of the more primitive opisthobranchs (C. Arneson photo).

The piscivorous *Conus purpurascens* devouring a small fish (J. Nybakken photo).

the less specialized forms are filter-feeders. *Vermetus*, however, can feed differently. *V. novae* is a species living in rough water on the edge of coral reefs and feeding in the traditional way. *V. gigas* on the other hand lives in very calm water and feeds by producing a mucous net over the surface of the water. This is secreted by glands on the foot, which is extended out into tentacles.

The advanced Neogastropoda include both rocky and sandy shore representatives. The least specialized belong to the whelk group and include genera such as *Buccinum*, *Busycon*, and *Nassarius*. The latter are scavengers feeding on dead and decaying material, but most whelks are carnivorous. This group also includes the Australian *Megalatractus* (= *Syrinx*), a giant among molluscs, reaching 60 cm in length. The Muricacea, murex or rock shells, show feeding adaptations connected with their predacious way of life. Like *Natica* they feed on living bivalves and utilize a variety of methods of penetration. *Murex* feed on oysters, holding the shell down and chipping away at the edge until it is finally opened. *Urosalpinx*, an oyster drill, drills holes into the shell of its prey, presumably by a combination of mechanical and chemical means. Once inside the shell the radula scrapes off the flesh. The whole process is very slow, and it may take over two days for a *Urosalpinx* to get inside a large oyster, but they are nevertheless important pests of this commercial shellfish.

Members of the genus *Conus* feed in a different way again. These essentially nocturnal snails come out to feed at night, worms being their main prey, although many feed on other gastropods or fish. When they attack prey a single hollow radular tooth is moved to the end of the proboscis and jabbed into the worm or snail, followed by a squirt of venom. The cone then engulfs the stunned victim.

The opisthobranchs are a more heterogenous group than the prosobranchs and may represent several convergent lines of evolution. Nevertheless they are all similar in the reduction of the shell and in the gradual trends of detorsion and gill loss. Their classification is based on the degree to which these trends have developed.

Group of eggs of an East Pacific squid, *Loligo opalescens* (D. Gotshall photo).

An attractive *Sepia* species. These animals are typical of the Mediterranean area and the cooler Indo-Pacific but are known to most people merely as "cuttlefish bone" bought for pet birds.

Squids are usually recognizable by their elongated body and long arms even without an arm count. Most are too active to be kept in aquaria. Above is *Loligo opalescens*, a very common East Pacific species (D. Gotshall photo); below is the small *Sepioteuthis*, one of the few squids likely to be found near tropical reefs.

The most primitive opisthobranchs are the bubble shells such as *Bulla*. These are large burrowing forms which retain the shell, although in some groups (for instance *Philine*) it is merely a thin internal plate. More usually, however, it is a very thin bubble-like growth somewhat smaller than that found in prosobranchs so that the bulloids present a streamlined shape for burrowing. This is aided by two large flaps laterally on the foot. These grow around the body, giving it a wedge-shaped appearance. In addition the head is large and shield-like. Primitive bulloids like *Acteon* are microphagous browsers, but the more advanced forms have developed a higher level of gut musculature and tend to be carnivorous; *Philine*, for example, feeds on young bivalves.

The next stage in shell reduction is shown by *Aplysia*, the sea hare. This has a tiny shell on the back which serves no protective purpose. Instead *Aplysia* relies on its cryptic coloration which is gained from the seaweed that it feeds on. When disturbed many species secrete a purple dye which acts as a smoke screen and allows them to escape either by crawling or swimming. An internal shell is found in the slug-like *Pleurobranchus*, a stage close to the nudibranchs, the true sea slugs.

The swimming pteropods consist of two groups, the shelled Thecosomata and the naked Gymnosomata. They are very light, colorless forms, sometimes known as sea butterflies, which spend all their life in the plankton. The Thecosomata are probably derived from the bulloids and like them retain the shell and mantle cavity. The most primitive (such as *Limacina*) are small forms with a transparent spiral shell. The broad foot is drawn out into two flaps known as wings. *Limacina* uses these as oars and swims on a broadly spiral course. By holding the wings together above the body, *Limacina* is able to sink. More advanced forms have lost the spiralling of the shell and are bilaterally symmetrical. In *Creseis* the shell is a long cone, and in *Clio* it is flattened dorsoventrally. In *Clavolina* it is swollen like a flask. The larger thecosomes are in the family Cymbulliidae. In these the true shell is replaced with a chitinous 'pseudoconcha' which forms a 'boat' in which the body lies. Swimming is again achieved by wings. In *Cymbulia* they are flapped like a butterfly or can be rowed backward and forward; *Corolla* has wings forming a fringe around the pseudoconcha.

The naked Gymnosomata are faster swimmers than the thecosomes. The body is bullet-shaped and the shell is entirely absent. Again wings are present, but here they are attached at a narrow point only. When swimming they are twisted at the joint, thus allowing a power stroke at the up and down beat.

The Pulmonata are such a terrestrial and freshwater group that the few primitive marine forms appear very insignificant. The Siphonariidae, which closely resemble true limpets, are entirely intertidal and may replace limpets on some tropical shores.

The gastropods form only one class of the enormous phylum Mollusca, but their importance in the intertidal environment cannot be over-emphasized, especially on rocky shores. Luckily because of their attractive shells they are well known to naturalists and shell collectors, and for the same reason there is a good fossil record. This shows that gastropods are at the height of their evolutionary expansion in present times, and this could justifiably be called the 'Age of the Snail.'

SEA SLUGS

The common sea slugs belong to seven groups that are all gastropod molluscs but are not otherwise closely related. Of these, the first are air-breathing pulmonates, the second (the Lamellariidae) are prosobranchs allied to the cowries, while the others are major opisthobranch groups consisting of the herbivorous sea hares or Aplysiomorpha, the side-gilled carnivorous Pleurobranchomorpha, the shallow-water herbivorous Sacoglossa, the minute sand-dwelling Acochlidiacea, and the common voracious Nudibranchia.

The evolution over many millions of years of sea slug

Acorn barnacles are a common sight almost anywhere at the tide line. Above is *Tetraclita squamosa* (Pacific; Y. Takemura and K. Suzuki photo); below is *Balanus cariosus* in typical setting of limpets and algae (East Pacific; K. Lucas photo at Steinhart Aquarium).

The specialized turtle barnacle *Chelonibia testudinaria* on the sea turtle *Chelonia mydas* (Pacific; L.P. Zann photo).

The delicate feeding legs of the barnacle *Balanus nubilus* are readily visible (center) in this mass of anemones (East Pacific; K. Lucas photo at Steinhart Aquarium).

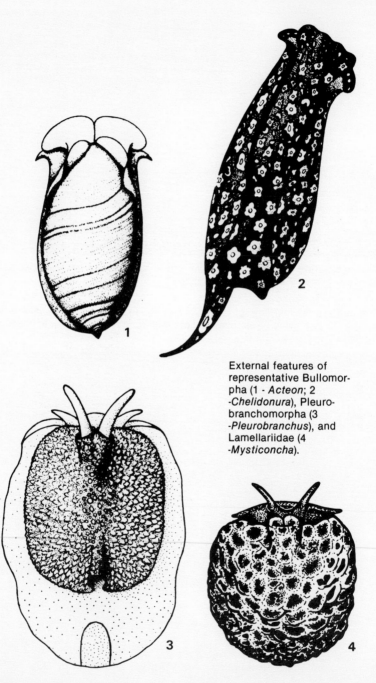

External features of representative Bullomorpha (1 - *Acteon*; 2 -*Chelidonura*), Pleurobranchomorpha (3 -*Pleurobranchus*), and Lamellariidae (4 -*Mysticoncha*).

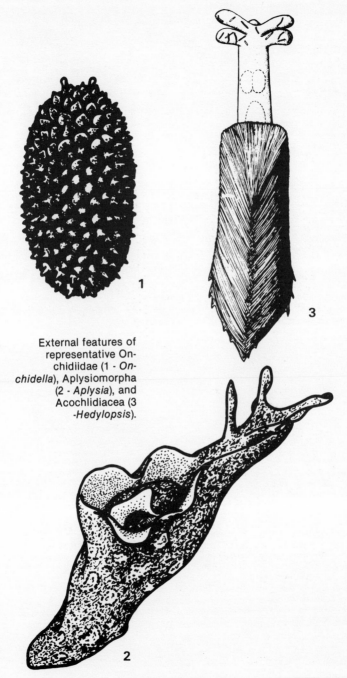

External features of representative Onchidiidae (1 - *Onchidella*), Aplysiomorpha (2 - *Aplysia*), and Acochlidiacea (3 -*Hedylopsis*).

Two types of stalked barnacles. Above, *Pollicipes polymerus,* with small plates on the stalk (East Pacific; K. Lucas photo at Steinhart Aquarium); below, *Lithotrya valentiana* (Caribbean; C. Arneson photo).

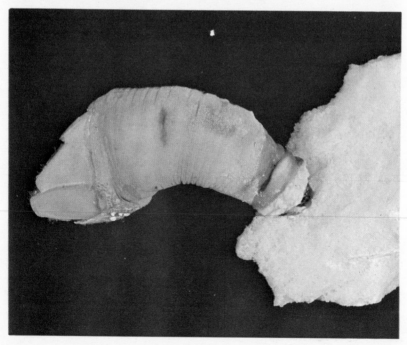

A common species of gooseneck barnacle, *Lepas ansifera* (cosmopolitan; A. Norman photo).

The unusual pelagic *Lepas fascicularis,* a widely distributed bluish species that produces its own float (T.E. Thompson photo).

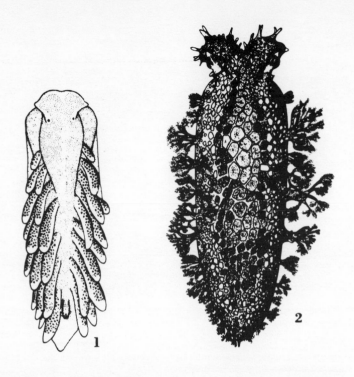

External features of representative Sacoglossa (1 - *Alderia*) and Nudibranchia (2 - *Marionia*). Below is a ventral view of the 'false' sea slug *Onchidella celtica* (3).

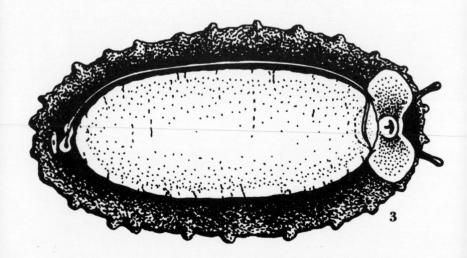

types from prosobranch gastropod ancestors which resembled the present-day top shells, abalones, and periwinkles has involved the gradual loss of the protection of the stout external shell with its opercular door. This characteristic molluscan organ of defense has, in the evolutionary lines that have led to the sea slugs of the present time, become replaced by diverse and (to the marine biologist) most exciting defensive adaptations. Dynamic chemical and biological methods of defense against sharp-toothed predators have taken over from the negative, passive method represented by the shell.

Once liberated from the necessity to retreat into a calcareous shell each time an alarm sounded in the prosobranch nervous system, these emancipated gastropods were able to evolve more vivid body ornamentation, extravagant epidermal color patterns, and varied types of movement behavior.

No doubt some of the early evolutionary experiments along these lines ended in failure, but many were partially or totally successful. We are fortunate that modern oceans contain more than three thousand species of gastropods which show intermediate stages in the general trend outlined above. Some of these still retain the shell and in others it has become greatly reduced and may be partially or wholly internal (covered by the mantle), but in the majority the shell is lost at the end of larval life in each generation and the adult is a sea slug and not a sea snail.

Sea slugs are among the most interesting and visually exciting of invertebrates; they are to the molluscs what the butterflies are to the insects or the orchids to the flowering plants. The slug-like level of organization has widened the evolutionary horizons of the stocks possessing it. They are no longer restricted in their range of behavioral activities, they are better able to employ tricks of coloration for defensive means, and they are able to pursue their prey (or seek shelter from their own enemies) in a multitude of crevices on the sea bottom which few hard-shelled molluscs could enter. So it is perhaps not surprising that the successful sluglike form has been independently acquired by descendants

Alepas parasita, an odd gooseneck barnacle found in the jellyfish *Cyanea* (Pacific; L.P. Zann photo).

The brightly striped *Conchoderma virgatum* can attach to large fish and may cause localized damage. Unless the feeding legs are seen, it is very hard to recognize this as a barnacle and not a parasitic copepod (Caribbean; C. Arneson photo).

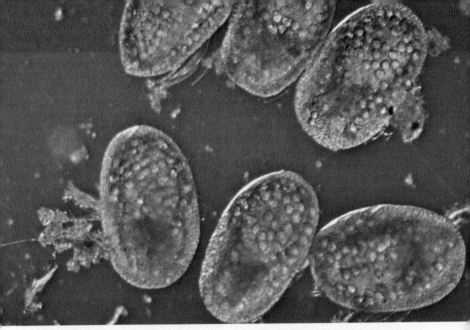

Newly hatched nauplii larvae of a dendrogastrid barnacle similar to those found in the body cavity of starfishes (Pacific; L.P. Zann photo).

Sacculinid barnacles are of economic importance as they cause slow growth and prevent sexual maturity in infected crabs. Shown here is *Sacculina granifera* under the abdomen of *Portunus pelagicus* (Pacific; L.P. Zann photo).

of many distinct stocks of gastropods.

Sea slugs are found in all the world's oceans and major sea areas. They feed on every kind of epifaunal animal material and play a significant part in some planktonic communities. Some sea slugs, such as the nudibranch *Cerberilla*, burrow into soft marine mud and sand in search of burrowing sea anemones. They owe their success to their unique maintenance, during their evolutionary history, of the delicate balance between the shell and the skin as defensive methods. They have agile, soft bodies and perfectly adapted chemosensory tentacles (rhinophores) and buccal apparatus, as well as virtual immunity to attack by fish and other predators. In many cases the searching larvae have an extraordinary ability to recognize and metamorphose upon the adult diet.

The pulmonate sea slug *Onchidella* and the larger nudibranchs are often easy to find between tide marks on rocky shores, sometimes in shallow water or in pools. Pulmonates of the genus *Onchidium* are characteristic of the swampy mud lying around mangroves in the tropics. The smaller sea slugs present considerable problems and can often be found only after patient and laborious searching through dredged hydroids, sponges, and bryozoans. Aqualung divers are privileged to see the sublittoral species in their natural habitats. In the Caribbean Sea with its tides of low amplitude, only divers will be able to find most sea slugs. Elaborate SCUBA equipment is fortunately not essential, and slug-hunting using basic equipment (snorkel, facemask, and rubber flippers or fins) can be very rewarding. But in other parts of the Atlantic Ocean and, better still, in Pacific and Indian Ocean localities, the receding tide often exposes a rich collecting ground. All the enthusiast will need here are a few jars and his rubber sneakers or hipboots.

The best localities are often the least accessible and the most hazardous. In temperate areas, rocky islands or exposed headlands in areas of rapid currents with water of high salinity are usually the most favorable. It oftens comes as a surprise to the student when he discovers or is told that the coral reefs of tropical seas are poor places to search for sea slugs. This is because coral polyps have tissues that are armored with nematocysts and the naked foot of a sea slug is repelled by these. A few species of nudibranchs have, despite the difficulty, mastered the coral reef environment, but these species are usually small, shy and inconspicuous. They can sometimes be found only by wrecking parts of the reef and sorting through the debris; this can be very laborious and it is of course ecologically most destructive and undesirable. It is better to turn attention to the lee or sheltered side of the reef and the quiet waters of the lagoon. The coral rubble commonly found in such places is a good place to hunt for nudibranchs, and the eel-grass or turtle-grass beds that characterize muddy lagoons from Trinidad to Brisbane usually give shelter to numerous aplysiomorphs, including the huge Indo-Pacific *Dolabella* (found partially embedded in the mud), the world's largest sea slug (weighing in air as much as two kilograms).

Spawn masses, sometimes taking the form of delicate white coils, often give a clue to the whereabouts of the adult sea slug which may be ' playing possum ' in the hope of eluding detection. Many sea slugs, however, are conspicuous anyway and no detective work may be necessary.

The responsible naturalist will never take more animals than are needed for his immediate purpose and, in fact, the greatest satisfaction may be obtained from simple ecological observations without the need to capture and kill the subjects.

With difficult individuals or obscure species, however, it is certainly not practical to attempt identification in the field. These molluscs can usually be transported safely back to the aquarium or the laboratory in clean sea water in a strong plastic bag or by using thermos bottles if the climate is extreme and the journey back to the homebase is long. Back at home or the laboratory the specimens should be transferred to clean sea water and never overcrowded. It is important to remember that some of the larger species are voracious when they get hungry and may attack delicate or injured individuals.

Pacific mantis shrimps. Above, the pastel *Odontodactylus japonicus* (Y. Takemura and K. Suzuki photo); below, *Echinosquilla guerini,* a species with erect spines on the telson (S. Johnson photo); opposite page, the brightly colored *Odontodactylus scyllarus* (M. Goto photo).

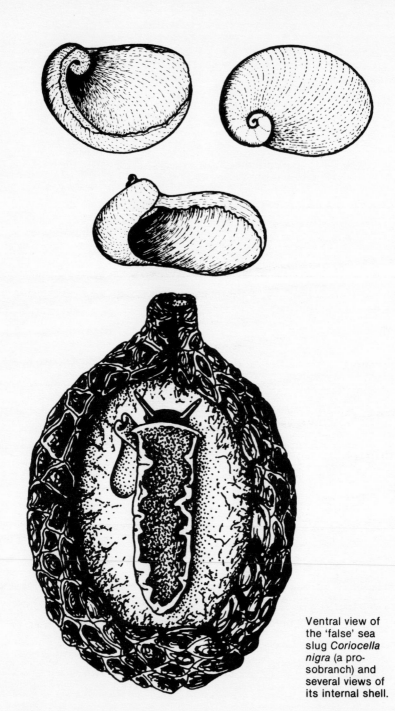

Ventral view of the 'false' sea slug *Coriocella nigra* (a prosobranch) and several views of its internal shell.

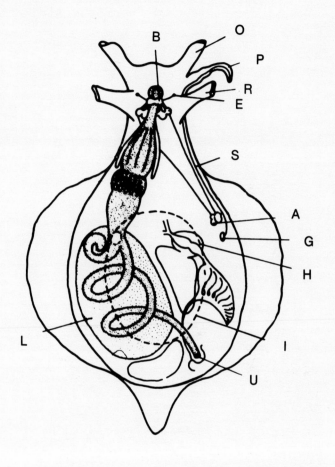

Anatomy of *Aplysia*. B = brain; O = oral tentacle; P = penis; R = rhinophore; E = eye; S = external seminal groove; A = ganglia; G = genital opening; H = heart; I = internal shell; U = anus; L = liver.

Three tropical East Pacific (Mexico) mantis shrimps. From top to bottom: *Squilla mantoidea, Parasquilla similis,* and *Hemisquilla ensigera.* Notice the variation in development of keels (carinae) on the abdomen, spination of telson, and size of the eyes (all A. Kerstitch photos).

Mysids are usually found in the plankton, but a few are reef-dwellers. *Mysidium* (above) occur in dense schools and are visible to the eye only as silver dots. The brightly striped *Heteromysis actinae,* on the other hand, is large enough to be seen easily (below); it is found only with the anemone *Bartholomea annulata* (both P. Colin photos).

Always take time to make observations from life; this is especially important with these soft and pliable animals. A good color photograph using close-up attachments and electronic ('strobe') flash can be helpful. It is important to stress that *photographs alone* (unaccompanied by full collecting data and a well preserved specimen or series) *are of little use*. No responsible scientific specialist will name specimens for you if you can give him only photographs; he knows how easy it is to be utterly wrong this way. So help him and help yourself by being systematic in your approach to the study of your sea slugs. Preservation should only follow careful noting of habitat, morphology, and colors. Make a large drawing and group your descriptive notes around this in the shape of labelled arrows. Such drawings can be added to and amended as more finds are made. Don't worry if at first your drawings look unprofessional—they will improve with practice.

Narcotization of sea slugs may be brought about with 7% aqueous magnesium chloride mixed in equal parts with sea water. The animals when relaxed can be preserved in 10% formalin and later be transferred to 70% ethyl alcohol, although this is not necessary (it just smells less offensive).

A detailed discussion of the internal anatomy of sea slugs is outside the scope of this chapter. Anatomical features which must be understood before we can address ourselves to the systematic zoology of sea slugs are shown diagrammatically in the drawings, which show the principal features of typical representatives of the main groups of sea slugs. Furthermore, it is necessary to study the radula for accurate identification because this organ is different in every mollusc and provides the patient investigator with a check on identifications made from external features alone.

Much remains to be discovered about sea slug radulae. We are still uncertain, for instance, about the functional significance of the various tooth patterns. We can see fairly well how the broad radulae of some of the nudibranchs or onchidiids are adapted to scraping up the two-dimensional sponge or other encrustations that they feed upon. But why are the radulae of those nudibranchs that feed upon sea anemones and other coelenterates usually so narrow? Why are the radulae of lamellariids and pleurobranchids (which both feed upon ascidians) so strikingly different? In such discussions, we must not forget that many sea slugs possess stout chitinous jaws in addition to the radula in the buccal bulb. The jaws usually serve to cut off slices from a three-dimensional food organism, the radula then acting to transfer these slices into the esophagus on its way to the stomach.

Within the gastropod molluscs, the zoologist recognizes two major categories. The first is the Streptoneura (or Prosobranchia), which usually have a heavy coiled external shell, live predominantly in the sea, and exhibit internally the effects of visceral twisting or torsion. The second category is the Euthyneura; they are usually more lightly shelled and the operculum, which was used to close the ' door ' of the shell in their remote ancestors, has almost invariably now been lost. Euthyneura is a technical term implying that the nervous system is untwisted and, indeed, most modern euthyneurans exhibit only slight traces of the ancestral torsional twisting.

Sea slugs have evolved from both Streptoneura and Euthyneura, and the resultant forms have come, through convergent evolution, to resemble one another to a striking extent. Acid secretion through the skin is a defensive mechanism that has evolved quite separately and independently in the streptoneuran Lamellariidae and the euthyneuran Pleurobranchidae. These two families are strikingly similar in many respects: both contain species that feed upon sea squirts, both have a substantial calcified shell completely roofed over and concealed by fleshy mantle lobes, both have a large bipectinate gill plume on the right side of the body (naked in the pleurobranchs, concealed in a mantle cavity in lamellariids), and both live in shallow hard-bottomed sea areas.

By far the most successful sea slugs belong to the gastropod orders Sacoglossa and Nudibranchia. Because

Although isopods are abundant in the sands and rocks of almost any coast, few are seen by general collectors. The parasitic ones such as *Anilocra* (above) are conspicuous when on the head or body of aquarium fishes (Caribbean; P. Colin photo), while other interesting isopods occur with invertebrates, such as *Cirolana lineata* (below), which lives in the anus of crinoids (Pacific; L.P. Zann photo).

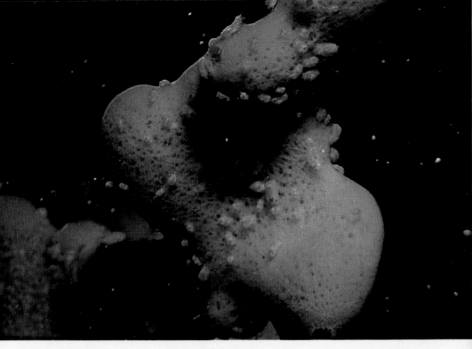

Most isopods are small, such as these numerous specimens of several species apparently feeding on a sponge (Pacific; W. Deas photo).

Lironeca vulgaris, a parasitic isopod of fish. The large brood of the species is visible on the belly of the female (East Pacific; A. Kerstitch photo).

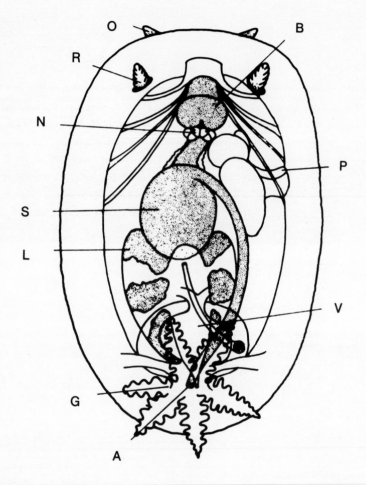

Anatomy of *Archidoris*. B = buccal mass; P = penis; V = ventricle; A = anus; G = gill circlet; L = liver; S = stomach; N = Brain; R = rhinophore; O = oral tentacle.

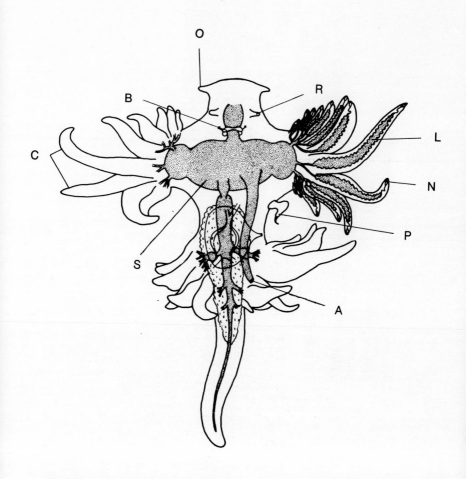

Anatomy of *Glaucus*. R = rhinophore; L = liver; N = cnidosac; P = penis; A = anus; S = stomach; C = cerata; B = brain; O = oral tentacle.

Skeleton shrimp, *Caprella,* an unusual amphipod adapted for camouflage of body form on marine vegetation (Pacific; L.P. Zann photo).

The incredible parasitic and commensal fauna of whales includes several highly modified barnacles and amphipods. Shown here are whale lice (flattened amphipods), *Cyamus scammoni,* and whale barnacles, *Cryptolepas rhachianecti,* removed from the California gray whale (K. Lucas photo at Steinhart Aquarium).

A euphausiid or krill shrimp. Krill occur in immense numbers in cooler waters and have been much talked about as a food source of the future.

these are included in the Euthyneura, it is necessary to look more closely at the evolutionary history of this great group, which comprises also the pulmonate slugs and snails of the land and fresh water.

In attempting to seek evidence about the origins and early history of the Euthyneura, one is forced to admit that hard-and-fast distinctions between the early opisthobranchs (the branch which led eventually to the Nudibranchia) and the early Pulmonata (the branch which led to the present-day land and freshwater slugs and snails) become difficult to define. Probably the earliest euthyneurans took their origin from some rather generalized, benthic, epifaunal, helically coiled, strong-shelled, marine prosobranch stock midway through the Paleozoic. Conceivably this stock was close in structure and habits to many present-day rissoaceans, which are prosobranch mesogastropods, although one authority seeks the origin of the Euthyneura among the early Archaeogastropoda. Some time later, the euthyneuran line split into two successful branches, one succeeding in the conquest of brackish and fresh waters and leading on to the invasion of purely terrestrial habitats; this became the Pulmonata. The other line led to the exploitation of the marine infaunal environment and produced by the late Paleozoic the earliest Opisthobranchia. It is interesting to note that these two major lines of descent allowed the Gastropoda access for the first time to fresh water and the land in the one case and to marine burrowing in the other. The ancestral prosobranch stocks had not succeeded in invading such habitats nor have their prosobranch descendants exhibited very much more success in these respects.

The two main lines of descent from the first euthyneurans have many features in common. In both lines the static, negative defensive mechanism represented by the operculum-shell complex came to be of reduced importance and was replaced by more dynamic biological and chemical methods of defense. The shells of euthyneurans are nearly always more frail than is the case in the prosobranchs, and the operculum is seldom present in pulmonates and exists in only a few of the most primitive living opisthobranchs. The effects of gastropod torsion have been progressively nullified then abolished in both pulmonate and opisthobranch lines of descent, and the most advanced members of both groups show external and internal bilateral symmetry of all systems except the reproductive apparatus (which usually remains predominantly on the right side of the body in connection with the pairing posture suitable for reciprocal copulation in such hermaphroditic animals).

The adult shells of the primitive opisthobranchs living today are often strongly developed and sometimes colorful, as in *Acteon tornatilis*, where a horny, snug-fitting operculum may be present and functional, but a more typical opisthobranch shell (as in *Micromelo undata* from the Caribbean Sea or *Hydatina physis* from the Indo-Pacific) of the present-day is fragile, inflated, egg-shaped, and has the spire more or less concealed; the operculum is absent after the veliger larval phase. In rather more advanced forms, such as *Umbraculum sinicum*, the external shell has a very wide gape, and in animals like *Berthella plumula* the widely gaping ' Chinaman's hat ' shell is wholly internal, covered by the mantle.

In the higher sacoglossans, pleurobranchomorphs and aplysiomorphs, some bullomorphs, and in all the nudibranchs, the true shell is completely lost after larval metamorphosis. In some cases an internal shell may persist for some months as a vestige before being completely resorbed, but in other cases the shell is cast away intact during post-larval metamorphosis. In extreme cases the shell may be obliterated completely from the ontogeny and even the shell gland may be lost from the developmental sequence.

The earliest opisthobranchs probably all became burrowing infaunal animals. Certainly the most primitive living opisthobranchs (such as *Acteon*) are burrowers. In these early forms the body became enlarged and anteriorly spatulate; the head became a flattened spade-like cephalic shield. The head tentacles, originally placed on top, were lost or became shifted to the sides or rear of the head. The

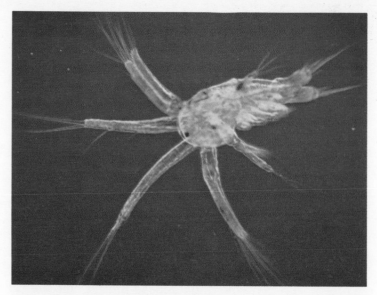

The nauplius larva of *Penaeus* is similar to that of most shrimp and crabs. There is a striking resemblance to the copepods, but the large paired appendages are antennae and mouthparts, not legs (Y. Takemura and K. Suzuki photo).

The genus *Penaeus* contains the most important commercial shrimp, with species is most warm countries. This is a Japanese food shrimp, *Penaeus latisulcatus* (Y. Takemura and K. Suzuki photo).

Penaeus plebejus, an important Australian commercial species. *Penaeus* spend the day buried in the substrate and are usually most active at night (W. Deas photo).

Deep-sea shrimp of several families tend to be rather spiny and often red in color. This *Acanthephyra* is bright red and has strong spines on the abdomen; it belongs to the family Oplophoridae.

mantle cavity, formerly placed anteriorly owing to the torsion exhibited by the ancestral forms, became more posteriorly situated and open to avoid clogging and fouling during the movements of burrowing. The shell became more streamlined in shape and elongated anteroposteriorly so as to permit subterranean locomotion, and the primitive coiling of the visceral hump was lost as the digestive gland, stomach, and ovotestis came to be accommodated as much within the foot as in the increasingly ineffectual shell. Great lateral or parapodial lobes of the foot gradually developed, and these, together with newly evolving posteriorly directed flattened lobes of the cephalic shield, helped to enhance streamlining and to exclude foreign particles from the shell and mantle cavities. Acteonid opisthobranchs (as well as a few strange species of *Retusa* brought up by a recent Soviet oceanographic expedition) still retain the operculum, but most other primitive opisthobranchs lost this passive organ of defense long ago and thus demonstrated a more complete commitment to the burrowing mode of life.

From the early burrowing opisthobranchs, with their growing emphasis on chemical and biological defense, and concomitant reduction in the importance of the shell, together with their decisive and irreversible loss of the operculum, arose a number of quite distinct lines of descent, the modern representatives of each of which constitute a group of ordinal rank. Each of the orders of living opisthobranchs shows groups of species, genera, and families that have taken up new, often rather restricted, dietary specialisms.

There are many groups of prosobranchs (especially among the Archaeogastropoda and the Mesogastropoda) which have slug-like forms; the shell has become reduced in importance. This tendency reaches its zenith in the Lamellariidae, which are mesogastropods in which the shell is fully internal (enclosed by the mantle), although it is large and spirally coiled like the shell of a cowry. The mantle cavity contains a well developed ctenidial gill, so they should not be mistaken for the onchidiids or the ' true ' opisthobranch sea slugs.

Lamellaria perspicua is found commonly in shallow waters in the northern Atlantic, usually on hard substrates where its prey, consisting of encrusting compound ascidians, abounds. The slugs move about little so long as food remains available. It has been noticed that the upper surface of the mantle may be modified in color and in texture so as to closely resemble the ascidian colony. The slug may actually eat its way into the prey in an attempt to effect further concealment. The mantle epithelium is equipped with acid glands that presumably function to dissuade an inquisitive fish which has been sharp-eyed enough to penetrate the visual disguise.

The color patterns of the dorsal mantle are extremely variable. The most interesting individuals are those in which the mantle has adaptations clearly designed to frustrate recognition. One such way is for the mantle to bear patches of red which resemble small colonies of red sponges, but the most remarkable case that has come to attention was the presence in two specimens of mantle processes colored and shaped exactly like intertidal acorn barnacles. It can be assumed that such perfect resemblances to their surroundings significantly benefit such individuals, but no experiments have been carried out using lamellariids and such potential predators as fish. Nor has it yet been established that such mantle patterns are inheritable.

The tropical lamellariids *Mysticoncha wilsoni* and *Coriocella nigra* are a great deal larger than *Lamellaria* and may reach a length of 6 cm. While *Coriocella* is well camouflaged, it appears that *Mysticoncha* exhibits vivid warning or aposematic coloration. Both are found occasionally on coral reefs in the Pacific Ocean. Nothing is known about their diet or their other habits.

Pulmonate sea slugs all belong to the Onchidiidae, a marine family of wide distribution (from Vancouver to Madagascar). The first onchidiid was described in 1800 from the swamps of Bengal (India). There are two genera, *Onchidium* (with principally tropical species inhabiting mangrove swamps) and *Onchidella* (comprising a small

Rock shrimp *(Sicyonia)* are close relatives of *Penaeus* but have much heavier and rougher bodies. Above is a nondescript Atlantic species (D. Faulkner photo); below is the interestingly patterned *Sicyonia penicillata* (East Pacific; A. Kerstitch photo).

Pandalus platyceros, a large pandalid shrimp of commercial importance from off California (D. Gotshall photo).

Two specimens of the snapping shrimp *Synalpheus stimpsoni* on an arm of their host crinoid; notice the resemblance of patterns between the shrimp and the crinoid (Pacific; L.P. Zann photo).

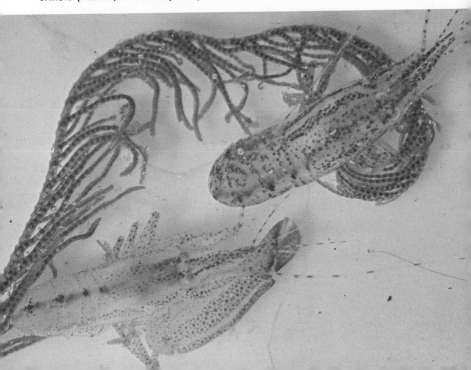

number of species living on rocky shores in more temperate regions). They are all air-breathing, having a small lung below the mantle rim at the rear of the body. Their cephalic eyes are situated on the tips of the head tentacles (one on each side of the head), indicating the affinity between the Onchidiidae and certain families of land-slugs. Plainly the onchidiids represent a line of evolutionary return into marine habitats by some of the descendants of the early terrestrial slugs.

Onchidiids are herbivorous and undertake periodic wanderings in search of food. According to some authorities, they have the ability to return to a ' home-site, ' which may be a particular crevice or indentation of the rock surface, after each excursion. If alarmed, an *Onchidium* may expel a poisonous stream from defensive glands around the mantle rim. In the Bermudan *Onchidella floridanum* such jets of venom may reach targets up to 15 cm away. The venom has been found in scientific tests to be repugnant to anemones, starfish, crabs, and fish. The mantle bears (in species of *Onchidium*) tall conical papillae, each of which bears a terminal eyespot and is highly contractile. These tiny pallial eyes are actually quite like those of some vertebrates, but we are uncertain how much they can discern. There may be as many as 100 of these eyes on the dorsal surface of a single *Onchidium*.

Although slug-like forms crop up in the Bullomorpha and Pleurobranchomorpha, and the Acochlidiacea contains numerous microscopic naked slugs, we shall have space only to deal with the three largest orders of opisthobranch sea slugs: the herbivorous Aplysiomorpha and Sacoglossa and the carnivorous ' true ' sea slugs of the Nudibranchia.

The aplysiomorphs form a compact group all of which feed upon the larger marine algae, especially Chlorophyceae, and are therefore restricted to the shallow continental shelves. The shell has become very much reduced and internal (roofed over by the mantle) and may be altogether lacking in some cases (such as *Dolabrifera*). The largest aplysiid, *Dolabella*, may reach 2 kg or more in live-weight, and aplysiids are certainly the largest living sea slugs. Despite this, many species (though not of the genus *Dolabella*) have acquired the ability to swim, correlated with their evolutionary reduction or loss of the shell. Most swimming aplysiids employ muscular movements of lateral parapodial lobes, but in *Notarchus* swimming is effected by a form of jet propulsion found elsewhere among the Mollusca only in the Cephalopoda.

Members of the order Sacoglossa have two common features: a herbivorous habit and the possession of a uniseriate radula in which the oldest, often broken, teeth are usually retained within the body and not discarded. These teeth are stiletto-like and are used to puncture individual plant cells so that the fluid contents may be swallowed. Dissociated chloroplasts from the prey may be retained and 'farmed' in the cells of the digestive gland, persisting in *Tridachia crispata* (Florida and the Caribbean Sea) for up to six weeks. Adult *Tridachia* do not need to ingest any more food, but live thereafter literally on sunshine and fresh air. Such a remarkable endosymbiosis is known to occur also in the northern Atlantic sacoglossan *Elysia viridis*, but this species has to carry on feeding throughout life; the products of endosymbiotic chloroplast photosynthesis are inadequate in turbid seas. An interesting exceptional case is the carnivorous sacoglossan *Olea hansineensis*, which feeds upon the eggs of other opisthobranchs and lacks a fully functional radula. *Stiliger vesiculosus* feeds on the eggs of the nudibranch *Favorinus branchialis* but retains the radula in order to slit open the egg cases.

It remains true, however, that the Sacoglossa are predominantly herbivorous; they are the plant-suckers of the Mollusca. Outside this nutritional uniformity, many other aspects of body form are extraordinarily diverse, more so than in any other comparable group of molluscs. Nowhere else but in the Sacoglossa can one find representatives with bivalve shells, with univalve shells, and with no shells at all; with a body form varying from smooth in some species to papillose in others; and with species living in habitats rang-

The red and white antennae of *Alpheus armatus* are a familiar sight to Caribbean skindivers. This species is often found under or near large anemones (D. Faulkner photo).

In the Indo-Pacific there are several commensal associations of burrowing blind snapping shrimp (such as this *Synalpheus* sp.) with gobies (here a species of *Cryptocentrus*). The shrimp's antennae are in contact with the fish, which detects danger, while the shrimp's large claws serve to hold off possible predators (Australia; L.P. Zann photo).

Snapping shrimp *(Alpheus* and *Synalpheus)* often make good aquarium animals. They vary from nearly plain species like the *Alpheus californiensis* above (East Pacific; A. Kerstitch photo) to the strikingly marked unidentified Philippine species below (A. Norman photo).

A sample of the variation in radular teeth of the Opisthobranchia.

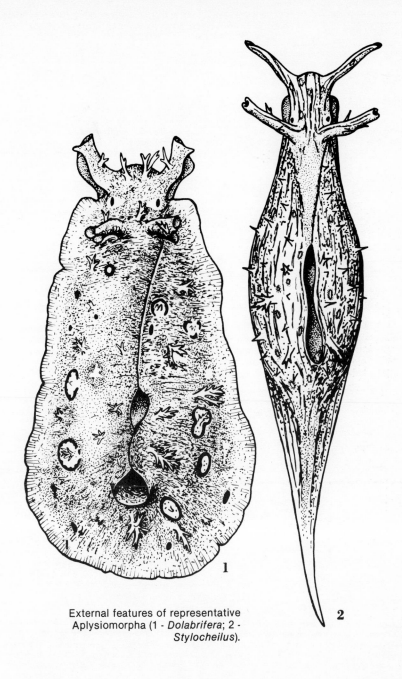

External features of representative
Aplysiomorpha (1 - *Dolabrifera*; 2 -
Stylocheilus).

Rhynchocinetid shrimp are unique in that their rostrum is hinged at the base and thus movable. Their large eyes and bright colors make them popular with aquarists when available. Above is a Hawaiian species of *Rhynchocinetes*, below a chain of many individuals of an Australian *Rhynchocinetes* (A. Norman photo above; R. Steene photo below).

The common nocturnal *Rhynchocinetes rigens* is widely distributed in the Caribbean (P. Colin photo).

Occasionally colorful and bizarre shrimps are imported but prove difficult for the aquarist to name. This unidentified little shrimp is certainly distinctive (Pacific; A. Norman photo).

ing from temperate salt marshes to tropical coral reefs!

In the primitive sacoglossans a relatively strong external shell may be present, and this may be univalve (*Oxynoe*) or bivalve (*Berthelinia*). A gill is present only in *Oxynoe* and *Lobiger*. In the shell-less forms the slug-like body may bear serially arranged dorsal tubercles or cerata, so that *Cyerce* and *Hermaea*, for instance, bear a strong external resemblance to some of the aeolidacean nudibranchs. This is, however, completely spurious, and the Sacoglossa undoubtedly took their origin separately from some early opisthobranch stock and have retained their phylogenetic integrity ever since.

In the Nudibranchia, the largest and most varied order of the opisthobranchs, there is some evidence of polyphyletic descent from a variety of long-extinct bullomorph stocks. In all those lines of nudibranchs which persist to the present day, the shell and operculum have been discarded in the adult form. In many of them the body has quite independently become dorsally papillate. In at least two suborders these dorsal papillae or cerata have acquired the power to nurture nematocysts derived from cnidarian prey, so as to use them for the nudibranch's own defense. In many other cases, such cerata have independently become penetrated by lobules of the adult digestive gland or may contain virulent defensive glands or prickly bundles of calcareous defensive spicules. In the Nudibranchia can be clearly seen the great advantages that accrue to gastropods which have exploited to the full the evolutionary loss of the constraints of passive mechanical defense represented by the shell-operculum system and have replaced this by active dynamic biological and chemical defensive adaptations.

The most obvious feature of a nudibranch is its soft pliability (with some exceptions like the tropical rock-hard *Notodoris*). This in turn permits swimming, an activity only possible in gastropods in which the shell has become flimsy, internal or, as in the nudibranchs, lost. Probably the pressure which led to the adoption of swimming in so many opisthobranchs was the need to escape from slow-moving, bottom-living, predatory carnivores. Many starfish and some crustaceans, for example, will eat virtually any motionless animal material. A number of species of opisthobranchs possess swimming escape reactions which may have evolved so as to escape just such dangers. The swimming mechanism may involve convulsive dorsoventral muscular contractions of the whole body (as in *Tritonia*), lateral undulations of the flattened body (as in *Bornella*), sometimes in an upside-down attitude, or it may be accomplished by the movement of certain parts of the body especially modified for the task. The North American *Cumanotus beaumonti* possesses dorsal cerata which can in unison perform up and down movements not unlike those of a pigeon's wings, pivoting at their base. The muscles that bring this about are of the striated type, but their functional arrangement has not yet been described. When the animal is abruptly disturbed, repeated downward sweeps of the cerata propel the body away from the sea bottom for minutes at a time.

The beautifully marked Indo-Pacific dorid nudibranch *Hexabranchus sanguineus* swims in a fashion that is unique among the gastropod molluscs and paralleled only by certain cephalopods. When at rest, the margins of the mantle are dorsally enrolled and slow locomotion then occurs by creeping on the broad pedal sole. In swimming, however, the mantle rim is spread out, dilated, and strong locomotor waves are propagated rearward from the anterior margins. At the same time the whole body undergoes great dorsoventral flexions. Swimming may continue for many minutes, and probably this is more than a simple escape reaction. Certainly a brilliant flash of color accompanies the unrolling of the mantle edge, and swimming may be the behavioral component of a warning display.

The aeolidacean nudibranchs *Glaucus atlanticus* and *Glaucilla marginata* contrive to stay all their adult lives in the surface layers of the ocean (where they feed upon the planktonic coelenterates *Velella* and *Porpita*, but especially on the Portuguese man-of-war, *Physalia*). They do so by a novel behavioral modification. Each aeolid gulps in air

Palaemon, a "typical" caridean or palaemonid shrimp. Numerous species and genera of this type of shrimp are common in brackish, fresh, and shallow marine environments.

Humpbacked shrimp, *Saron marmoratus,* are common reef shrimp of the Indo-Pacific. They are easily recognized by the distinctively ringed legs and small circles on the body, plus the tufts of bristles. Females (shown here) have short front legs with many bristles, while males have extremely long legs (A. Norman photo above; T.E. Thompson photo below).

from above the water surface and holds a bubble inside the stomach. These glaucid nudibranchs are thus passively planktonic, buoyed up in an upside-down posture. The skin of the body is delicately countershaded by blue and white pigment so as to give camouflage, but because of the head-over-heels posture, it is the foot which is darkest blue and the back or dorsum which is silvery.

All nudibranchs are carnivorous, and each family contains species which feed on broadly similar types of prey. For example, the Coryphellidae, Dendronotidae, Eubranchidae, Facelinidae, Heroidae and Lomanotidae contain species which feed principally upon Hydrozoa. Many species of the Aeolidiidae feed on Actiniaria (although the American *Phidiana pugnax* feeds chiefly on other aeolid nudibranchs), and the Glaucidae all attack chondrophores and siphonophores. Members of the Tritoniidae all feed on alcyonarians. The Antiopellidae and Onchidorididae contain species that attack bryozoans; *Onchidoris bilamellata*, however, feeds upon acorn barnacles.

A feeding type which has proved successful in members of several opisthobranch families is the egg-consumer. This habit crops up in the aeolids, such as *Favorinus branchialis* and *Calma glaucoides*, as well as in the sacoglossans *Stiliger vesiculosus* and *Olea hansineensis*. Molluscs like these are often very pale in color and pass unnoticed among cream-colored or white spawn. Whereas *Calma* feeds on the eggs of cephalopods and teleost fishes, *Favorinus*, *Stiliger*, and *Olea* take the eggs of other opisthobranchs. On examining such predators, it is usual to detect not only ova and young cleavage stages in the stomachs, but also late veligers of the molluscan prey. In *Calma* the anal opening is closed off in postlarval stages, presumably because there are few solid waste materials remaining after the digestion of such yolky food. *Olea* and *Favorinus* are less firmly committed to a diet of embryos, and in these genera the anus remains functional throughout free life. The West African *Favorinus ghanensis*, moreover, does not feed on eggs but instead crops the polyps of the bryozoan *Zoobotryon verticillatum*.

Another highly successful specialization has evolved independently in the Fionidae (*Fiona*) and the Glaucidae (*Glaucus* and *Glaucilla*). It involves attacking planktonic cnidarians (*Porpita, Velella, Physalia*). *Fiona* is said to also attack stalked barnacles, and a recent report described *F. pinnata* feeding on *Lepas anatifera* in Californian waters. These prey organisms are extremely common in the oceans, and glaucid and fionid nudibranchs are found in all warm seas. Both *G. atlanticus* and *F. pinnata* are circumtropical in distribution.

Many dorid nudibranchs are well adapted for feeding upon encrusting marine sponges and possess dorsoventrally flattened bodies, a more or less oval outline, and a broad flattened foot. Such a body form is also exhibited by other nudibranchs which browse on flattish encrustations (for example *Alcyonium*, bryozoan and ascidian colonies, and barnacle aggregations). Their radula is usually broad and they lack strong jaw plates. In their natural surroundings they are frequently well camouflaged through their shape, coloration, and behavior. Mysteriously, these remarks do not apply well to the chromodorid nudibranchs of warm seas. These feed upon encrusting sponges and possess a broad radula and feeble jaws, but they are often extraordinarily vividly marked. Nor do they apply to the tiny *Okadaia elegans*, which is a dorid that in Hawaii drills through the calcareous tubes of serpulid polychaetes so as to consume the worms within, or to the dendrodorid nudibranchs of warm seas which feed upon sponges by the novel method of sucking up the sponge tissues through a strongly developed proboscis; the radula is totally lacking.

Nudibranchs which feed upon arborescent, more erect organisms (for example hydroids, actiniarians, and some bryozoans) are usually of smaller size and of more elongate shape with a long narrow foot adapted for clinging to this type of prey. Nudibranchs that fall into this category are active voracious forms in which cannibalism may often be seen in laboratory aquaria. Jaws are often well developed and the radula shows a tendency toward reduction in

Bumblebee shrimp are small, usually colorful gnathophyllids often associated with sea urchins or other echinoderms. Above is the brightly spotted *Gnathophyllum panamense* (East Pacific; A. Kerstitch photo); below is the more familiar *Gnathophyllum americanus* (tropical; Savitt and Silver photo).

Although it belongs to the family Gnathophyllidae along with the bumblebee shrimp, *Hymenocera picta* bears little obvious resemblance to its relatives. The bright pattern, large flattened claws, and adaptability make it a familiar aquarium species (Indo-Pacific; A. Norman photo).

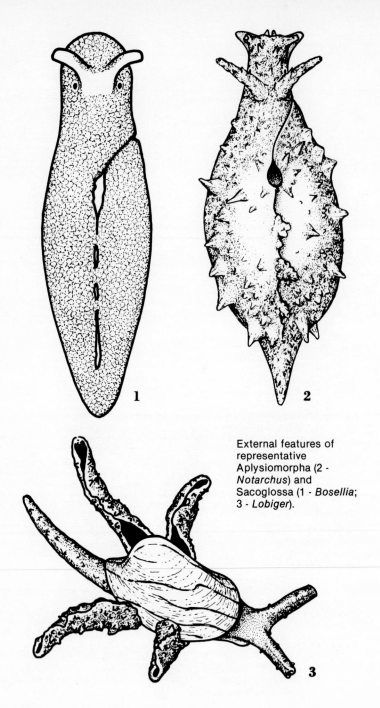

External features of representative Aplysiomorpha (2 - *Notarchus*) and Sacoglossa (1 - *Bosellia*; 3 - *Lobiger*).

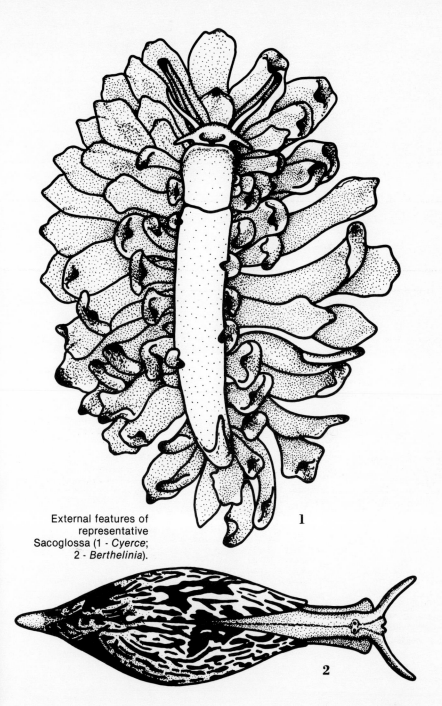

External features of
representative
Sacoglossa (1 - *Cyerce*;
2 - *Berthelinia*).

1

2

Hymenocera picta feed on the tubefeet of starfish and other echinoderms, the relatively small shrimp being able to turn over annd kill a large starfish (K. Lucas photo at Steinhart Aquarium).

Lebeus grandimanus, a brightly colored cold-water hippolytid from the East Pacific (T.E. Thompson photo).

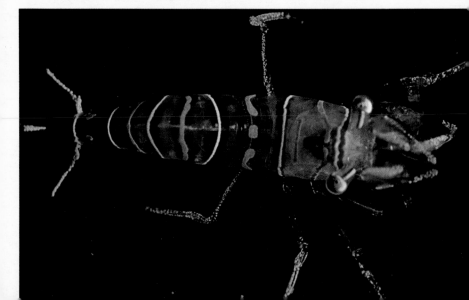

Most species of *Lysmata* are attractively but inconspicuously colored, often with red and white stripes like this East Pacific species (A. Kerstitch photo).

Lysmata grabhami, however, is an exception in the genus as it is vividly striped with red and has nacreous white antennae; it is one of the two most widely known cleaner shrimp and is found throughout the tropics.

width. In extreme cases the radula may exhibit only one (central) tooth in each row (*Aeolidia, Doto*).

So much for the feeding activities of the nudibranchs; but what are their enemies, animals that may prey upon them? Aquarium experiments showed that nudibranchs were almost invariably refused as food by a variety of hungry fishes. Occasionally a fish would take a proffered nudibranch into its mouth, but nearly always rejected it violently and immediately.

We are beginning to understand the functioning of some of the chemical and biological defensive mechanisms employed by these naked gastropods. First of all, it is important to note that any individual species may include in its repertoire several different defensive adaptations. For instance, *Discodoris planata* spends most of its life on sponge-encrusted boulders sublittorally, on which it may be almost impossible to see it. But if nosed out by a hungry fish it can in a crisis expel quantities of sulfuric acid from multicellular acid glands in the mantle. It has been known for many years that teleost fishes have an intense dislike of anything tasting acidic, and many species of opisthobranchs (as well as the prosobranch *Lamellaria*, mentioned earlier) have capitalized upon this.

It happens that sulfuric acid is relatively easy to identify chemically, so we know a good deal about this phenomenon of acid secretion in gastropods. Other naked forms produce other, quite different, defensive fluids, the nature of which has not so far been analyzed. In all the nudibranchs adequately studied skin glands have been found, the position and function of which indicate that they must be defensive. These glands are usually most common in those areas of the skin which would first be encountered by an inquisitive predator. In species like the North American *Polycera tricolor*, where dorsal papillae are present, these glands are usually concentrated there. Papillae of this kind are usually non-retractile unlike the gills and rhinophores and can be rapidly regenerated if damaged. The defensive glands vary greatly in structure; they may be superficial or below the skin, unicellular or multicellular; their secretion may be exclusively fluid or may contain glistening concretions. The secretion may be drab or colorful, and it may or may not taste unpleasant to the human tongue. More than one kind of gland may of course be present in a single species.

Although the chemical make-up of the majority of nudibranch non-acidic defensive secretions is almost completely unknown, it seems likely that they will be shown to be proteins. Some of them have proved in laboratory tests to be dramatically poisonous to other animals.

A biological defensive mechanism has been found in the aeolidacean nudibranchs (of which *Aeolidia papillosa* is a familiar American example), where the nematocysts of the anemone or other coelenterate prey are utilized for the defense of the predator. We now know that nematocysts are separated in the aeolidacean alimentary canal from the nutrient parts of the meal and passed through special ducts to storage sacs at the tips of the dorsal papillae or cerata. These sacs are placed below external pores at the ceratal tips, so if one of the cerata is nipped the nematocysts burst free. Once they reach the seawater medium they discharge, with the probable consequence that the hungry fish abandons the attack.

A similar biological defensive mechanism occurs independently in the dendronotacean family Hancockiidae, but here the nematocysts are stored not only in the projecting cerata but also beneath the skin on the sides of the body.

When attacked an aeolidacean nudibranch usually holds the cerata erect and may muster them together aimed toward the enemy. In other words, attacks are invited toward the ceratal tips and directed away from the vital head and body. These defensive adaptations may not be infallible, but they seem to function adequately.

There has been much argument about the significance of coloration in nudibranchs. The cerata of the aeolids and the dorsal papillae of many dorids or dendronotaceans are frequently patterned in such flamboyant ways as to foster the

Although it lacks the large claws of *Stenopus,* the other common cleaner shrimp, *Lysmata grabhami* is an effective cleaner that is tolerated by reef fishes. Shown on this and the facing page are cleaners working on the butterflyfish *Chelmon rostratus.*

theory that in nature these serve to attract the attention of a visually searching predator away from the less conspicuous (but more fragile and precious) head and visceral mass.

In many cryptically marked nudibranchs (as well as in some sacoglossans and aplysiomorphs) the resemblance in color pattern to the normal surroundings is so perfect that it would appear to disguise these molluscs. But against this it can be argued that many opisthobranchs appear to live predominantly in situations where there is little light (for instance in deep waters or under boulders) and their color patterns cannot therefore be perceived by potential enemies. The truth is that we have very little accurate information about the method of selection of food by predatory fish in different kinds of marine habitats.

ORDER SACOGLOSSA

Characters of the order:

Shell external (univalve or bivalve), internal, or absent in the adult phase; operculum absent in postlarval stages; body shape may be limaciform, aeolidiform (with dorsolateral cerata), or flattened and leaf-like; if cerata are present, they lack cnidosacs (sacs of nematocysts); head bearing digitiform, auriform, or enrolled rhinophoral tentacles (occasionally absent); mantle cavity and gill usually lacking; foot elongated, usually closely united with the head and visceral mass; gizzard plates lacking; potash-resistant jaws absent; radula narrow, uniseriate; abraded or broken teeth are not always discarded from the radula ribbon, and many or all of the worn teeth may be retained in a special sac (the ascus); pharynx often with a muscular buccal pump; digestive gland often much-branched, with tributaries in the head, tentacles, foot, and cerata (where present); hermaphroditic reproductive system usually lacking an external seminal groove; penis often armed with a stout stylet; impregnation often hypodermic; central nervous system euthyneurous, forming a ganglionic ring around the foregut, behind the buccal mass. In habit, most sacoglossans are herbivorous, feeding principally (but not exclusively) upon green algae.

SUBORDER I. CYLINDROBULLACEA

This is the most primitive suborder of living Sacoglossa and contains the family Cylindrobullidae with the single genus *Cylindrobulla* Fischer, 1857. There is an external univalve shell and an external seminal groove on the right side of the head. *Cylindrobulla* is an interesting link between the ancestral bullomorphs and the remainder of the Sacoglossa.

SUBORDER II. VOLVATELLACEA

Members of the single living family Volvatellidae have an external univalve shell (terminated fore and aft by a siphonal ' spout ') but, lacking an external seminal groove along the side of the head, possess an internal vas deferens. There is a single genus, *Volvatella* Pease, 1860.

SUBORDER III. JULIACEA

These are the famous ' bivalved gastropods ' found alive for the first time only recently (by the Japanese workers Kawaguti and Baba), feeding on the green alga *Caulerpa*. They all fall readily into a single family, the Juliidae, and there is little reason to utilize for them more than one generic name, *Berthelinia* Crosse, 1875. The body is laterally compressed and enclosed in a bivalved shell having a dorsal hinge and a conspicuous spiral protoconch adherent to one valve.

SUBORDER IV. OXYNOACEA

Members of this suborder resemble the bivalved gastropods anatomically but differ in that the oxynoaceans have a univalve shell. This shell is inflated and very fragile. The two best known genera are *Oxynoe* Rafinesque, 1819 and *Lobiger* Krohn, 1847. They are able to cast off (' autotomise ') parts of the mantle and foot if abruptly disturbed.

Periclimenes is a diverse genus with numerous species, but most are colorless non-cleaners and commensals of little interest to aquarists (Caribbean; P. Colin photo).

At the other extreme are colorful *Periclimenes* that, from the white antennae, are probably cleaners. The identification of this species is uncertain.

Periclimenes imperator is a commensal that lives on the large "Spanish dancer" nudibranchs and also on some sea cucumbers. It presumably feeds on small bits of host tissue and detritus (Indo-Pacific; A. Norman photos).

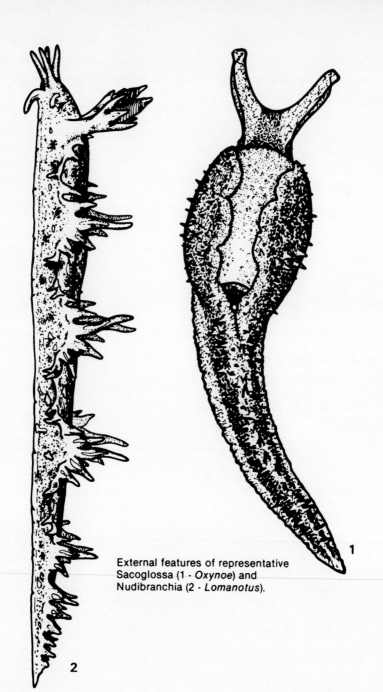

External features of representative
Sacoglossa (1 - *Oxynoe*) and
Nudibranchia (2 - *Lomanotus*).

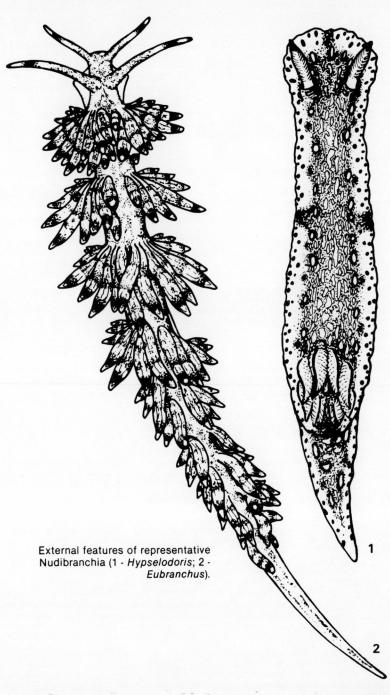

External features of representative
Nudibranchia (1 - *Hypselodoris*; 2 -
Eubranchus).

1

2

427

Contrasts in *Periclimenes*. Above, the brilliantly spotted *Periclimenes colemani* lives in pairs on sea urchins. Below, the slender *Periclimenes cornutus* has the knack of matching the colors of its host (Pacific; L.P. Zann photos).

Periclimenes brevicarpalis, a free-living species often associated with large anemones, is sexually dimorphic. The small, nearly transparent male (above) also differs in form from the large, more brightly colored female (below) (Indo-Pacific; Dr. H.R. Axelrod photo above, K.H. Choo photo below).

SUBORDER V. POLYBRANCHIACEA

This suborder contains a number of shell-less forms which are usually flattened and leaf-like in shape but frequently bear dorsolateral papillae or cerata. These cerata contain tributaries of the digestive gland and, sometimes, branches of the reproductive organs.

The family Polybranchidae is characterized by the possession of cerata which are flattened and of head tentacles which are large, complex, and enrolled. Typical genera are *Polybranchia* Pease, 1860, *Cyerce* Bergh, 1871, *Caliphylla* Costa, 1867, *Olea* Agersborg, 1923 (an unusual form which feeds on the eggs of other opisthobranchs, up to 20 eggs per minute), and *Bosellia* Trinchese, 1890 (a subtropical and tropical genus containing species which are very flattened and live upon ' leaves ' of the calcareous alga *Halimeda*).

The other family, the Hermaeidae, contains sacoglossans in which the head tentacles are simpler (sometimes the oral tentacles are reduced or absent) and the cerata are finger-like, not flattened. The best-known genera are *Hermaea* Loven, 1841, *Stiliger* Ehrenberg, 1831, and *Alderia* Allman, 1845.

SUBORDER VI. ELYSIACEA

Although some species may bear a felting of tiny papillae, the dorsal surface in the elysiaceans lacks dorsolateral ceratal tubercles. There are three families, each of which is best known by a single successful genus.

The Elysiidae, characterized by the world-wide genus *Elysia* Risso, 1822, contains forms which have flattened leaf-like bodies and expanded lateral foot-lobes or parapodia (occasionally used for swimming). In *Tridachia* Deshayes, 1857, these parapodial lobes are much folded so as to increase the surface for insolation of the endosymbiotic chloroplasts.

Placobranchus Hasselt, 1824, the only genus of the Placobranchidae, is flattened, with longitudinal ridges over the dorsum.

The remaining elysiaceans belong to the family Limapontiidae, with the single genus *Limapontia* Johnston, 1836. These are the most drab of the sacoglossans, having dark coloration, smooth, featureless bodies, and extremely small size (usually less than 3-4 mm in overall length).

ORDER NUDIBRANCHIA

Characters of the order:

Shell and operculum absent in the adult phase; calcareous spicules may be present in the skin; body shape may be smooth and limaciform, aeolidiform (with dorsolateral cerata), or flattened; if cerata are present, they may contain cnidosacs (sacs of nematocysts); head often bearing both oral and rhinophoral tentacles, the latter often wrinkled, lamellate, or branched and sometimes retractable into elaborate pallial sheaths; mantle cavity absent; gill may take the form of a simple set of folds along the sides of the body or may be a crescent or a ring of structurally complex branchial appendages situated at the rear of the body; foot elongated, occasionally broad, closely united with the head and visceral mass; calcareous gizzard plates lacking, horny plates rare; potash-resistant jaws often present; radula very variable, uniseriate to broad, with or without the median tooth in each row; abraded or broken teeth are discarded, not retained in a special sac; pharynx sometimes with a muscular buccal pump; digestive gland compact or much-divided, with tributaries from the head, foot, and cerata (where present); hermaphrodite reproductive system lacking an external seminal groove; penis sometimes armed with a stout stylet; impregnation never hypodermic; central nervous system euthyneurous, forming a ganglionic ring around the foregut.

SUBORDER I. DENDRONOTACEA

This suborder contains ten families of nudibranchs, all possessing a midlateral anal papilla and distinct, often elaborate, sheaths for the rhinophoral tentacles to be pulled into after an alarm. They feed upon coelenterates as varied

The violet-spotted *Periclimenes pedersoni* was one of the first cleaners studied in any detail. The female above is ovigerous (in berry). Note the strong contrast of the white antennae, the typical sign of the cleaner, against the background (Caribbean; P. Colin photo above, D. Reed photo below).

Periclimenes yucatanicus is another common and brightly colored Caribbean species, but it seems that it does not normally clean fishes (C. Arneson photo).

as medusae (in the case of the Mediterranean *Phylliroe bucephala*), hydroids and anemones (in European and North American species of *Lomanotus, Hancockia, Dendronotus* and *Doto*) or soft alcyonarian and gorgonian corals (preferred by the world-wide species of *Tritonia* and *Marionia*). Horny jaws may or may not be present, and the radula shows a great deal of variation in the different families, from broad and multiseriate (having many rows) (Tritoniidae) to uniseriate (*Doto*), resembling the radula of some of the aeolidaceans. The body has a dorsolateral ridge on each side of the body, frequently bearing arborescent processes which function as gills. Sometimes these gills contain tributaries of the alimentary canal and a dendronotacean such as *Doto* can be mistaken for a true aeolidacean by the unwary student (the presence of rhinophore sheaths in *Doto* is decisive evidence for its retention in the Dendronotacea). Finally, mention must be made here of *Tethys* and *Melibe*, which occur on American coasts and lack both jaws and radula. They feed by casting about in muddy eelgrass and other rich areas for small crustaceans which are captured with the aid of a dilated fimbriated buccal hood and swallowed whole.

The families of the Dendronotacea are as follows (with representative genera in brackets): Tritoniidae (*Tritonia, Marionia, Tochuina*), Marianinidae (= Aranucidae) (*Marianina*), Lomanotidae (*Lomanotus*), Dendronotidae (*Dendronotus*), Hancockiidae (*Hancockia*), Bornellidae (*Bornella*), Dotoidae (*Doto*), Tethyidae (= Fimbriidae) (*Tethys, Melibe*), Scyllaeidae (*Scyllaea*), and Phylliroidae (*Phylliroe, Cephalopyge*).

SUBORDER II. DORIDACEA

This is the largest nudibranch suborder, and in it are placed about 25 families. Most of these families are distributed throughout world seas, but some of them (Platydorididae, Asteronotidae, Baptodorididae, Actinocyclidae, Bathydorididae, and Doridoxidae) are absent or rare in cooler northern hemisphere waters, while others are quite decidedly tropical in their preferences (Chromodorididae, Hexabranchidae). The Corambidae, Goniodorididae, Onchidorididae, Polyceridae, Cadlinidae, and Archidorididae are families that seem to be centered chiefly in the cool northern temperate regions.

The doridaceans often possess rhinophoral cavities into which the tentacles may be retracted, but they only rarely have external sheaths surrounding their bases. The anal papilla is nearly always situated midposteriorly, under the mantle rim in *Corambe* and *Doridella* (= *Corambella*) (and, incidentally, in all dorids at an early developmental stage), but situated dorsally in all the typical forms, such as *Doris, Archidoris, Polycera, Adalaria, Hexabranchus*, and many others. A variable number of retractile foliaceous gill plumes emerge from the mantle surface close to the anal papilla. In typical dorids the gill plumes are arranged in a circlet and may be retracted in a coordinated way into a dorsal pocket on alarm in *Archidoris, Discodoris, Jorunna, Chromodoris*, and *Asteronotus* (and many other dorid genera). This pocket may, in *Asteronotus cespitosus* from the South China Sea, be distinctly crenulate so that the lips interlock and give greater resistance to an inquisitive fish. In less advanced dorids, such as *Laila, Polycera, Crimora, Adalaria*, and *Hexabranchus*, no such pocket is present and each gill is contracted separately down to the mantle surface on alarm. Sometimes large spiculose mantle papillae give extra protection to the gills and rhinophores, often containing elaborate defensive glands and luminescent or colored lures to draw the attention of a predator away from the fragile vital parts. Such papillae never contain lobes of the digestive gland in the Doridacea, and the liver usually forms a more or less compact single mass close to the stomach. The radula is usually rather broad and certainly never uniseriate. One successful family of dorids, the Dendrodorididae, contains species which lack the radula and jaws and ingest sponges by an ingenious sucking modification of the muscular buccal mass. Sponges, bryozoans, acorn bar-

Periclimenes lucasi, an East Pacific relative of *Periclimenes pedersoni* (A. Kerstitch photo).

A pair (female larger and sitting on gills) of the commensal shrimp *Conchodytes meleagrinae* in the cavity of an opened black-lipped pearl oyster, *Pinctada margaritifera* (Pacific; L.P. Zann).

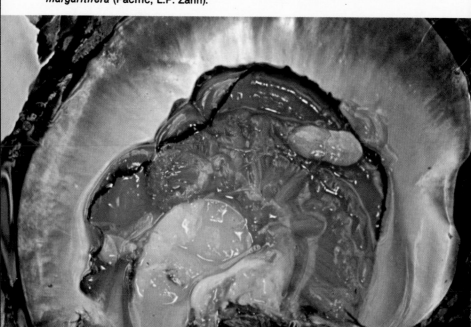

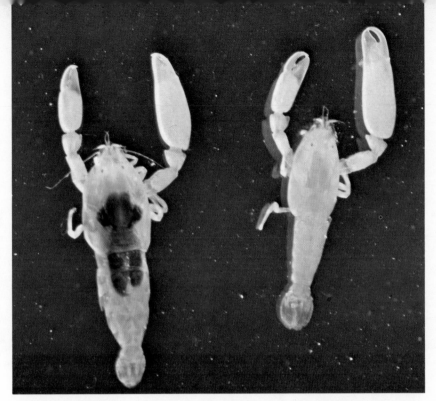

A pair of commensal shrimp, *Pontonia pinnae,* from pen shells *(Pinna);* notice the unequal development of the claws and the relatively stouter abdomen of the female (East Pacific; A. Kerstitch).

The golden cleaner shrimp, *Stenopus scutellatus,* is seen less often than its banded cousin but it is also an effective cleaner (Caribbean; K. Lucas photo at Steinhart Aquarium).

External features of representative Nudibranchia (1 - *Phyllidiopsis*; 2 - *Trapania*; 3 - *Hexabranchus*).

438

nacles, compound sea squirts, and tube-dwelling polychaete worms form the diet of most dorid nudibranchs.

The families are as follows (with representative genera): Doridoxidae (*Doridoxa*), Bathydorididae (*Bathydoris*), Corambidae (*Corambe, Doridella* (= *Corambella*)), Goniodorididae (*Goniodoris, Okenia, Ancula, Trapania, Hopkinsia*), Onchidorididae (*Onchidoris, Adalaria, Acanthodoris*), Triophidae (*Triopha, Plocamopherus, Crimora, Kalinga*), Notodorididae (*Notodoris, Aegires*), Polyceridae (*Polycera, Thecacera, Laila, Limacia*), Gymnodorididae (*Gymnodoris, Nembrotha*), Vayssiereidae (*Vayssierea, Okadaia*), Hexabranchidae (*Hexabranchus*), Cadlinidae (*Cadlina*), Chromodorididae (*Chromodoris, Hypselodoris, Casella, Miamira, Ceratosoma*), Actinocyclidae (*Actinocyclus, Hallaxa*), Aldisidae (*Aldisa*), Rostangidae (*Rostanga*), Dorididae (*Doris, Austrodoris, Alloiodoris*), Archidorididae (*Archidoris, Trippa, Atagema*), Homoiodorididae (*Homoiodoris*), Baptodorididae (*Baptodoris*), Discodorididae (*Discodoris*), Kentrodorididae (*Kentrodoris, Jorunna*), Asteronotidae (*Asteronotus, Sclerodoris, Halgerda, Aphelodoris*), Platydorididae (*Platydoris, Hoplodoris*), Phyllidiidae (*Phyllidia*), and Dendrodorididae (*Dendrodoris*).

SUBORDER III. ARMINACEA

This suborder is more difficult to define and certainly more difficult to recognize from external features alone. The nine families currently contained here will certainly not fit into any other suborder, which (to a certain extent) explains our conviction that a separate suborder is needed for them. Species of *Armina* found in the northern Atlantic have externally some of the features of a primitive doridacean, such as *Phyllidia* or *Corambe*, with external pallial leaflets forming a respiratory series beneath the mantle edge, but internally *Armina* proves to be very unlike a dorid for it has a much-divided liver, a laterally placed anal papilla and strong jaws.

Similarly, *Hero*, *Antiopella*, and the Pacific Ocean

The banded cleaner shrimp, *Stenopus hispidus,* is familiar to most divers and aquarists. The large claws, the long arms, and the red and white banded pattern are very distinctive. The species is found throughout the tropics (P. Colin photo).

Recently another stenopid cleaner shrimp has been imported from the Pacific, the ghost cleaner. The brilliant red stripe down the abdomen distinguishes it from other known species. This shrimp was recently described as *Stenopus pyrsonotus* (Philippines; A. Norman photo).

The copper lobster or furry lobster is solid bright orange, flattened, and lacks spines on the bristle-covered carapace. *Palinurellus gundlachi* is known from the Caribbean, but a very similar or identical lobster is now being taken in Hawaii (Caribbean; P. Colin photo).

Close-up of the mouthparts of a spiny lobster, *Panulirus*. The mandibles are hidden by the bristly maxillipeds. Notice the small stalked barnacles attached near the mouth (Pacific; L.P. Zann).

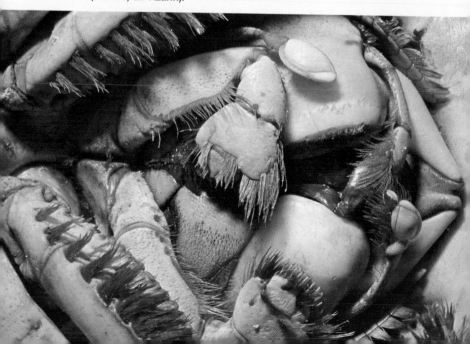

Dirona have a strong resemblance to the aeolidacean nudibranchs which will be dealt with shortly. But their dorsal ceratal processes lack cnidosacs with contained nematocysts, a fact which instantly separates them from the aeolids.

The arminacean nudibranchs do, on the other hand, have certain features in common. Their rhinophoral tentacles do not possess external protective sheaths. The anal papilla is usually situated rather far forward, either dorsally or laterally, on the right side. The radula may be narrow, but it is never uniseriate (with a single tooth in each row). Oral tentacles are usually lacking.

The diets of the arminaceans are varied. *Hero formosa* certainly feeds upon naked (gymnoblastic) hydroids, and *Armina*, too, is said to attack coelenterates (the alcyonarian sea pansies in the case of the best known species, *Armina californica*). But *Antiopella* and *Proctonotus* feed upon encrusting and erect bryozoans, while *Dirona albolineata* is known to devour a wide variety of shelled molluscs and other invertebrates.

The families of Arminacea are as follows: Heterodorididae (*Heterodoris*), Doridomorphidae (*Doridomorpha*), Arminidae (*Armina, Linguella, Dermatobranchus*), Madrellidae (*Madrella*), Dironidae (*Dirona*), Antiopellidae (*Antiopella*), Gonieolididae (*Gonieolis*), Charcotiidae (*Charcotia*), and Heroidae (*Hero*).

SUBORDER IV. AEOLIDACEA

This is a very compact group containing 20 families of delicate, beautifully formed and patterned species. They all bear clusters, groups, or rows of elongated finger-like smooth dorsal ceratal processes. These cerata contain the much-divided lobes of the digestive gland, together with, at their tips, the defensive cnidosacs. The cerata are usually vividly marked in a characteristic way in each species. These cerata are extremely short in the sand-burrowing aeolid *Pseudovermis* but in the majority of the species are the most conspicuous feature of the body. The rhinophoral tentacles are never retracted into pallial sheaths, while the oral tentacles are often very long and graceful and the propodial extremities are sometimes extended so as to form a third pair of anteriorly placed sensory processes.

Jaws are usually well developed, but the radula is reduced to a very narrow ribbon and sometimes bears only a single file of teeth (the uniseriate condition). Aeolidaceans (such as *Aeolidia, Eubranchus,* and *Catriona*) are usually active hunters feeding upon coelenterates. *Glaucus* and *Glaucilla* are planktonic aeolids preying upon siphonophores and chondrophores. The Pacific American species *Phidiana pugnax* attacks other opisthobranchs and even some aeolids as its normal prey, while *Calma* and some species of *Favorinus* devour only the eggs of their molluscan and fish prey. *Favorinus ghanensis* feeds, however, on the bryozoan *Zoobotryon verticillatum* in western Africa. *Fiona pinnata* is a very unusual aeolid of wide geographical distribution which occurs occasionally in American waters and attacks stalked barnacles such as *Lepas.*

The aeolidacean families are listed below, with representative genera in brackets. Notaeolidiidae (*Notaeolidia*), Coryphellidae (*Coryphella*), Nossidae (*Nossis*), Protaeolidiellidae (*Protaeolidiella*), Caloriidae (*Caloria*), Phidianidae (*Phidiana, Moridilla*), Facelinidae (*Facelina, Hermissenda*), Favorinidae (*Favorinus*), Myrrhinidae (*Myrrhine, Phyllodesmium*), Glaucidae (*Glaucus, Glaucilla*), Pteraeolidiidae (*Pteraeolidia*), Herviellidae (*Herviella*), Aeolidiidae (*Aeolidia, Aeolidiella, Cerberilla*), Spurillidae (*Spurilla, Berghia*), Eubranchidae (*Eubranchus, Egalvina*), Cumanotidae (*Cumanotus*), Flabellinidae (*Flabellina*), Pseudovermidae (*Pseudovermis*), Cuthonidae (*Precuthona, Cuthona, Phestilla, Catriona, Tenellia, Tergipes, Embletonia*), Fionidae (*Fiona*) and Calmidae (*Calma*).

CLASS BIVALVIA

Members of the molluscan class Bivalvia include many familiar molluscs such as the oyster, scallop, and mussel.

Justitia longimanus, the longarm spiny lobster, is an uncommon reef lobster with a distinctive color pattern. The first legs are very large and end in a very strong curved claw (Caribbean; P. Colin photo above, Savitt and Silver photo below).

The European spiny lobster, *Palinurus elephas. Palinurus* species are found only in the eastern Atlantic from Europe to South Africa and shouldn't be confused with the similarly named *Panulirus* species. *Palinurus* have very short flagellae on the second antennae, while *Panulirus* have very long flagellae.

Panulirus interruptus, one of the few common spiny lobsters of the East Pacific (D. Gotshall photo).

The class is characterized by the presence of a double shell, made up of two halves or valves, and a lateral compression of the body. The head is greatly reduced in size while the foot varies in size according to its function in each particular species. In general, however, its flattened shape gives rise to the name Pelecypoda, 'hatchet foot,' which is sometimes used to describe bivalves. The mantle cavity is very large in bivalves as the greatly enlarged gills take on the function of feeding organs in addition to that of gaseous exchange.

Much of the general biology of bivalves depends on the gills, and a brief description of their anatomy will illustrate the important role that they play. A bivalve breathes by pumping water in through the incurrent siphon, through the gills and out through the exhalent siphon. The siphons are extensions of the mantle protruding between the valves as the posterior end. Their relative lengths vary according to the depth at which the bivalve lives and the way in which it feeds. The water current is produced by numerous bands of cilia which lie all along the length of the gills around tiny perforations. By rhythmic beating they draw water into the mantle cavity. There are two gills, each consisting of a double plate with a water channel between the plates. The passage of water across the gill is as follows. When water enters the mantle cavity it is drawn through the holes in the gills into the internal water channel from where it flows upward to a larger canal running along the top of the gills. These canals carry the water back to the posterior end of the animal where they unite and discharge into the exhalent siphon.

When the bivalve feeds it secretes a mucous layer over the surface of the gills through which the water current must pass. As it does so any suspended material is trapped in the mucus. This is subsequently collected together and passed to the mouth.

The bivalve shell is another characteristic feature. Although all are basically of the same plan, no two bivalves have exactly the same type of shell, and it is usually possible to identify bivalves solely by examination of the shell.

Typically the bivalve shell consists of two like valves which are broadly oval in outline, convexly curved, and attached to one another dorsally. As the shell grows, new material is secreted by the mantle around the edge of the shell so that the concentric lines so often found on bivalve shells are really growth rings. This basic shell type is modified as each species adapts itself to its environment and, while very many show no change at all, others are drastically altered from the original pattern.

There is also a great variation in size of shell, ranging from only a few millimeters to over a meter. The giant clams, *Tridacna*, are coral-dwelling species which may reach over 150 centimeters in length.

Variation within a species may occur, usually reflecting differences in the substrate. Thus individuals living in smooth light sand tend to have smoother larger shells when compared to members of the same species taken from heavy sticky mud where the shells are rounded and small.

From the above descriptions it becomes obvious that bivalves are relatively sedentary animals, and some are indeed completely sessile. Most are burrowers in sand and mud, the foot being the main burrowing device used. Because of this it has been greatly modified in shape so that it is greatly compressed and blade-like. The actual mechanism of burrowing is a highly coordinated sequence of events causing the protraction of the foot and its penetration into the sand. Here it acts as an anchor for the rest of the body to draw down upon and the whole process is repeated. The actual movement of the foot is achieved by muscular activity together with changes in blood pressure. Many clams can dig very quickly. The razor clams (*Ensis, Solen*, etc.) have a very long extensible foot which enables its owner to shoot rapidly down into the mud. Anyone who has come across the opening of a razor clam burrow on a sandy beach and has tried to dig out the occupant will know just how rapidly they can dig. Some clams, such as *Macoma*, move the body backward and forward as they dig so that the cutting action of the shell can assist in burrowing.

447

The commercial spiny lobster of the Caribbean, *Panulirus argus*. Although several other species occur in the Caribbean, it is the one most likely to be taken by the casual lobster hunter. Its exploitation is carefully regulated in most areas (P. Colin photo).

One of the most attractive, if not the most attractive, spiny lobsters is the Indo-Pacific *Panulirus versicolor*. The starkly black and white carapace and abdominal patterns, with the light antennae and tail fan, set it off from all other species (A. Norman photo).

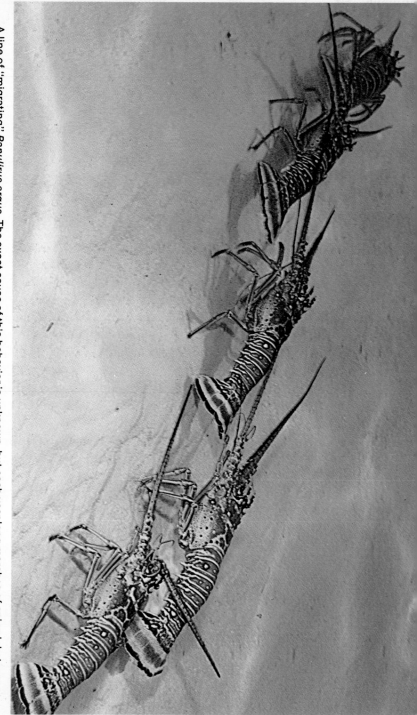

A line of "migrating" *Panulirus argus*. The exact cause of this behavior is unknown, but each year long marches of spiny lobsters are seen in the Caribbean.

Although most bivalves inhabit soft substrates such as sand or mud, a few have invaded clays, rocks, and even wood, boring rather than burrowing into their chosen substrate. Obviously the shape of the body is modified for this new activity. Generally the shell is elongated and the edges of the shell are serrated so that they will help to cut the rock. Boring is achieved by rubbing these roughened edges over the substrate, although different species do this in different ways. In a few genera, such as *Lithophaga*, the date mussels, boring is a chemical process. This bivalve bores into limestone and secretes an acidic mucus from around the edge of the mantle.

Some species, notably the shipworms such as *Teredo*, bore into wood. *Teredo* is a highly modified worm-like bivalve with a greatly reduced shell which is used for drilling. Penetration of the wood takes place immediately after larval settlement, the burrow being enlarged as the bivalve grows. The long thin body, which may reach 30 centimeters in length, is naked. It fits snugly into its burrow, which it lines with a calcareous secretion produced by the mantle. Shipworms can do very extensive damage to wooden pilings, wharfs, and ship bottoms, large pieces of wood becoming like paper as they are honeycombed by burrows. There has been much research into the development of special coatings to prevent larval settlement by these pests.

Not all bivalves are burrowers. Many are sessile animals attaching themselves to a hard substrate. Most sessile clams have a reduced foot. Oysters, for example, have one valve fused onto the substrate. Another group, including the mussels, attach by means of byssus threads which are strong horny threads secreted by a gland in the foot. Other species use a byssus thread to help them move around. *Kellia*, for example, attaches by a single byssus thread. When it moves it throws out a new thread ahead of itself and breaks the old one so that it can draw itself up.

Many clams move around over the surface by extending the foot and then contracting it suddenly so that the body is thrown forward. Razor clams can move over 35 centimeters in this manner, and even the much heavier cockles (Cardiidae) can move in this way.

The ultimate in movement is shown by those bivalves that can swim. This is a feature of the scallops (*Chlamys, Pecten,* etc.). The swimming is a jerky kind of action achieved by water jets. The valves are rapidly closed and water squirted out through openings near the hinge.

Most bivalves are filter-feeders collecting suspended material on the gills as described above. Once the food particles are entangled in mucus, they are passed across the gill filament to a ciliated food groove. The food particles are carried along this groove to palps near the mouth where they are sorted according to size. Smaller particles are retained for ingestion and periodically fed into the mouth as the palps are wiped across it. Larger particles are rejected by the palps and fall onto the foot from where they are eventually ejected through the inhalent siphon. Periodically the valves are closed and water is forced out of the inhalent siphon, taking the waste material with it.

In species such as *Macoma nasuta* the inhalent siphon acts like a vacuum cleaner, collecting the surface layers of detritus for ingestion. Such bivalves could be termed deposit-feeders although the actual feeding process is still one of filtering.

A few highly specialized species have evolved different feeding mechanisms. The shipworms, for example, feed on wood pulp, gaining nutritional value from the cellulose. At this stage it is not known whether the shipworms produce their own enzymes for the breakdown of this substrate or whether they rely on enzymes produced by symbiotic bacteria.

Tridacna, the giant clams, has a very unusual feeding mechanism. As well as feeding in the usual way, it actually grows its own unicellular algae as a food source. The algae, known as zooxanthellae, live in the fleshy tissue of the very large siphons which protrude between the valves. The clam is oriented so that the gape is directed upward. This means that the algae receive the maximum amount of light for

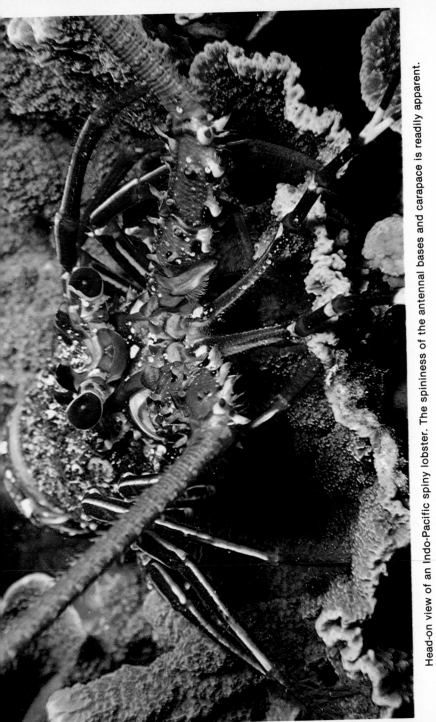

Head-on view of an Indo-Pacific spiny lobster. The spininess of the antennal bases and carapace is readily apparent.

Slipper lobsters or "shovelnosed" lobsters are related to spiny lobsters but have the antennae modified into paddles useful for burrowing. In *Scyllarides,* the most common genus, the front and sides of the antennae are not cut into large teeth. Above is the Indo-Pacific *Scyllarides martensi* (S. Johnson photo); below is the East Pacific *Scyllarides astori* (K. Lucas photo at Steinhart Aquarium).

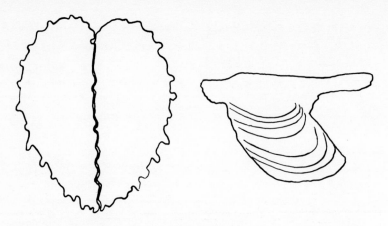

Two unusual bivalves: at left, the heart cockle, *Corculum*, family Cardiidae; at right, *Pteria*, a winged oyster, family Pteriidae.

photosynthesis. In addition some algae can live deeper in the tissue as special crystalloid vesicles in the mantle diffuse the light to a greater depth. At periodic intervals amoeboid cells in the clam will ingest some of the zooxanthellae.

A few bivalves are commensals living in association with other animals. *Pseudopythina*, for example, lives attached to the underside of the sea mouse *Aphrodita* or on the abdomen of a shrimp, and *Phlyctaenachlamys* lives in the burrows of coral-dwelling shrimp. This clam has become so very specialized that its shell is practically absent.

The majority of commensal bivalves live in association with, rather than directly attached to, their hosts. *Cryptomya* lives in the burrow of certain echiuroids such as *Urechis*. *Cryptomya* itself has very short siphons which would normally restrict it to a surface existence, but by housing the siphon openings in another animal's burrow it is able to live at much greater depths where it is protected from predators. *Cryptomya* causes no inconvenience to its host since it feeds independently and is unlikely to take food away from the *Urechis*.

Mytilimeria lives embedded in a compound ascidian. The latter offers protection to the clam so that its shell becomes very thin and fragile.

454

Taking the association a stage further, *Pholadidea* bores into the shell of abalones. This can result in the production of a pearl when the clam actually pierces the shell as the abalone lays down an extra nacreous layer over the irritation. A few clams actually parasitize other animals. *Entovalva*, for example, is parasitic in the gut of a sea cucumber.

Reproduction in bivalves is essentially a very simple process. In most cases the sexes are separate and eggs and sperm are discharged into the seawater, where fertilization takes place. A few bivalves are hermaphrodites and some, such as *Ostrea*, the oyster, change sex. In the oyster fertilization occurs within the parent, the young larvae undergoing development in the gill chamber.

Larval development is similar to that of the gastropods. The egg gives rise to a free-swimming trochophore which in turn develops into the typical molluscan veliger. Unlike the case in gastropods, torsion never takes place.

Each adult may produce millions of eggs in a season. These large numbers reflect the high fatality levels of the larvae. Many are eaten by predators while they are in the plankton and others die at the settlement stage. There has been a lot of research into larval settlement in the commercial shellfish such as *Ostrea* and *Crassostrea*, and it is known that settlement is induced by the presence of certain chemicals in the seawater, sometimes pheromones from adults of the same species.

A few species of bivalve brood their young. The shipworms are an example of this, and in their case the advantages are obvious. By keeping the larvae with the parent they ensure a suitable substrate for the young to bore into.

From this brief review of general bivalve characteristics it should now be possible to look at some individual species in more detail.

Anyone who has walked across a sandy beach will have passed across several species of bivalve, probably without realizing it. To most of us bivalves are the shells washed up on the shore line which the beachcomber collects. We do

Parribacus antarcticus (above) and *Arctides regalis* (below) both have the antennae strongly spined at the margin, but in *Parribacus* the sides of the carapace
are also flattened and cut into large spines. *Parribacus antarcticus* is found in
the Indo-Pacific and Caribbean (A. Power photo); *Arctides regalis* is Pacific (S.
Johnson photo).

Homarus, the true lobsters, differ from other clawed lobsters in having the claws asymmetrically developed, the left with crushing knobs, the right with sharp cutting edges. Shown is the western Atlantic *Homarus americanus,* but the European *Homarus gammarus* is virtually identical (K. Lucas photo at Steinhart Aquarium above; R. Abrams photo below).

A giant clam, *Tridacna crocea*, with the mantle exposed and showing the siphonal opening in the middle and the crystalloid vesicles (Dr. L. P. Zann photo).

not think of the animal living inside the shell. Yet most common beach shells represent animals actually living beneath the beachcombers' feet.

The typical sand-dwelling bivalve of northern waters is represented by *Glycymeria*, which is almost circular in outline, equivalve, and equilateral. This is thought to be the approximate condition of the ancestral bivalve. Such a shape is only suitable for a relatively inactive animal though. A triangular outline is more common in those bivalves which need to shift their position, such as *Cerastoderma* and *Cardium*, the cockles. The more streamlined shape is advantageous if the cockle needs to re-embed itself after storm action has exposed it. Consequently there are many triangular rounded bivalves found intertidally.

Other burrowers have been modified further so that the valves are no longer equilateral and instead the posterior end is produced. This means that the all-important siphons are nearer the surface of the sand.

The shape and form of the siphons also reflect the way of life of a bivalve. The family Veneridae contains many specialized shallow burrowers. *Venerupis pullastra* occurs in patches of sand under stones. In such a shallow-living species the inhalent and exhalent siphons are fused throughout most of their length.

The deeper burrowers such as *Tellina*, *Abra*, and *Scrobicularia* have larger siphons. These are very mobile, and the inhalent siphon plays an important part in the collection of food. In *Macoma*, for example, it sucks up the top layer of detritus and passes it to the gills for filtering. The siphon is only flexible over a small area, however, and the bivalve rapidly exhausts this food source. It is then able to move through the substrate so that the siphon can work over a fresh surface. When the tide is in *Macoma* feeds in the normal filter-feeding manner.

In very deep burrowers such as *Mya* the siphons are protected by a horny sheath which surrounds them.

An extreme example of adaptation to the environment is shown by the razor clams *Solen* and *Ensis*. These have short

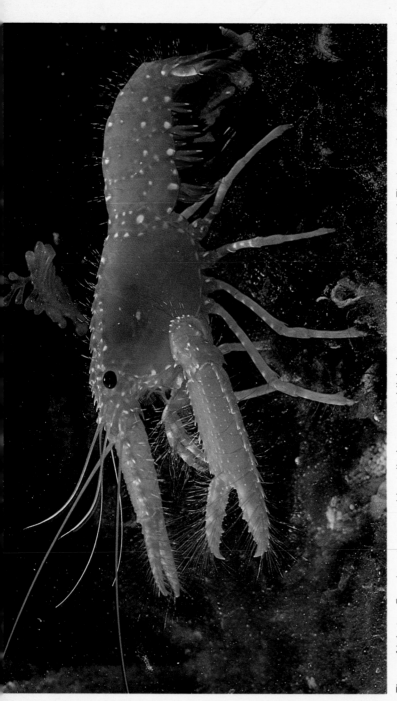

The reef lobster, *Enoplometopus occidentalis*, is now established as a popular aquarium animal. The large claws, long bristles, and bright red coloration make it hard to confuse with anything else. It and a few similar species are found on deep reefs in the Indo-Pacific.

Between the true lobsters (family Nephropidae) and the mud shrimp (family Callianassidae) fall a few small families with intermediate characters. Above is *Axius vivesi* (East Pacific; A. Kerstitch photo) of the family Axiidae. Below is an unidentified axiid-like lobster from Hawaii (S. Johnson photo).

fused siphons for feeding near the surface, but their stream-lined shape and powerful foot enable rapid retreat to deeper layers in the event of danger. The openings of the siphons in *Ensis* and in many other burrowers are frilled to prevent un-wanted debris wafting down and possibly blocking them. A further protective adaptation allows some razor clams to shed the tips of the siphons if they are by chance caught by a predator. Like the lizard losing its tail, this enables the bivalve to escape.

Some inhabitants of muddy shores retain their byssus threads. Originally these were a feature of very young bivalves only, as they enabled the metamorphosing animal to attach to sand grains until the animal could successfully burrow into the substrate. During the course of evolution some bivalves have retained this behavior in the adult stage. Thus *Modiolus*, the horse mussel, attaches onto small stones on the surface of muddy shores. The somewhat flattened shape of the shell may help to prevent its sinking into the mud. The pen shell, *Pinna*, is similarly attached in muddy shores in the tropics.

Attachment by byssus threads is commonly seen in species living on rocky shores. The retention of this postlarval feature may be associated with small size. The minute *Lasaea* is a good example of this. A tiny bivalve with a red-dish shell, *Lasaea* is found in great numbers in rock crevices high in the splash zone. It is possible that its small size could be related to its unfavorable environment rather than to its means of attachment.

Mytilus, a mussel, is the best known bivalve attaching in this way. It occurs in dense colonies on exposed rocky shores, and its general shape may reflect adaptation to these conditions. The posterior end with the siphons is the best developed, while the relatively unimportant anterior end is tucked away.

Reduction of the anterior end has been carried even fur-ther in the pectens, which have completely lost the anterior adductor muscle. Small byssal threads are extruded near the hinge, and *Chlamys*, for example, is found attached on its right side. Members of the clam genus *Lima* have carried the production of byssal threads to extremes. *Lima lima* is quite normal, lying attached by a few threads to the underside of boulders, but *L. hians* lies in a nest of byssal threads. If disturbed it can leave the nest and swim briefly. The byssus threads are used to consolidate a gravelly substrate rather than to attach the bivalve. In saddle oysters the byssus consists of a single calcified column which apparently perforates the lower shell.

Attachment to the rock surface by cementation has occurred several times in bivalve evolution. The best known example is probably the oyster. Oysters are of considerable commercial importance because of their value as a food source, and oyster fisheries are important enterprises. Research and development have led to oyster farms where spat is allowed to grow and settle out under controlled conditions free from predators.

The pearl oysters belong to a different group living mainly in the South Pacific. A pearl is formed when the bivalve secretes thin layers of calcium carbonate around an irritation, the thinness of the layers causing the iridescent colors. Pearls may be induced to develop by artificially placing irritants inside the shell of the bivalve.

Rock boring may have originated in burrowing molluscs which invaded a harder environment or could have resulted from byssally attached bivalves sinking downward. Whatever the method, rock boring has certainly resulted in great specializations. As described above, most species penetrate the rock by the abrasive activity of the valves, *Lithophaga* being the main exception. This genus bores chemically into limestone and is particularly common on coral reefs. *Rocellaria* also bores into coral boulders. Although its shell is fairly thin and appears fragile, it does not use chemical means to eat into the rock, but uses only its shell. It can presumably do this because coral rock is comparatively soft, abrasion being achieved by the rapid opening and shutting of the shell.

Members of the family Pholadidae are well adapted for boring. In these the shell is able to rock about a vertical axis

Two mud shrimp, *Callianassa,* showing the characteristic narrow abdominal base and short eyestalks. The Pacific species (above) is unusually brightly colored (S. Johnson photo), with white being more typical, like the Caribbean species below (C. Arneson photo).

Hermit crabs are common almost from pole to pole. This long-armed hermit, *Pagurus longicarpus,* is one of the most abundant species on the eastern coast of North America (A. Norman photo).

In a very few hermit crabs the abdomen is not curved to fit inside shells. These egg-bearing *Paguritta* females live in old worm tubes in coral heads, thus the nearly straight abdomens. The heavily "feathered" antennae help gather plankton as food (Australia; L.P. Zann photo).

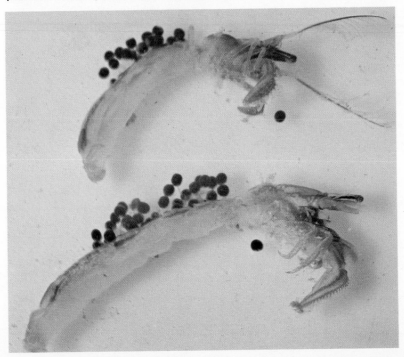

Larger bivalves of interest to shell collectors. 1) *Lopha cristagalli*, the cock's-comb oyster. 2) *Hippopus hippopus*, a small relative of the giant clams. 3) *Tridacna squamosa*, the ruffled giant clam. (G.v.d. Bossche photos.)

Larger bivalves of interest to shell collectors. 1) *Ostrea hyotis*, a scabrous oyster. 2) *Spondylus regius*, a spiny oyster. 3) *Chama lazarus*, a jewel-box clam. (G.v.d. Bossche photos.)

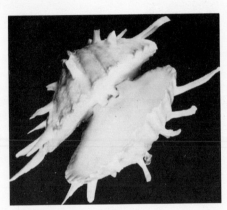

The deep-water hermit *Pylopagurus varians* in a coenoecium of hydrocoral (East Pacific; A. Kerstitch photo).

A demon hermit crab, *Aniculus maximus*. This species seems to be uncommon in the Pacific and has been incorrectly called *Dardanus punctulatus* (Hawaii; A. Norman photo).

White-backed demon hermits, *Aniculus strigatus*, are popular with aquarists but are seldom available. The glossy white back and greatly depressed form allow it to utilize many shells avoided by most other hermits (Pacific; R. Steene photo).

Paguristes cadenati, a brilliantly colored little Caribbean hermit. If two hermit families are recognized (Paguridae and Diogenidae) for the aquatic species, the previous species belong to the Paguridae, with *Paguristes* and following genera in the Diogenidae (D.L. Ballantine photo).

which enables it to achieve a better rasping action against the rocks. Several accessory shell plates have been evolved which provide additional protection for the delicate mantle in this rough environment.

Wood boring is really little different from rock boring, and certain members of the Pholadidae occur in wood (for example *Martesia*). However, *Martesia* does not exploit its habitat for food the way the shipworms do and, as mentioned earlier, it is shipworms that are the champions in specialization to a wooden home.

A variety of unrelated bivalves have a completely free-living unattached life on the surface, the pectens being a good example with representatives on several different substrates. Thus *Pecten maximus* and *Chlamys distorta* are common on firm substrates in deep water, while *Chlamys varia* occurs on similar substrates in shallow water and *C. septemradiata* prefers muddy substrates in deep water. Many species of *Chlamys* can swim by jet propulsion and use this behavior as an escape mechanism from such predators as starfish.

One of the best swimmers is *Amusium pleuronectes*. This shows an interesting adaptation to its way of life. It always rests with the left shell valve uppermost. This is colored a rich chestnut brown and camouflages well with the substrate. The right valve, however, is white in color. This valve is only visible when viewed from beneath when the bivalve is swimming, and its pale coloration, paralleled by similar markings in many fish, makes it less visible against a watery background.

But although *Amusium* is so well adapted to its way of life, it must remain an oddity. To us the typical bivalve is *Tellina* or *Macoma*, hidden beneath the sand surface where only the wading birds can reach it and feeding at high tide with siphons wafting in the overlying water. Yet these hidden animals are readily accessible to anyone who will take the time to spend a while on a beach digging where they see the tell-tale holes in the surface marking the openings of the bivalve burrows.

CEPHALOPODA

It may come as a surprise to some readers to learn that cephalopods—octopuses, cuttlefish, squids, and nautilus—are grouped by zoologists with the familiar seaside cockles, limpets, and periwinkles in the phylum Mollusca. The apparently obvious difference between, say, a cuttlefish and a periwinkle is in fact only superficial: they are linked by a number of basic similarities. Both have an external skin-like covering, called a mantle, open at one end. This encloses the viscera and secretes the shell, which is external for the periwinkle but internal in the cuttlefish. Both have a symmetrical arrangement of horny rasping teeth, the radula, in the mouth and breathe by means of feather-like gills, the ctenidia, within the mantle cavity; in both animals the respiratory pigment in the blood is hemocyanin.

Comparison of the cuttlefish with the animal of the pearly nautilus shows further close affinities. Both have, in addition to the radula, a pair of horny mandibles, like a parrot's beak, situated in the center of a circle of tentacular grasping arms. Both have a chambered shell for controlling buoyancy. A striking similarity is the presence of a funnel beneath the head which circulates water in the mantle cavity during respiration and doubles as an organ of locomotion—the famous ' jet propulsion ' of the cephalopods.

Once the zoological affinities of the nautilus and the cuttlefish are recognized, comparison of the externally-shelled nautilus and periwinkle shows that they are, after all, not so very different. Remaining doubts are dispelled by comparing cephalopods with the rest of the invertebrates, for they have nothing at all in common with the pentaradiate echinoderms, the articulated arthropods, the brachiopods, annelids, corals, or the simple sponges.

The name Cephalopoda was proposed by the French zoologist Cuvier in 1798 for a class of the phylum Mollusca typified by the nautiloids, cuttlefish, squids, octopods, and the extinct ammonoids and belemnoids. The name, derived from Greek, means simply ' head-footed. ' All living cephalopods have a ring of tentacular arms surrounding the

Dardanus is a common genus of large reef hermits with enlarged left claws. Above is the Caribbean *Dardanus venosus* (P. Colin photo); below is the somewhat more brightly colored Pacific *Dardanus lagopodes* (S. Johnson photo).

The very large size and pattern of black-ocellated white spots on red make *Dardanus megistos* recognizable at a glance. Widely distributed but seldom common, it is a very popular species with aquarists (Pacific; K. Gillett photo).

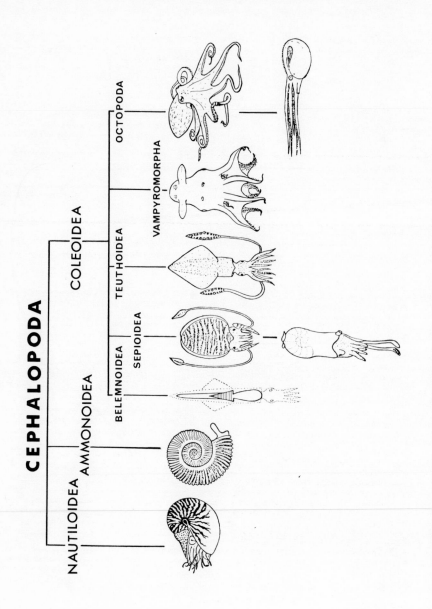

CEPHALOPODA

NAUTILOIDEA

AMMONOIDEA

COLEOIDEA

BELEMNOIDEA

SEPIOIDEA

TEUTHOIDEA

VAMPYROMORPHA

OCTOPODA

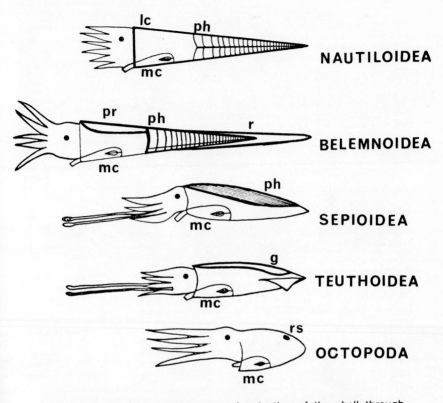

NAUTILOIDEA

BELEMNOIDEA

SEPIOIDEA

TEUTHOIDEA

OCTOPODA

Above: Diagrammatic representation of reduction of the shell through cephalopod evolution, freeing the mantle cavity for development of muscular tissues and the fins for locomotion. Abbreviations: lc = living chamber; ph = phragmacone; mc = mantle cavity with gills; pr = proostracum; r = rostrum; g = gladius or pen; rs = residual shelly plates in Octopoda. Shell indicated by thicker lines.

Facing page: The usually accepted classification of the Cephalopoda. Nautiloidea have four gills in the mantle cavity, Coleoidea have two, and the gills of Ammonoidea are unknown. The Sepioidea, Teuthoidea, and Vampyromorpha have ten arms with pedunculate suckers, the Octopoda have eight arms with sessile suckers, and the arms of the Belemnoidea are uncertain.

Species of the genus *Clibanarius* tend to be dull in color, have spoon-shaped claw tips, and are typical of cool shallow water. Above is the Japanese *Clibanarius virescens* (Y. Takemura and K. Suzuki photo); below is the eastern North American *Clibanarius vittatus* (P. Giwojna photo).

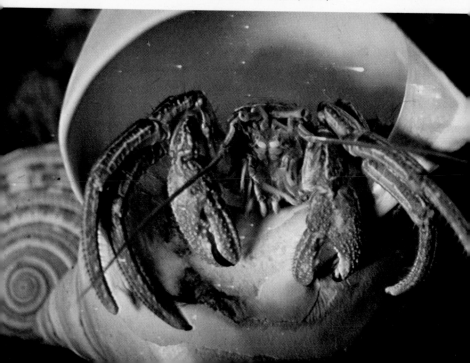

Calcinus also has spoon-shaped claw tips, but its species tend to be very color-ful and typical of tropical reefs. Above is *Calcinus latens;* below is *Calcinus elegans* (Hawaii; S. Johnson photos).

mouth, and no good reasons exist for doubting the extinct forms were constructed on the same plan. The arms of coleoid cephalopods are numbered 1-4 from the dorsal midline and designated left and right. Owing to the folding of the alimentary tract in cephalopods, the terms dorsal and ventral have a different connotation from their use in relation to the other molluscs.

About 650 species of Cephalopoda live in the oceans and seas of the world, extending from within the Arctic circle to the Antarctic continent. All are carnivorous and most are active predators feeding on fish and crustaceans at depths of between a few meters and nearly 5,000 meters. All are exclusively marine; none survive in fresh water. Among the Cephalopoda are the largest invertebrates ever known, the giant squid *Architeuthis* which measures up to 17 meters overall. In contrast, the tiny *Sepiola* and *Ideosepius* are usually less than 5 centimeters long.

Cephalopods have a long geological history, beginning in the Cambrian period and flourishing exceptionally during the Mesozoic era. The main evolutionary trend seems to have been the substitution of the shell by other means of achieving neutral buoyancy. Octopods have lost the shell entirely and squids retain only a horny strip for muscular attachment, but cuttlefish and *Spirula* still have a chambered and buoyant internal shell. *Nautilus* alone has an external coiled shell.

The Ammonoidea and Belemnoidea are wholly extinct, but five other orders have living representatives.

The Nautiloidea are represented by four species of *Nautilus*. The genus has a planispiral shell coiled above its head and divided by septa into about 36 closed chambers pierced by a calcareous tube, the siphuncle. The animal occupies the last chamber, which can be closed by a leathery shield or operculum. Within a circle of about ninety retractile arms is the mouth with its horny beak, which in *Nautilus* has a coating of calcium carbonate. As in all cephalopods the upper mandible closes within the lower and behind it is the radula formed of chitinous teeth. The arms are without suckers or hooks but have encircling annulations. The two simple eyes have no lens or cornea. Beneath the head two muscular flaps overlap ventrally to form a funnel which circulates water in the mantle cavity for both respiration and locomotion. There are four gills at the back of the mantle cavity but no ink sac. A pair of adductor muscles is attached to the inside wall of the living chamber; the funnel is attached to the underside of the head but is free anteriorly and protrudes 3-4 centimeters beyond the edge of the shell. When the animal is jet-swimming the funnel can be adjusted with great delicacy so that precise movements are possible.

The rest of the living cephalopods have only two gills in the mantle cavity, either eight or ten arms with suckers or hooks, and well developed eyes with cornea and lens; the shell, if present, is internal.

The Sepioidea include the ten-armed shallow-water cuttlefish; *Sepia officinalis*, the common European cuttlefish, may be taken to represent this order. It has an internal minutely chambered shell (familiar as the 'cuttle bone') which makes the animal buoyant. The mantle cavity is muscular and the funnel is an entire rounded tube. Eight grasping arms have suckers all along the inner surface; two long tentacular arms, normally tucked into pouches between the third and fourth arms, have bunches of suckers only on their club-like tips. In both cuttlefish and squids the suckers are furnished with a serrated horny ring. The body is flattened, shield-shaped, and the mantle has numerous pigmented cells which can produce a dazzling array of colors and patterns for sexual display or camouflage. The ink sac is very large and is used very rapidly by irritated or frightened cuttles. Swimming is powered either by the funnel or by means of the two lateral flange-fins which almost encircle the body up to the mantle opening.

The deepwater *Spirula* is included with the sepioids in nearly all classifications. *Spirula* is anomalous in having a squid-like body with an internally spirally-coiled shell divided, like *Nautilus*, into 25-37 gas-filled chambers. It is small, up to 7.5 cm long, with two terminal fins on the

The land hermits or tree crabs belong to the family Coenobitidae, genus *Coenobita*. The common Caribbean *Coenobita clypeatus* is a popular terrarium pet at the moment. The large claw closes the aperture of the shell much as did the operculum of the original snail occupant (H.R. Axelrod photo above, K. Lucas photo at Steinhart Aquarium below).

The distinctly flattened inner surface of the legs allows *Coenobita perlatus* and other land hermits to "seal" the body and help prevent loss of water from the gills (Australia; K. Gillett photo).

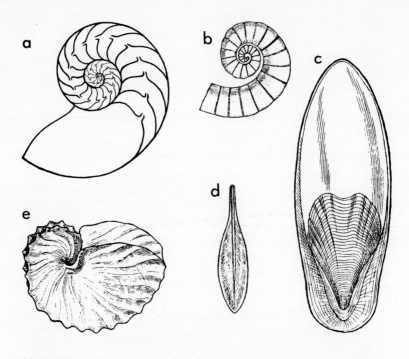

Above: The shells of living cephalopods. a: *Nautilus pompilius*, median section to show the septa; b: *Spirula spirula*, whole shell; c: *Sepia officinalis*, ventral view; d: Gladius of *Loligo vulgaris*; e: 'Cradle' of *Argonauta hians*.

Facing page: The cuttlefish *Sepia officinalis* with the mantle cut down the mid-dorsal line and pulled back to expose the shell in position (G. Marcuse photo).

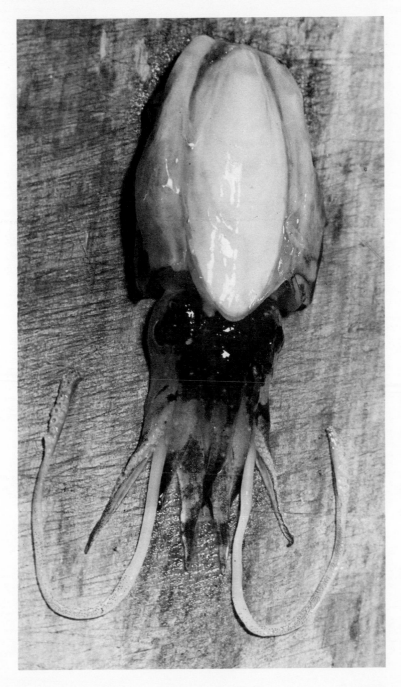

483

The gigantic coconut crab, *Birgus latro*, is the only hermit that does not require protection for the abdomen. Adult coconut crabs have an exoskeleton thick enough to prevent water loss and avoid minor predation (Pacific; S. Johnson photo).

Lobster krill, *Pleuroncodes planipes*, may occur in immense swarms off Baja California and are very important items in the diets of whales and large sea fishes (A. Kerstitch photo).

Squat lobsters, family Galatheidae, are intermediate between the true crabs and the hermit crabs and allies. Above is *Munidia quadrispina* from the East Pacific (T.E. Thompson photo). Below is *Galathea elegans,* a species commensal on crinoids in the Indo-Pacific; it is usually striped to somewhat resemble its host (L.P. Zann photo).

short, dumpy body. *Spirula* lives in nearly all the tropical and sub-tropical seas at depths between 180 and 900 meters.

Teuthoidea are represented by the ten-armed squids of the open seas. The essential difference between a squid and a cuttlefish is that the squid has a reduced horny strip—the gladius—in place of the cuttle's broad, chambered internal shell. As a result the cuttlefish is broad, flat, and shield-shaped, while the squid is long, round, and torpedo-shaped. The fins of squids are usually confined to the posterior third or half of the body and tend to be spear- or arrow-shaped. In both squids and cuttlefish the suckers are pedunculate (raised on muscular stalks) and strengthened with a serrated horny ring.

Between the squids and the octopods a single species, *Vampyroteuthis infernalis*, constitutes the order Vampyromorpha. It lives between a third of a mile and two miles deep in tropical and subtropical waters. Its color is a deep purplish black and it has the consistency of a jellyfish. The eight arms are joined by webs reaching almost to the tips of the arms. A pair of reduced worm-like arms can be completely withdrawn into two pockets lying between the first and second arms. They are thought to have a sensory function but cannot be homologous with the tentacles of cuttlefish and squids since these occupy a place between the third and fourth arms.

The fifth extant order of cephalopods is the Octopoda. Octopods have only eight arms—no tentacles— and no shell apart from two small slivers embedded in the mantle. Unlike the flattened cuttlefish and the cylindrical squids, the octopods resemble a rounded sack furnished with eight arms and two prominent eyes at the opening. They live mostly in shallow waters, feeding on crustaceans, mainly crabs, but some are found at great depths, *Eledonella* at 5278 meters and *Octopus* at a maximum of 3412 meters for example. In octopods the suckers are sessile and without a horny ring. The argonauts are octopods with the first pair of arms modified for secretion of the ' egg-nest ' shell, and the male is very reduced in size.

Nearly all molluscs lay eggs and a few retain the eggs within the mantle cavity until they are hatched. The majority of marine bivalves and gastropods have a free-swimming larval (or veliger) stage before settling on the bottom where metamorphosis takes place and the juvenile begins growth and development. Cephalopods lay relatively large yolky eggs and omit the veliger stage; the young hatch out as miniature adults, ready to engage directly in an active predatory life. Young octopods and cuttlefish are remarkably precocious and begin jet-swimming and hunting small crustaceans almost as soon as they are hatched; some newly-hatched cuttlefish will even squirt their ink at a source of fright or irritation.

The female oviduct or male duct opens into the mantle cavity beside the anus; the ducts may be either paired or single. Perhaps the most extraordinary element of cephalopod reproduction is the male sex organ. One of the arms is modified to serve the function of a penis. The male produces elongated cylindrical packets of sperm (spermatophores) which are pulled from the genital opening by the genital arm and deposited on various parts of the female or even introduced directly into the mantle cavity. The specialized male arm is called a hectocotylus. In the argonauts the hectocotylized arm is completely detached from the male and left in the mantle cavity of the female where it has a short independent existence. This organ was first described by Cuvier as a parasite and given the name *Hectocotylus octopodis*. Only after much bitter dispute was the real nature of the sucker-bearing ' parasite ' understood and the name perpetuated for the arm. In fact Aristotle had recognized the sexual nature of the hectocotylus more than 2,000 years ago.

Copulation in *Loligo* takes place after the male has swum alongside the female and made a sexual display by spreading the arms and turning dark red. They may then copulate head to head or parallel, with the ventral sides in contact, while the male grasps the female firmly by the head. Spermatophores are placed inside the mantle cavity by the hectocotylized fourth left arm.

Porcelain crabs are now placed in several genera, but we will call all of ours *Petrolisthes* for convenience. Porcelain crabs are filter-feeders often associated with anemones, sea urchins, or hermit crabs. Above is *Petrolisthes coccineus* (Pacific; S. Johnson photo); below is the much more colorful *Petrolisthes maculatus* (Pacific; R. Steene photo).

The relatively smooth and finely spotted *Petrolisthes ohshimai* is usually associated with *Stoichactis* and other giant anemones (Indo-Pacific).

King crabs, such as this *Paralithodes rathbuni* from the East Pacific, are actually relatives of the hermit crabs. There are only eight large legs visible, the fifth pair being small and tucked under the abdomen (K. Lucas photo at Steinhart Aquarium).

In the octopods the modified male arm is usually the third on the right. The female is caressed by the male at arm's length, the tip of the hectocotylus inserted in the mantle cavity, and the sperm deposited near the opening of the oviduct. In cuttlefish the male places spermatophores in the region of the mouth of the female, and in *Nautilus* four arms on the right side are modified to form a sleeved projection called a spadix.

The number and appearance of eggs laid by cephalopod females are variable. *Octopus* lays about a hundred, *Eledone* nearer sixty. *Loligo* produces sausage-shaped egg-masses attached at one end and forming radiating clusters. The eggs of *Sepia* are black like bundles of small grapes and are given a coating of ink as they are laid.

Octopus exhibits a great deal of parental care, brooding over her eggs, cleaning them with the tips of her arms, and jetting water from the funnel at them. However, the prize for parental care goes to the female argonaut. She produces a beautiful paper-thin, often elaborately ornamented shell in which she lives brooding over her thousands of small eggs which are arranged on branching egg-strings, like strings of beads. The female is not attached to the shell, but if she is forcibly deprived of it she dies. This remarkable shell is without partitions and, though partly coiled, is open at the apex. It is secreted by the two topmost arms which are large and flattened, embracing the sides of the shell and secreting new material as it grows; it is therefore in no way homologous with the mantle-secreted shell of the nautiloids and the ammonoids. There nevertheless remains a disturbingly coincidental similarity between the shells of argonauts and those of some Cretaceous ammonites. The argonaut shell is literally an egg-nest to house both mother and eggs. Sober zoologists, perhaps too frequently confronted by the extraordinary, are not given to lyricism or playfulness, but two of them, perhaps delighted by this elaborate and beautiful structure for parental care, have been moved to refer to it as a ' cradle ' or more wittily as ' not a house but a perambulator. '

Cephalopods are rated by zoologists as 'advanced' molluscs, usually as the 'most advanced.' The elaborate behavior of these creatures, resulting from a highly developed nervous system and a centralized brain, their parental care and intelligent hunting technique, and the undoubted intelligence exhibited under experiment by *Octopus* confirm this assessment. But if 'advancement' is rated on general distribution or occupation of available habitats, cephalopods fail to equal even the simple headless bivalve, passively filtering particles suspended in water. Both scaphopods and cephalopods have remained conservatively in the sea, neither having penetrated even the littoral zone as a regular habitat. Chitons, bivalves, and gastropods all vigorously occupy the littoral zone, but the bivalves and gastropods leave the chitons behind when it comes to penetrating into fresh water and the gastropods entirely outstrip the bivalves when it comes to occupying terrestrial habitats. By this rating gastropods dominate all other molluscs, being herbivorous and carnivorous, marine and freshwater, terrestrial, arboreal, and even subterranean, occupying all the continents up to heights of 5486 meters and all the seas down to depths of 5200 meters. They have, like the cephalopods, abandoned the shell in the marine nudibranchs and the terrestrial slugs, and they show no sign of decline even in the face of man's increasing pollution and destruction of habitats.

Nevertheless, the slow gait of the dull garden snail cannot compare with the flashing speed, the scintillating color changes, and the brilliant bioluminescence of the marine squids. Cephalopods are advanced animals which have thoroughly exploited one habitat—the sea. They have successfully challenged the vertebrates in their repeated returns to the sea, the reptiles in the Mesozoic and the mammals in the Cenozoic, and have retained a place in the hierarchy of marine animals second perhaps only to the cetaceans and extending in time nearly 500 million years.

The first need of any swimming animal is to achieve near neutral buoyancy so that little or no energy is wasted in

The Alaskan king crab, *Paralithodes camtschatica,* one of the most heavily fished commercial crustaceans of the North Pacific. It is found from near Korea through the Bering Sea to British Columbia (D. Gotshall photo).

Lopholithodes foraminatus, the box crab, is a king crab that can draw the legs very tightly to the body (East Pacific; D. Gotshall photo).

The small but extremely spiny *Acantholithodes hystix* is related to the other king crabs and also found close to the Arctic Circle (Y. Takemura and K. Suzuki photo).

Mole crabs, such as this East Pacific *Emerita* species, are small, usually elongated burrowers related to the hermit crabs. The antennae are usually feathery and project from the burrow to collect detritus (A. Kerstitch photo).

keeping afloat. Nautiloids and the extinct ammonoids achieved buoyancy by means of a chambered and gas-filled external shell, conical, coiled, or straight and divided from the apex to near the aperture by septa into a series of chambers which could be filled with gas to rise or liquid in order to sink. This part of the shell is the phragmacone and the opening is the living chamber. During the Palaeozoic and Mesozoic eras the chambered shells of the nautiloids and ammonoids enabled the animals to rise into the teeming life of the plankton fields, freeing them from the restrictions of bottom-living. As early as the Mesozoic era the belemnoids and teuthoids freed themselves from the unwieldy shell. The belemnoids kept the phragmacone at the back, balancing it with a calcareous rostrum—the familiar fossil belemnite—and retained only the dorsal part of the living chamber, the proostracum. The squid-like teuthoids did even better and abandoned all but the proostracum, which they kept for the attachment of the muscles. How they lightened the body is not know.

During the Jurassic cuttlefish shifted the phragmacone over the back, abandoned the counter-balancing rostrum, and placed the center of buoyancy over the center of gravity. One result of all these innovations was that the mantle, freed from the enclosing shell, became muscular and evolved into a powerful means of jet-propulsion. There nevertheless still remained a place for the ' old fashioned ' external shell of *Nautilus*, whose long-term success is confirmed by the four living species found abundantly in the southwestern Pacific.

The internal shell of *Spirula* causes it to float head down at depths between 180 and 900 meters, where the pressure on the shell must be in the region of half of a ton.

Squids, having abandoned pneumatic means of buoyancy, have developed other methods. In the cranchiid squids the coelomic space between the viscera and the body wall is filled with a liquid in which a high proportion of ammonium chloride replaces sodium chloride. As ammonium chloride solution is lighter than sea water, its presence gives the animal neutral buoyancy.

Giant squids have an even more remarkable solution to the problem. A large stranded *Architeuthis* in New Zealand was towed soon after it was dead into the water where it floated, and local fishermen are reported to have given up using the flesh of *Architeuthis* as bait since it caused the hooks to ' rise to the surface. ' Numerous reports exist of dying squids floating at the surface of the sea, and during the decade 1870-80 many dead and dying giant squids were reported floating at the surface off Newfoundland. *Architeuthis* has no coelomic cavity filled with ammonium enriched liquid and no buoyant material is known from any other part of its anatomy; one can only conclude that the solid tissues themselves have lightened and become buoyant. This must be the ultimate step in the emancipation of the cephalopods from the calcareous shell.

The idea that cephalopods anticipated the human discovery of jet-propulsion by about 600 million years is well known, but these versatile animals have developed three types of propulsion: jet-swimming, fin-swimming, and web-swimming.

Jet-swimming, the most spectacular method, is most highly developed in the squids, particularly in *Loligo*. The mechanism is extremely simple. Water entering the mantle cavity through spaces between the sides of the head and the mantle passes rearward in two streams to the feather-like gills at the back of the mantle cavity. There respiration takes place; the two streams of water then unite, change direction, and pass forward past the anus and ink sac to the funnel, to be forcibly ejected by muscular contractions of the mantle. A system of flap-valves in the funnel and on either side of the head ensures that water circulates in one direction only.

Respiration is the primary function of the circulation of water in the mantle cavity and it continues as long as the animal lives, whether it is active or at rest. The process is under the control of the pallial nerves, which are a little over 1 mm in diameter—about 50 times thicker than the nerves in other animals. A fear-inducing stimulus interrupts

Close-up of the "front" of a typical crab. The eye is of course the round red object, with the antenna placed next to it; the second antenna or antennula is just to the side of the midline.

Ventral view of *Petalomera lateralis,* a Pacific sponge crab. The sponge covering plus the tendency to "play dead" when disturbed make this crab very difficult to see in nature (T.E. Thompson photo).

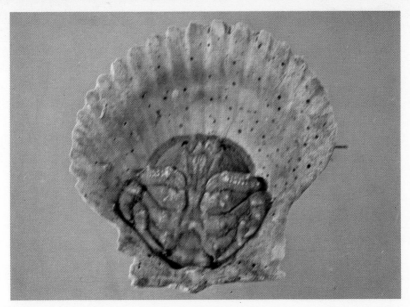

Not all sponge crabs carry sponges. Some use ascidians or coelenterates, while a few, such as this *Hypoconcha lowei,* manage to fit into old clam shells (here a scallop) (East Pacific; A. Kerstitch photo).

Dromidiopsis species are normal-appearing sponge crabs except for the brilliant silver eyes under certain conditions. This Hawaiian species is small, only a couple of centimeters in carapace length, but some species exceed 15cm in length (S. Johnson photo).

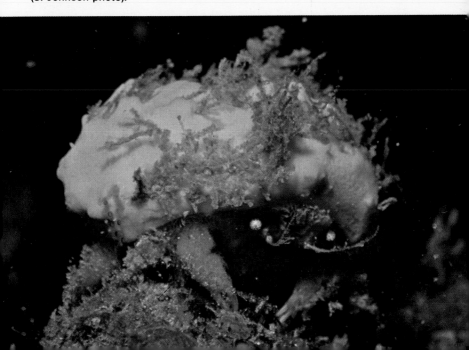

the normal respiratory cycle with a powerful impulse that travels swiftly down the 'giant' pallial nerves to activate the mantle muscles into rapid and violent motion. Water is expelled from the mantle cavity, together with a cloud of ink, and the squid darts backward several feet. The very rapid darting movement of *Loligo*, controlled by its 'giant' nerves, has earned it the description 'a little squirt with a big nerve.'

(In passing, it might be mentioned that the long-held view that cephalopod ink functions like a visual smoke-screen has been challenged recently by a suggestion that it is the strong smell of the ink cloud that attracts and holds the predator, rather than the black opaque cloud hiding the retreating cephalopod.)

Loligo, though mainly pelagic, occasionally rests on the sea bottom; it is confined to the continental shelf. Since it lacks the buoyant phragmacone of *Sepia*, having only a thin but rigid horny strip for the attachment of muscles, it is heavier than water and must therefore swim continuously. Captured squids cruise back and forth, day and night, from wall to wall of their tank, not by jet-swimming (although the respiratory cycle continues) but by means of the broad arrow-shaped fin extending from the posterior tip to about half-way down the body. At the end of each run the action of the fin is reversed and the squid changes direction without turning around: there seems to be no preference between backward and forward swimming. If an emergency arises and jet-swimming is resorted to, the fin is wrapped close around the body, producing a streamlined shape. Escape movements are always backward in squids and cuttles.

In *Sepia* two lateral fins almost encircle the body and undulations pass in either direction for, like squids, cuttles travel backward and forward with equal ease. All cuttlefish and squids use fin-swimming for general migratory movements; jet-swimming is reserved for serious occasions—toward prey or away from danger—but the gentle ejection of water from the funnel during respiration probably contributes to the overall movement during fin-swimming.

Web-swimming occurs in all the cirrate octopods, in two incirrate genera, in *Vampyroteuthis*, and in two squids, *Histioteuthis* and *Cirroteuthis*. The eight arms are interconnected by triangular webs which in some forms reach almost to the tips of the arms. This umbrella-like arrangement opens out slowly and closes a little faster, the movement resembling the pulsing contractions of the medusoid jellyfish. The funnel is usually weak and probably serves only the respiratory needs of all these sluggish web-swimmers. Near-neutral buoyancy is achieved by the presence of gelatinous tissue in the bodies of these strange cephalopods, which have a strong superficial resemblance to jellyfish.

In the cranchiid squids jet-swimming is used for escape movements only; respiration takes place by means of peristaltic movements of the coelomic fluid. The mantle cavity is divided by a septum on each side of the body, producing four chambers. A hole in each septum allows water to pass from the anterior to the posterior chamber, while the gill itself acts as a valve ensuring one-way circulation. Contraction of the anterior coelomic sac increases the volume of the anterior mantle chamber, so that it fills with water drawn through a restricted passage between the side of the head and the collar valve. Dilation of the anterior coelomic sac forces water through the opening in the septum, across the gill, and into the posterior mantle chamber, pushing the water already there out through the funnel. During this respiratory cycle the mantle cavity remains distended and the muscles play no part.

Octopods can be viewed at many marine aquaria. They are sluggish animals in comparison with squids, moving with a graceful, almost comical, slow motion. One octopod, *Eledone cirrhosa*, I watched at the Plymouth Marine Aquarium grew tired of one side of its tank and moved to the other. It kicked off by means of its eight arms in an obvious swimming movement, but once water-borne it closed the arms behind it and used the funnel in a series of fairly rapid, but not powerful, jets, resulting in a jerky copepod-

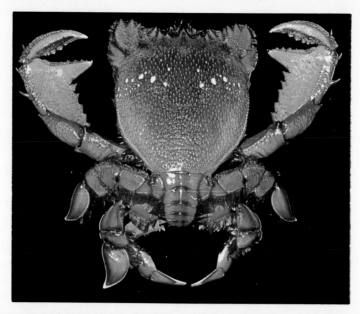

Frog crabs are burrowers recognizable by their unusual shapes and odd claws. Above is the giant *Ranina ranina* with a carapace length of up to 12.6cm (Pacific; K. Gillett photo). Below is the much smaller and anteriorly narrowed *Lyreidus stenops* (Pacific; Y. Takemura and K. Suzuki photo).

Although flattened swimming legs are typical of the portunid crabs, they occasionally occur in unrelated families. This is *Matuta banksi,* a strikingly patterned calappid or box crab with swimming modifications (Pacific; A. Norman photo).

The calico crab of the West Atlantic, *Hepatus epheliticus,* is one of the most familiar box crabs (N. Herwig photo).

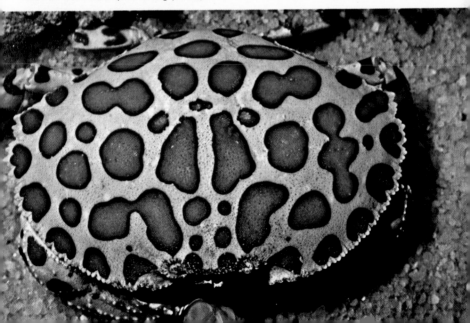

like motion. At the other side of the tank jet-swimming ceased and the arms spread out to make a balletic eight-point landing. There the octopus rearranged its arms and continued to regard the humans on the other side of the glass. The presence of a crab will stimulate different behavior: the octopod then moves rapidly with a mixed swimming and scrambling motion over sand and rocks, no less graceful but more purposeful.

The feeding habits of *Nautilus* are less well known than those of other cephalopods, although the remains of bottom-living crustaceans are found in those caught. *Nautilus* inhabits depths down to 610 meters.

Octopods and cuttlefish, being easy to keep in marine tanks, have probably been studied more than any other cephalopods. Octopods hunt and eat crabs, which they appear to relish as much as humans do. The crab is captured and held by the sessile suckers lining the inside of the arms. *Octopus vulgaris* has about 240 suckers arranged in two rows on each arm, totaling nearly 2,000, so that a crab once caught stands no chance of escape but is turned over and bitten through the soft ventral side by the octopod's horny beak. Behind the beak is a salivary gland secreting a poison which attacks the nervous system, so that a crab with strength enough to struggle against the suckers is soon rendered helpless. The radula resembles that of the gastropods but is probably used to assist in swallowing shelly pieces of crustacean.

The rest of the octopod's alimentary tract consists of esophagus, stomach, cecum, pancreas, liver, and intestine. In *Octopus vulgaris* digestion takes place in about 18 hours. The alimentary tract follows the usual folded cephalopod plan, like a U lying on its side; the mouth lies at the end of one arm of the U, the anus at the other. An Antarctic octopod, *Grameledone setebos*, was found to have pieces of seaweed in its stomach, and the common American squid often bites off and swallows pieces of eelgrass which are found undigested in the intestine. Since the cephalopod digestive system seems unable to deal with vegetation, this would appear to be aberrant behavior.

Most reports indicate that octopods hunt either at dusk or at night and, unless very hungry, are inactive during the day. Their favorite food is undoubtedly crustaceans, particularly crabs, but they will if necessary eat fish. Octopods certainly open bivalve shells in the same manner as starfish, and *Octopus vulgaris* is known to pull abalone (*Haliotis tuberculata*) from rocks in the Channel Islands.

In addition to the well known octopod cannibalism, they have been seen to eat some extraordinary food. Near Singapore an *Octopus* was observed eating a shore-living species of spider. Another Indo-Pacific species is locally believed to lie in wait at night for rats which come down to the shore to scavenge. A captured octopod was kept alive in a tank on a diet of hard-boiled eggs, the shell perhaps resembling its normal crustacean diet.

The two tentacles and the funnel of the cuttlefish are its main hunting weapons. The long reach of the extensible tentacles, coupled with their lightning speed, allows the cuttle to strike at prey while it is still sufficiently far away for the predator's presence to be unsuspected. They seem to employ two hunting techniques, active hunting and lying in ambush. The first relies upon the buoyancy of the phragmacone, which allows the cuttle to hover a few inches above the sea floor. Undulations of the lateral fin propel it forward while jets of water are directed from the funnel at the sandy bottom. Eventually a shrimp is uncovered and, apparently unaware that its sandy color and nearly transparent body render it virtually invisible, the shrimp begins to cover its head with sand. The movement betrays it to the cuttlefish which darts down and seizes it.

The ambush of a shrimp by a cuttlefish makes fascinating viewing. The cuttle reduces its buoyancy and sinks to the sea floor, covering its back with sand by means of the lateral fin and burrowing down a few centimeters. Only the top of the head, covered with sand, remains above the sea floor: the lidded eyes watch and wait. At the first sight of a shrimp the cuttle turns to face it, marking its scuttling progress until it comes within the range of the tentacles which by now

Calappids are called box crabs because of their shape and shame-face crabs because the very large curved claws hide the entire front of the body when desired. As can be seen in the small species of *Calappa* above, the fingers are asymmetrical in shape, one with a large knob at the base. Below is the Indo-Pacific species *Calappa lophos* (D. Faulkner photo above; Y. Takemura and K. Suzuki photo below).

The oddly shaped purse crabs (*Leucosia obtusifrons* above, *Leucosia anatum* below) are often recognizable by the narrow, projecting front of the carapace. The crabs below are starting to mate; notice the projecting gonopods in the male (foreground) (Pacific; Y. Takemura and K. Suzuki photos).

are almost out of their pouches and ready to strike. There is a sudden movement and a settling cloud of sand. Much too fast for the human eye to follow, the tentacles have shot out, seized the shrimp with the terminal suckers, and hauled it back to be held by the eight arms until it is eaten. Digestion in cuttles is completed in about 12 hours, compared with 18 hours in octopods and 6 hours in squids.

Squids, except for the Loliginidae and a few other small groups, live away from shore lines and at various depths. Some have been taken at depths of two miles. Most squids found in shallow water are myopsids, the eyes being covered with transparent skin, perhaps a protection against slit. Off-shore squids usually have the eyes naked and exposed to salt water; they are commonly known as oegopsids.

They eat a variety of foods, but the majority hunt and eat fish. *Loligo*, with its torpedo-shaped body, has a remarkable turn of speed and can catch fast-swimming mackerel which it kills by a swift bite in the neck. Sometimes they run riot among a school of fish, like foxes in a chicken coop, attacking more than they can possibly eat in a frenzy of killing. The ferociously aggressive Humboldt Current squids, *Ommastrephes gigas*, weigh up to 350 pounds and measure up to 3.5 meters overall. They are known to attack tuna and eat everything except the head.

In squids the radula seems only to assist in swallowing; the digestion of these energetic animals, with their high metabolic rate, is usually completed in 4-6 hours.

In the cephalopod molluscs the sexes are always separate. Elaborate courtship rituals and parental care testify to the advanced nature of these animals. Sexual dimorphism is apparently common, the male usually being smaller than the female; this characteristic has been convincingly demonstrated in Jurassic ammonites.

Phylum Arthropoda

Without doubt the arthropods are the dominant form of animal life on earth. Of the million or more species, 95% are insects, with over two-thirds of these beetles. Of course everyone recognizes an insect on sight, and most other arthropods are just as easily recognizable. The phylum is characterized by jointed legs or other appendages, these often greatly modified for functions such as feeding and sensing the environment. The body consists of three major sections (head, thorax, and abdomen) variously fused or developed, the whole covered with a thick or thin chitinous exoskeleton often hardened by calcium carbonate. Although the majority of species are beetles less than 5 mm in length and terrestrial, a large number of arthropods are large, common marine animals such as crabs and shrimp, the larger crustaceans. Our chapter will discuss only the crustaceans and mention the other major marine groups, horseshoe crabs and sea spiders, so the following classification includes the non-marine members as well.

SUBPHYLUM ONYCHOPHORA: *Tropical thin-skinned terrestrial worm-like animals with unsegmented bodies and paired appendages. They form a connecting link between annelid worms and typical arthropods and are often placed in a separate phylum. Commonly called peripatids.*

SUBPHYLUM PENTASTOMIDA: *Elongated, small parasites of the nasal cavities of mammals and reptiles; the circulatory, respiratory, and excretory systems are absent.*

SUBPHYLUM TARDIGRADA: *Microscopic arthropods usually with four short legs ending in large claws; heavily*

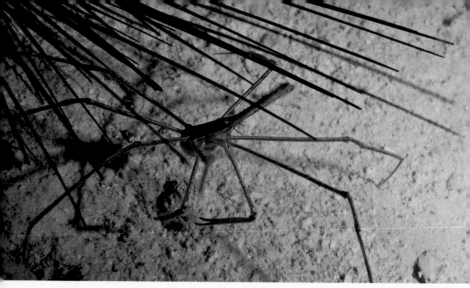

Stenorhynchus seticornis, the arrow crab, is a bizarrely elongated yet adaptable spider crab often seen in aquaria (Caribbean; D. Reed photo).

Seldom are specimens of the small spider crab *Pugettia producta* as clean as the specimen illustrated; usually they are heavily covered with algae and bryozoans (East Pacific; K. Lucas photo at Steinhart Aquarium).

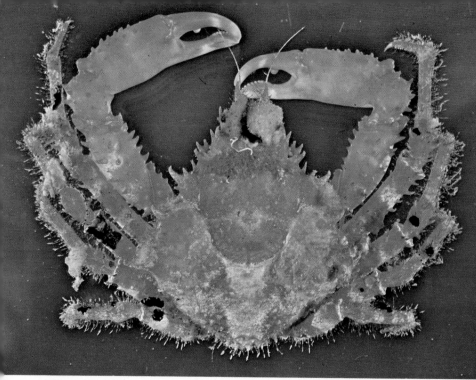

Mithrax, a common genus of "typical" spider crabs, contains both small, rather colorful species (above) and giants such as the two-kilogram, 20-centimeter long *Mithrax spinosissimus* (below) (both Caribbean; C. Arneson photo above, D. Reed photo below).

armored and able to survive long periods of desiccation; systems except for digestive system all reduced. Terrestrial in moist situations, freshwater, or (few) marine. The 'water bears' of popular literature.

SUBPHYLUM CHELICERATA: Mostly moderately large arthropods with four or more pairs of legs and the anterior feeding appendages not modified as jaws, but usually ending in a large claw. The spiders and their allies.

Class Merostomata: Horseshoe crabs, easily recognized by the shape of the cephalothorax and the terminal spine. Marine. Also called xiphosura.

Class Pycnogonida: Sea spiders. Mostly small marine animals with five or seven pairs of very long multi-segmented legs. The abdomen in most species is very small, almost absent.

Class Arachnida: Spiders, scorpions, ticks and mites, and their relatives. Antennae and jaws absent, the anterior appendages or chelicera well developed. Body consisting of a cephalothorax and abdomen without appendages; four pairs of legs. Many species, virtually all terrestrial, in nine orders.

SUBPHYLUM MANDIBULATA: Small to rather large arthropods with the anterior appendages modified as jaws or mandibles. Usually three to five pairs of legs, with distinct antennae. Sometimes highly modified or microscopic and not closely resembling typical arthropods.

Class Crustacea: A very large group of mostly marine and freshwater animals with two pairs of antennae and usually with at least some of the appendages with two branches. The body is often heavily covered with a calcified carapace joining the head and thorax, the abdomen large and multi-segmented in many species. Microscopic to very large. This is the only class of arthropods of real interest to the marine aquarist, as among its 25,000-30,000 species are found the familiar shrimp and crabs as well as dozens of minor groups including the copepods, barnacles, and isopods.

Classes Pauropoda, Diplopoda, Chilopoda, and Symphyla:
Elongated, rather slender terrestrial animals with a single pair of antennae and more than five pairs of legs. Millipedes, centipedes, and their allies.

Class Insecta: *Insects, which need little definition. Three pairs of legs, one pair antennae, often with wings and specialized larval stages. The one million or so insects are divided into a large number of orders, most entomologists (insect specialists) recognizing between 25 and 30. Most insects are beetles, moths, and flies. Although many insects are found along the shore, few are truly marine except for a few genera of flies and true bugs.*

Next to the sea anemones, crustaceans, especially shrimp, are the most popular group of marine invertebrates for aquariums. They have many advantages, such as larger size, bright colors, adaptability to moderate changes in temperature, and generally broad feeding habits that allow them to adapt to aquarium foods. The smaller crustaceans of several groups can be easily cultivated for food for larger aquarium inhabitants, while a few crustaceans occasionally appear as parasites of aquarium fish. Requirements for keeping crustaceans have been given in many books and articles, but as general rules: 1) keep crustaceans of different sizes and habits segregated to avoid predation; 2) avoid fish in the same tank; 3) keep temperatures and salinities constant; 4) feed a varied diet that will not foul the water; and 5) remember that many shrimp have short lifetimes and will not survive over six months in aquariums regardless of care. Many species of shrimp and crabs are available to aquarists either through purchase or wise collecting, and the variety of colors, forms, and habits is almost endless.

Identification of even the larger crustaceans is often complicated because of the large number of species, but in many cases general form and pattern are quite distinctive. There are numerous local guides to the larger crustaceans of many areas, but none cover all the species likely to be encountered by either collectors or aquarists. Fortunately there are seldom major differences between genera of similar appearance as regards feeding habits and life in the aquarium.

"Decorator crabs" are usually some type of spider crab. This specimen has covered itself not only on the carapace but over the legs as well (Pacific; W. Deas photo).

Herbstia camptacantha, one of many interesting East Pacific spider crabs (A. Kerstitch photo).

Spider crabs may camouflage themselves either by a resemblance to the usual habitat (such as *Xenocarcinus* on gorgonians above) or by attaching bits of the background directly to the carapace as in the crab below covered with bits of *Dendronephthya* (both Pacific; R. Steene photo above, L.P. Zann photo below).

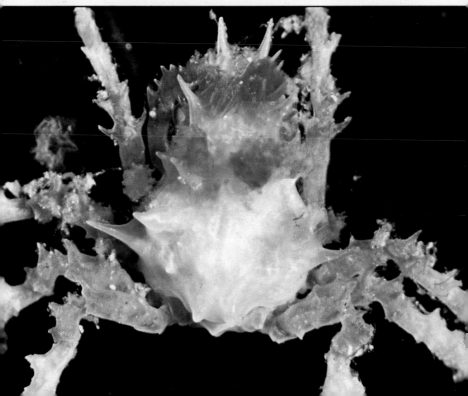

The arthropods (jointed-legged animals) have established themselves in nearly every habitat on the Earth's surface. Well over three-quarters of the known animal species belong to the phylum Arthropoda, the major classes of which are the Insecta (insects), Crustacea (water fleas, shrimp, crabs), and Arachnida (spiders, mites); these, along with three minor classes, the Pycnogonida (sea spiders), Merostomata (horseshoe crabs), and Tardigrada (water bears), have representatives occurring in the sea.

The class Insecta contains the most numerous of all arthropod species that have colonized land and freshwater habitats, but insects have never successfully invaded the sea and their occurrence there is confined, more or less, to surface waters of rock pools and similar habitats. By comparison, members of the class Crustacea abound in the open sea; some have invaded fresh water and a limited number occur near the shore. Members of the class Arachnida, with the exception of a few marine mites, are also absent from the sea. Of the minor classes, the Pycnogonida and Merostomata are exclusively marine, and only a very few species of the class Tardigrada occur in the sea. Only two other classes of arthropods need be considered with respect to the marine environment. These are the Diplopoda (millipedes) and Chilopoda (centipedes) that, along with insects, some crustaceans, and arachnids, are temporary or permanent inhabitants of the region just above high tide where many species scavenge among decaying vegetation of the "strand line."

CLASS CRUSTACEA

Crustaceans are predominant members of the marine fauna. They compete only with the molluscs in numbers of species and subspecies that successfully occupy diverse habitats that the sea offers. This class contains the familiar crabs, lobsters, hermit crabs, shrimp, krill shrimp, sea slaters and their relatives (isopods), sea scuds and shore hoppers (amphipods), barnacles, brine shrimp, and a multitude of less familiar forms that have no popular names but that are described or mentioned in the following account.

Although crustaceans show great diversity of body shape and some demonstrate considerable variation in this during their life cycle, all have segmented bodies and limbs at some stage of their lives and either have gills or breathe through the surface of their bodies. Similar to insects, they have an integument (skin) composed chiefly of a complex nitrogenous polysaccharide often strengthened by calcium salts and known as chitin. This chitinous layer may vary from an extremely thin flexible covering to a thick layer the hardness of bone. Crustaceans also show extremes of size from the giant Japanese spider crab that can span 3.6 m from claw to claw, to the microscopic copepods, some species of which may be less than 1.0 mm long.

The familiar lobster will serve to introduce the general features of crustaceans. The body is composed of a rigid head fused with the thorax (cephalothorax) and a flexible segmented abdomen. This body plan has evolved from an ancestral one that consisted of numerous articulated segments, many of which bore pairs of segmented limbs. The fusion of these body segments from the head backward and the changes in limbs (alteration of shape, reduction in size and number) have occurred independently within the various subclasses. In the lobster some of the thirteen anterior segments composing the cephalothorax can be seen imperfectly fused together, while the dorsal part of the fifth segment has been greatly developed into a shield (carapace) that overhangs on either side to protect the delicate gills. Anteriorly the carapace is produced as a rostrum between the eyes.

The abdomen is composed of six shelly rings articulating one with the other. The last has joined to it a tail fan (telson) that, along with the part bearing the eyes, is not considered a true segment as these are the parts remaining after all the segments have been budded off during embryonic development. Although all the nineteen segments of the lobster's body are not easily evident, the actual position of each and

515

The Japanese giant spider crab, *Macrocheira kaempferi*, reaches a carapace length of over 30 centimeters and a leg span of over three meters (Y. Takemura and K. Suzuki photo).

Elbow crabs are typically oddly-shaped brown lumps like the East Pacific *Parthenope* species below (A. Kerstitch photo), but few are more distinctive, like the striped *Zebrida adamsi* (above), a commensal of the venomous sea urchin *Salmacis* (Pacific; L.P. Zann photo).

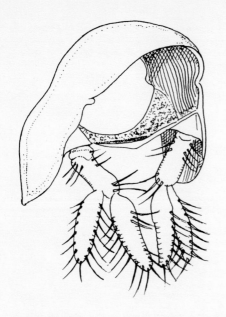

A shelly 'ring' of a lobster's abdomen.

Lateral view of the northern lobster, *Homarus*. A1 = first antenna; A2 = second antenna; R = rostrum; CP = carapace; T = telson; SW = swimmerets; ThL = thoracic limb.

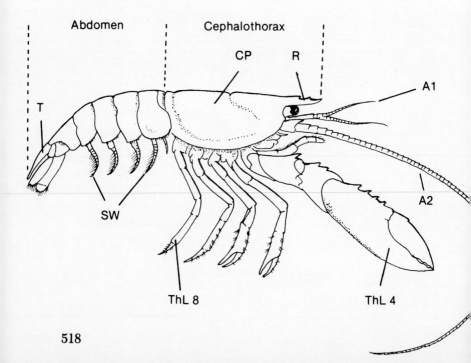

518

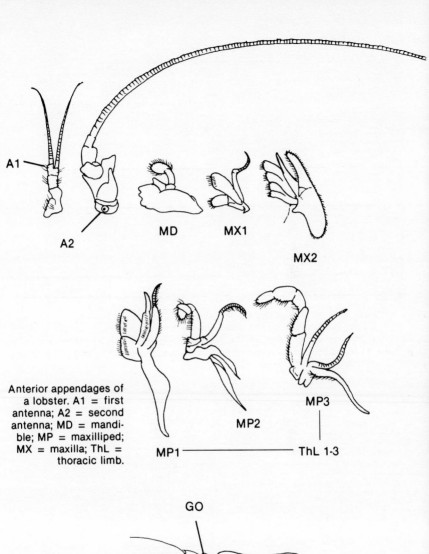

Anterior appendages of a lobster. A1 = first antenna; A2 = second antenna; MD = mandible; MP = maxilliped; MX = maxilla; ThL = thoracic limb.

A1
A2
MD
MX1
MX2
MP1
MP2
MP3
ThL 1-3

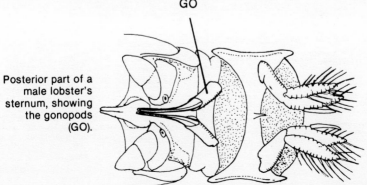

Posterior part of a male lobster's sternum, showing the gonopods (GO).

GO

519

Cancer species, the edible box crabs or rock crabs, are familiar as the origin of the astrological "sign of the crab" and are also sold as food. They are typical of cool shallow water. Above is *Cancer anthonyi:* below is the commercially important *Cancer magister* (both East Pacific; D. Gotshall photo above, K. Lucas photo at Steinhart Aquarium below).

Cancer amphioetus, a rather colorful East Pacific box crab (A. Kerstitch photo).

Telmessus cheiragonus, a Bering Sea species of the small family Atelecyclidae (Alaska; D. Gotshall photo).

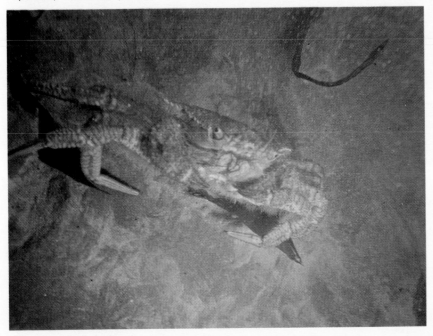

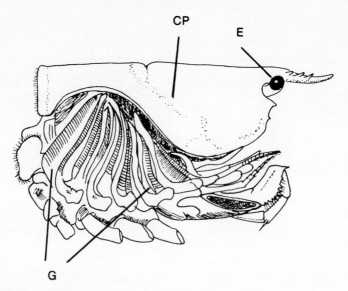

Lobster's carapace cut away posteriorly to show the gills. E = eye; CP = carapace; G = gills.

View of the ostracod *Conchoecia elegans* with the right valve removed to show the body; 1.3 mm long. A1 = first antenna; CP = carapace; MD = mandible; ThL = thoracic limb.

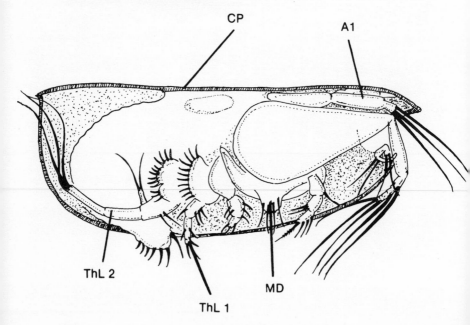

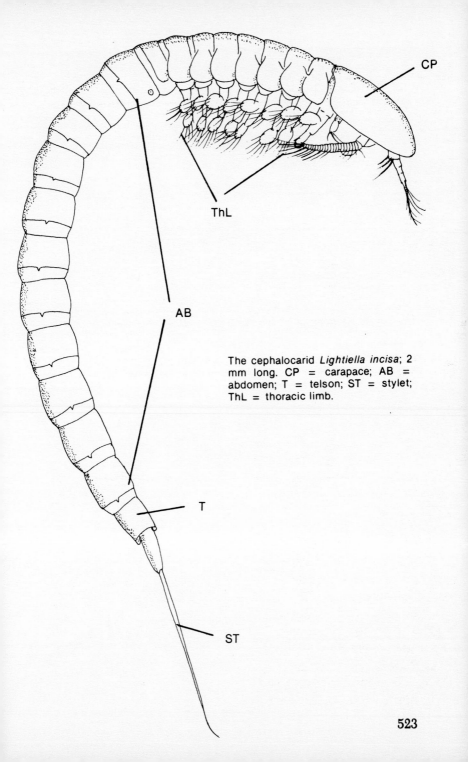

The cephalocarid *Lightiella incisa*; 2 mm long. CP = carapace; AB = abdomen; T = telson; ST = stylet; ThL = thoracic limb.

Swimming crabs, Portunidae, are familiar and often aggressive species recognized by the paddle-shaped hind legs and long, slender claws. Above is *Portunus ordwayi;* below is *Portunus sebae* (both Caribbean; P. Colin photos).

Blue crabs, *Callinectes*, are often distinguished by blue claws. *Callinectes marginatus* (above) is a tropical Caribbean version of the edible blue crab *(C. sapidus)* of the eastern American coast (D.L. Ballantine photo).

A pair of *Cronius ruber,* the tropical red crab (Caribbean; C. Arneson photo).

the total number can be verified by noting where the paired limbs arise. The first five segments are fused to form the head. The paired limbs attached to the first and second segments are the first and second antennae respectively. These, along with the eyes and hairs (setae) on the body and limbs, are the lobster's chief sense organs.

The eyes, although prominent, are probably more efficient at detecting motion than forming detailed images. Similar to insects, each is composed of some 14,000 sensitive individual units; each unit is capped externally by a corneal lens and each transmits its own image to the optic nerve.

The basal segment of each first antenna has a small cavity (statocyst) lined with minute sensitive setae; the cavity also contains sand grains (statoliths). These respond to gravity by falling downward as the lobster turns, displacing the sensitive setae by resting upon them and telling the lobster of its position with respect to gravity. They are thus the organs of balance.

The short whip-like flagellum of each first antenna is invested with numerous setae, some of which serve to detect chemicals in the water while similar setae on the long flagella of the second antennae probably function as touch receptors enabling the lobster to detect prey, sex partners, and obstacles.

From the third head segment arise a pair of foot-jaws (mandibles) that cut into smaller pieces food already shredded by the paired appendages of the following segments, particularly the first pair of maxillae of the fourth segment. The paired second maxillae of the fifth segment also aid feeding, but in addition function as ' gill bailers. ' Rapid vibration of the large oval paddle (scaphognathite) draws a respiratory water current over the gills.

The segments that follow the head belong to the thorax. The limbs arising from the first to third thoracic segments are modified as large foot-jaws, more properly maxillipeds; these also assist food ingestion. The third pair are frequently seen seizing large pieces of food but are also used to clean debris from the antennae. The paired thoracic legs of the fourth to eighth segments are the most conspicuous. The fourth to sixth pairs have their distal segments modified as grasping claws. The claws of the fourth pair are greatly enlarged, provided with strong muscles, and are dissimilar in the lobster. One claw is adapted for crushing, the other for cutting; they are also used for defense and during mating. The remaining thoracic limbs are used for walking although the small claws of the fifth and sixth pairs sometimes hold food and, along with the distal segments of the seventh and eighth legs, are used to 'preen' various parts of the body, removing unwanted debris and attached organisms.

Each second innermost segment of these thoracic legs bears a furrow where the leg may voluntarily break off when grasped by a predator, enabling the lobster to escape. This furrow marks the point where the internal cavity of the leg is crossed by a transverse partition through which nerves and blood vessels, but not leg muscles, pass. The rapid muscular contraction at this junction results in the limb breaking off. The wound heals quickly and rapid growth of new tissue in that region gives rise to a new limb that reaches the size of its predecessor after a few molts.

The six segments of the abdomen each bear a pair of appendages. The first pair of the female are short, slender, and unbranched, while those of the male are stout and modified as twisted rods. The following pairs (swimmerets) are similar in both sexes and are two-branched; the second pair of the male has a small lobe on the inner branch. The last pair of abdominal appendages are laterally placed and expanded into broad stout two-branched blades (uropods) that, together with the telson, form the tail fan.

The lobster's body is enclosed in a hard-shelled articulated covering that cannot expand to accommodate body growth. This shell must be periodically shed (molted) to allow growth to take place. Molting is both exhausting and hazardous, and preparations for it begin with many changes within the body tissues. The essentials of a new shell are formed beneath the old. Glycogen, previously stored in the digestive gland, is the chief component of the

Eriphia sebana, the red-eyed crab (Indo-Pacific; K. Gillett photo).

Typical xanthid or mud crab defensive posture as exhibited by *Eriphia spinifrons* (S. Frank photo).

The colorful xanthid *Lybia tessellata* uses anemones *(Triactis* or *Actinia* among others) both as defense and to get food. It drags the anemones along the bottom and then eats the fine detritus that adheres to the tentacles. Hawaiian specimens (below) are sometimes called *L. edmondsoni* (Indo-Pacific; B. Carlson photo above, S. Johnson photo below).

horny chitin composing this new shell. Parts of the old shell are first weakened by the removal of lime, particularly from the midline of the carapace. This lime is accumulated as lens-shaped concretions of calcium carbonate ('crab eyes' or gastroliths) about 5 mm thick on the sides of the gizzard between the old and new shell and, in the lobster, are resorbed.

Molting begins with part of the shell between abdomen and thorax splitting across its width, as this area is always thin and uncalcified. By repeatedly flexing the body the lobster enlarges this opening and slowly pulls its body out of the old shell. At the same time each side of the carapace bends outward along the weakened midline and the limbs are withdrawn along with the complex internal skeleton that includes the lining of both fore and hind gut, gills, and statocysts. The shell of a newly molted lobster is soft and allows the animal to swell and increase in size by taking in water. During this soft-shelled stage the lobster is vulnerable to predators and hides away; the shell gradually hardens over some six to eight weeks in lobsters or just a few hours in smaller crustaceans.

During late summer newly molted female lobsters are receptive to advances of hard-shelled males, and mating then occurs. The male turns the female onto her back and deposits sperm packets (spermatophores) into a wedge-shaped pocket between the bases of her seventh and eighth thoracic limbs. The male uses his modified first abdominal limbs to accomplish this task. The female spawns chiefly in late summer. She lies upon her back and curls her flexible abdomen, forming a pocket into which the eggs stream from the oviduct openings situated at the base of the sixth pair of thoracic limbs. These eggs, as they are liberated, pass into a jelly-like liquid in which they are fertilized by sperm released from the spermatophores. The eggs then attach to the setae on the basal parts of the swimmerets and are carried for some nine to ten months before they hatch into minute larvae in late spring and early summer. These larvae swim freely in middle and surface waters, passing through a number of molts to become lobsterlings that sink to the sea bed where they live in similar situations to the adults.

This account of the lobster summarizes the more important features of the class Crustacea. The variations of body shape and segmentation, of life histories, behavior, and habitats displayed by its members will become apparent from the descriptions that follow of selected marine representatives belonging to the various subclasses and orders. These representatives are discussed in an ascending order of complexity.

SUBCLASS CEPHALOCARIDA

Cephalocarids are small (2-3.7 mm long) blind colorless crustaceans that live in fine bottom sediments from just below low tide down to about 300 m. Members of this subclass were unknown until 1955, when the first of the few known species, *Hutchinsoniella macracantha*, was described from Long Island Sound, New York, living in organic-rich silt-clay sediments on which it feeds. The similar *Lightiella serendipita* occurs on subtidal silty sand, while *Lightiella incisa* lives among the root-trapped sediments of turtle grass (*Thalassia*).

The anterior part of the cephalothorax of *Hutchinsoniella macracantha* is horseshoe-shaped and the carapace protrudes backward. There are eight recognizable thoracic segments bearing limbs and an eleven-segmented abdomen with a telson to which is attached a long pair of stylets.

Hutchinsoniella moves over the sea bed by beating its limbs. This movement begins at the hind end and passes forward from limb to limb as a smooth wave. The claws of the thoracic limbs and the paddle action of its components (outer branch and false gill) as well as the second antennae also assist forward movement. *Hutchinsoniella* is a hermaphrodite. Two large eggs are carried in egg sacs attached to the limbs of the first abdominal segment. The first larval stage (metanauplius) is not unlike the adult but differs in limb structure. The final shape is acquired after a number of molts, during which segments and limbs are added.

Domecia acanthophora, a small xanthid crab that causes galls in the coral *Acropora* (Caribbean; P. Colin photo).

Another anemone-utilizing crab. The teddy bear crab, *Polydectus cupulifer,* uses small *Telmatactis decora* for protection (Indo-Pacific; S. Johnson photo).

The large, heavily sculptured Indo-Pacific mud crab *Zosymus aeneus* is one of the few crabs directly responsible for human deaths; apparently the flesh is at least sometimes poisonous (A. Power photo).

The brilliant little xanthids of the genus *Trapezia* are commensals on corals, with the color pattern of the various species closely matching their host. Shown is the Indo-Pacific *T. wardi* (S. Johnson photo).

SUBCLASS REMIPEDIA

Described in 1981, Remipedia consists of a single blind 20 mm species, *Speleonectes lucayensis*, from a cave in the Bahamas. Somewhat similar in shape to cephalocarids, it has a many-segmented trunk (31-32 segments) with each segment bearing laterally directed biramous swimming appendages. The antennae are both two-branched. Virtually nothing is yet known of the natural history of *Speleonectes*.

SUBCLASS BRANCHIOPODA

True sea-dwelling branchiopods are limited to a number of water flea species, although the brine shrimp inhabits land-locked salt water lakes and tadpole shrimp are found in saline and brackish water pools; a few species of clam shrimp also occur in brackish water. The water fleas and brine shrimp are the only members considered here as truly marine.

Branchiopods ('gill-footed') typically have flattened leaf-like or lobe-shaped limbs for swimming, breathing, and sieving food from the water. Such development is clearly seen in brine shrimp but is less obvious in the water fleas, in which these limbs are enclosed between two halves of a well developed carapace.

Adults and larvae (nauplii) of brine shrimp, *Artemia salina*, inhabit strongly saline water bodies. At certain times in the year their presence in large numbers turns the water deep red. Brine shrimp and their eggs are used in aquaculture for feeding young fish and crustaceans. The eggs, which can withstand prolonged drying, are hatched by placing them in a salt solution.

A common marine water flea is *Evadne nordmanni*, which occurs in plankton of European coastal waters. As in many similar species, the female for most of the year produces young that can grow from unfertilized eggs. These young develop within a brood pouch that they eventually leave to swim freely in the water. At the end of the year males appear and sexual reproduction occurs. The females then produce fertilized resting eggs that can withstand adverse conditions, even drying. *Evadne* lives just below the sea surface and feeds upon unicellular plants and minute larvae of other invertebrates. The powerful raptorial limbs are used to catch prey. Another widely distributed water flea is *Penilia avirostris* of tropical coastal waters. This species occurs as far apart as the Peruvian coast and the Mediterranean. It lives in swarms in sounds and bays but sometimes occurs offshore. The female produces two resting eggs during the period of sexual reproduction.

SUBCLASS OSTRACODA

Ostracods occur from shore rockpools to some 2800 m in oceanic waters. They have unsegmented or at most very indistinctly segmented bodies with the carapace expanded on either side into a shell hinged along the dorsal midline to enclose the body organs and limbs. In this respect they resemble small bivalved molluscs and are not infrequently mistaken for them. The limbs can be withdrawn into the shell to give some protection when closed.

The shape of limbs and texture of shell surfaces reflect, to some extent, the habits and habitats of ostracods. Species that swim for much of their time, such as *Cypridina norvegica*, have long setae on the first and second antennae that form efficient paddles; the carapace is generally thin and smooth, providing reduced weight combined with streamlining. Species that burrow, such as *Macrocypris minna*, have the antennal setae much reduced and often replaced with claw-like spines used to obtain purchase on sand grains and detritus when burrowing. The thoracic limbs are often similarly modified, and the carapaces of these sediment burrowers are generally smooth and elongated, adapted for penetrating the substrate.

Ostracods that spend much of their life crawling over bot-

Cycloxanthops sp., a small but colorful mud crab from the tropical East Pacific. Notice the resemblance to the shell of the false cowry *Jenneria* from the same area (A. Kerstitch photo).

The Caribbean coral crab, *Carpilius corallinus,* is one of the most gaudily colored coral-reef crabs. (P. Colin photo).

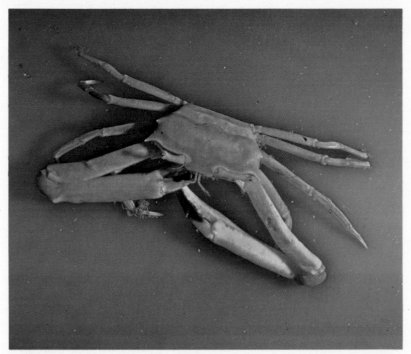

Notice the greatly elongated eyestalks of this male *Goneplax rhomboides;* goneplacids are among the few crabs that construct burrows in mud below the low-tide level (East Atlantic; T.E. Thompson photo).

Female *Xanthasia murigera,* a pea crab or pinnotherid commensal in the giant clams, *Tridacna* (Indo-Pacific; L.P. Zann photo).

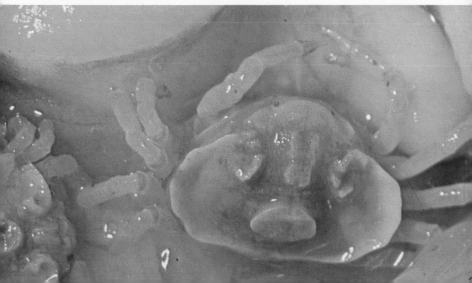

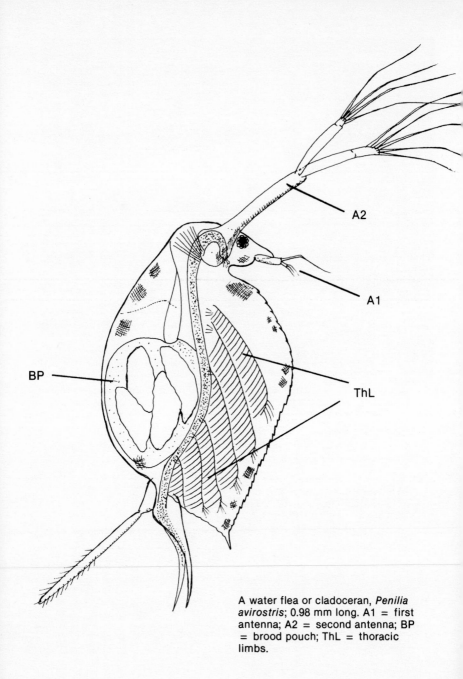

A water flea or cladoceran, *Penilia avirostris*; 0.98 mm long. A1 = first antenna; A2 = second antenna; BP = brood pouch; ThL = thoracic limbs.

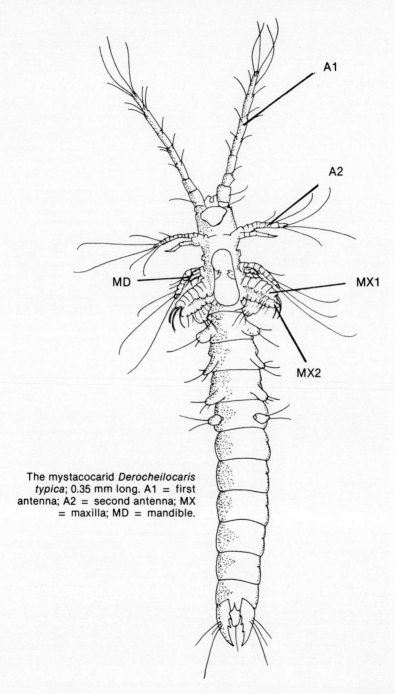

The mystacocarid *Derocheilocaris typica*; 0.35 mm long. A1 = first antenna; A2 = second antenna; MX = maxilla; MD = mandible.

A1

A2

MD

MX1

MX2

Two divergent types of shore crabs. Above, the greatly depressed and mud-crab-like *Gaetice depressus* from the Japanese Sea; such a shape is not typical of grapsoid crabs (Y. Takemura and K. Suzuki photo). Below is the more typically shaped *Pachygrapsus crassipes* from the East Pacific (K. Lucas photo at Steinhart Aquarium).

Planes minutus, the sargassum crab, is one of many pelagic invertebrates associated with floating masses of sargassum and other debris; the color is quite variable in this virtually cosmopolitan grapsoid crab (South Africa; T.E. Thompson photo).

A brightly colored shore crab, probably of the genus *Hemigrapsus,* from the East Pacific (K. Lucas photo at Stainhart Aquarium).

tom sediments often have highly ornamented carapaces. The valves are thick and modified in some species into ventrally or laterally projecting keels. Species living in coarse sediments may have spined or reticulated patterns on their carapace, while smooth forms tend to inhabit the finer substrate layers.

As fossils, ostracods are well represented in many types of rocks such as limestones, shales, and marls. Some building stones, such as Cyprus Freestones of the English Upper Jurassic, are composed almost entirely of their shells.

Reproduction in ostracods is still imperfectly understood. Some podocopid ostracods mate frequently during the warm season. The male brings the ventral edges of his now opened valves into contact with those of the female and transfers sperm into the female sperm receptacle. Mating may last from thirty seconds to a few minutes, and a single mating may suffice for the fertilization of a large number of eggs. The larvae that hatch from these eggs already have a bivalved carapace. The number of larval stages varies from species to species; there may be as many as eight, and new limb buds and limbs are added at each molt. The thoracic limbs are generally fully developed before the last but one molt.

Some species of myodocopid ostracods mate in the surface waters. The males of *Philomedes globosa* normally live near the surface, but females dwell on muddy bottoms. When adult females undergo their final molt they acquire setae on their limbs that enable them to swim. During darkness at certain times of the year these females migrate to the surface from depths of over 180 m; there they mate with the males. After mating they again descend to the bottom, cut off their swimming setae using the claws of the first pair of thoracic legs, and condemn themselves to a permanent bottom existence.

Many ostracods are continuously migrating from one water level to another, rising to shallow waters during darkness and descending during the day. *Conchoecia spinifera*, for example, is chiefly concentrated in depths between 500-570 m during the day, but at night the largest numbers occur in depths around 150-220 m.

Although many species of ostracods are quite small, varying between 0.4 mm to 1.5 mm long, there are a few large marine species. The largest is *Gigantocypris agassizi*, reaching 23 mm in length. It is probably a sedentary species that floats or swims slowly, waiting for its prey to come within reach before catching it.

Some ostracods luminesce or eject clouds of luminous particles into the water. This is a particular feature of many species of *Cypridina*. When stimulated by light from a lamp, *Cypridina serrata* produces bright blue luminous clouds that last for about 1-2 seconds. Another species, *Cypridina hilgendorfi*, lives in sand, coming out at night to feed; if disturbed, it produces a blue luminescent cloud. The glands responsible for this light production are situated within the valves; *Conchoecia daphnoides* retains the luminescence within glands located in various extensions of the carapace.

SUBCLASS MYSTACOCARIDA

Mystacocarids are very small crustaceans of not more than 0.5 mm body length. They are thought to be related to the copepods but have primitive features that also link them to cephalocarids and branchiopods.

Derocheilocaris typicus has forwardly directed first antennae that are probably sensory structures, while the mandibles and second antennae, along with the maxillules, assist in locomotion as well as feeding. *Derocheilocaris* hatches as a larva (metanauplius) bearing only a superficial resemblance to the adult. Through progressive molts the abdominal segments are added and limb segments developed.

Mystacocarids live in spaces between sand grains (interstitial habitat) of intertidal marine beaches; very few occur in offshore sands. They are collected by digging holes in the sand near the water line. Groundwater flowing into the hole will carry the animals into it, where they are then scooped out with a fine mesh net.

Pachygrapsus marmoratus, one of the common intertidal shore crabs often called "Sally Lightfoot" crabs (Atlantic; S. Frank photos).

Although not strictly marine, potamid or land crabs occasionally appear for sale in pet shops as marine animals. Above is a parathelphusid-like species from Sri Lanka that ventures into brackish water (R. Jonklass photo). Below is an unidentified geothelphusid-like species from Thailand (A. Norman photo).

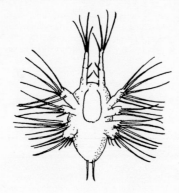

Left: Nauplius larva of *Calanus*; 0.27 mm long. **Below:** Copepodid larva of *Calanus*.

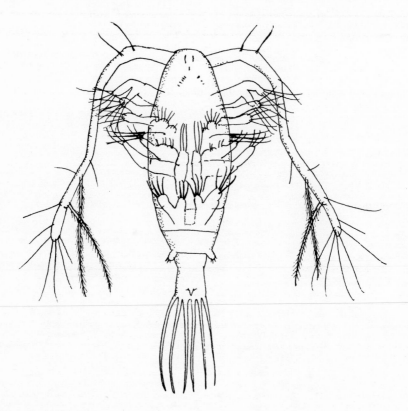

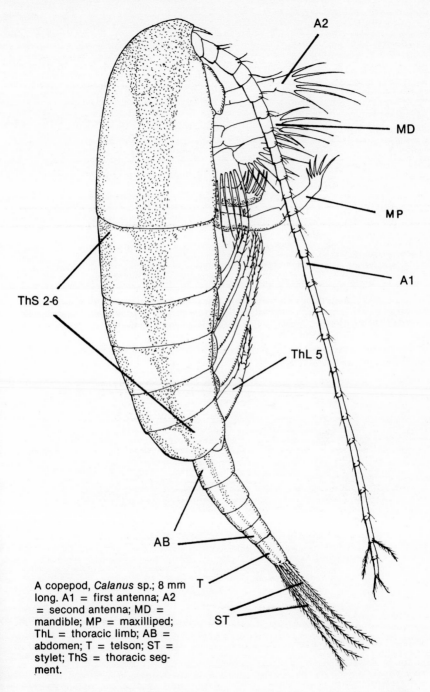

A copepod, *Calanus* sp.; 8 mm long. A1 = first antenna; A2 = second antenna; MD = mandible; MP = maxilliped; ThL = thoracic limb; AB = abdomen; T = telson; ST = stylet; ThS = thoracic segment.

547

True land crabs, family Gecarcinidae, are often incredibly colorful. What other marine invertebrates could match the colors of this *Cardisoma latimanus* from the East Pacific? (K. Lucas photo at Steinhart Aquarium).

Ocypode ceratophthalma, a ghost crab with unusual eyestalk projections (Indo-Pacific; K. Gillett photo).

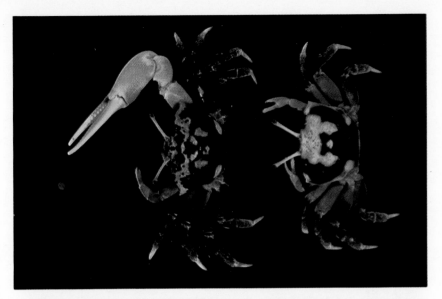

Fiddler crabs *(Uca)* are familiar inhabitants of muddy or sandy shores around the world. Above is a colorful species from Fiji (B. Carlson photo); below is a male *Uca coarctata* from northern Australia (W. Deas photo).

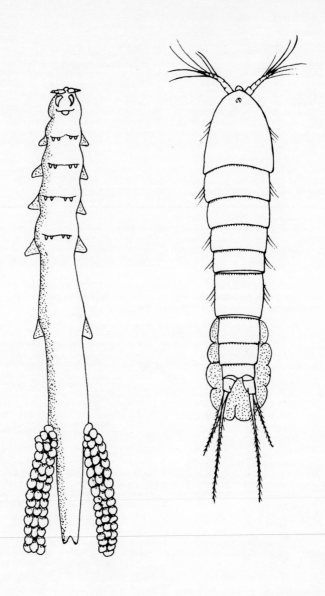

Left: Female cyclopoid copepod, *Mytilicola porectus*; 3 mm long, ventral
view. **Right**: Female harpacticoid copepod, *Mesochra lilljeborgi*; 0.5 mm
long.

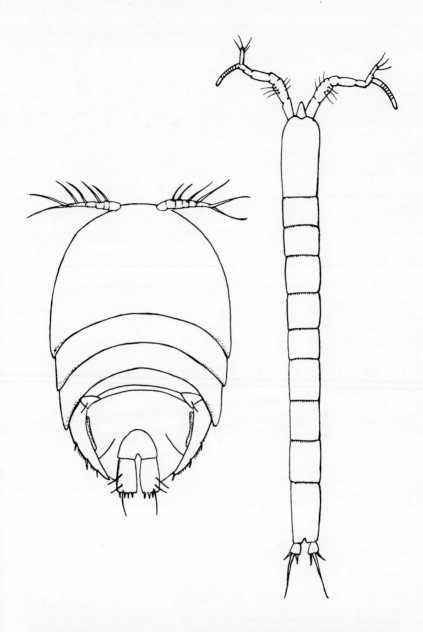

Left: Female harpacticoid copepod, *Porcellidium echinophilum*; 0.6 mm long. **Right**: A sand-dwelling harpacticoid copepod, *Stenocaris arenicola*; 0.6 mm long.

Mictyris longicarpus, the Indo-Pacific soldier crab, masses in large "armies" on tidal flats at low tide. The inflated carapace can be completely sealed by the mouthparts to allow conservation of water over the gills (K. Gillett photo).

Coral gall crabs, family Hapalocarcinidae. Above, *Cryptochirus coralliodytes* forms a pit gall in massive corals. Below, galls of *Hapalocarcinus marsupialis* at the end of branches of *Seriatopora* (both Indo-Pacific; L.P. Zann photos.). These are the most un-crablike of the known crabs.

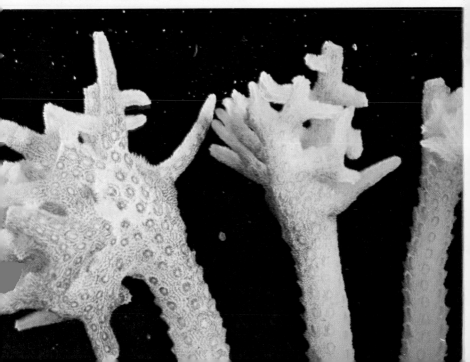

The few known species of mystacocarids closely resemble each other, and each has a wide geographical range. *Derocheilocaris typicus* occurs in both subtropical (Florida) and subarctic (Massachusetts) waters, while *D. remanei* is found off the Atlantic coast of France, in the Mediterranean, and from the Cape of Good Hope to Durban in South Africa.

SUBCLASS COPEPODA

The copepods (' oar-footed ') occur in a variety of habitats in salt, brackish, and fresh waters as free-living and parasitic forms. Many species are permanent inhabitants of the oceanic zooplankton, of which, at times, they form the predominant bulk. There are well over 7500 known species of copepods that vary in size from the large *Bathycalanus sverdrupi* of deep waters, measuring some 17 mm in length, to the small *Sphaeronellopsis monothrix* that parasitizes ostracods; the males of this species grow to only 0.11 mm. Free-living copepods are fairly uniform in shape, but commensal and parasitic species may have their bodies and limbs drastically modified so that some are hardly recognizable as copepods.

Calanus hypoboreus is a free-living planktonic species. The carapace is not developed and the first thoracic segment is fused with the head; the following five thoracic segments are of similar size. The four-segmented abdomen is much narrower than the thorax, and the telson bears a pair of long stylets. Each of the five pairs of thoracic limbs are interlocked and serve to propel the animal through the water while the long first pair of antennae bear many sensory setae and, with the help of the well developed setae of the telson stylets, prevent the animal from sinking too rapidly when it momentarily ceases to swim.

Species belonging to *Calanus*, *Temora*, *Arcartia*, and related genera are widespread throughout the Atlantic, Pacific, and Arctic oceans, where they often dominate zooplankton communities. These calanoid copepods are said to be the most numerous animals in the world that ' graze ' upon the biggest plant crop of the world—the flagellates and diatoms that occur in a superabundance in the surface waters of the ocean. In the northern Atlantic *Calanus finmarchicus* occurs in such enormous numbers as to turn the water red during the summer, making it the most important food of the herring. In the Barents Sea, for a part of the year this species of *Calanus* provides some 65 % of the food of this fish. The large *Calanus hypoboreus* female grows to some 8 mm long and feeds chiefly by filtering suspended microscopic algae from the plankton. Swimming slowly, it creates currents that sweep small particles toward the setae investing the thoracic limbs. This efficient filter can trap organisms as small as 10 microns. *Calanus* will also catch and feed upon smaller copepods or their immature stages.

The female *Calanus* lays eggs in batches of 200-300 at intervals of about 10-14 days. The number of eggs produced is related to the quantity of food available. The eggs hatch as minute larvae (nauplii), each with three pairs of unspecialized thoracic limbs used for food gathering and propulsion. After a series of molts each nauplius becomes a metanauplius when the body acquires segments and the limbs appear. This stage is followed by a number of copepod stages that, after each molt, begin to resemble more and more the adult but are still without their full complement of limbs. At the final molt to adult the full number of body segments and limbs is acquired, the second pair of antennae become reduced, and the first pair is greatly enlarged.

Free-living cyclopoid copepods are less abundant in plankton than calanoids. One such species is *Oithona nana*, the female of which grows to a length of 0.9 mm. Many cyclopoids are predators, feeding upon smaller planktonic animals and their larvae. A parasitic cyclopoid is *Mytilicola intestinalis*. The female grows to 8 mm and the red body is modified for living within the gut of the mussel *Mytilis*; this species also occurs in oysters and other bivalves. The early larval stages are free-living; the second (copepodid) stage invades the gut of the mussel and continues to develop there. A number of cyclopoids are associated with other marine in-

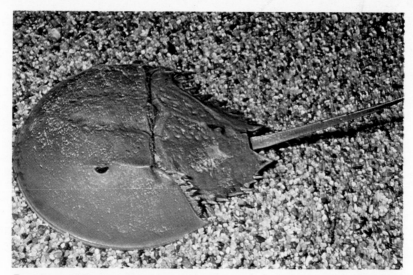

Two views of the horseshoe crab, *Limulus polyphemus*, from the West Atlantic. This living relict is more closely related to the arachnids than to the crustaceans (K. Lucas photos at Steinhart Aquarium).

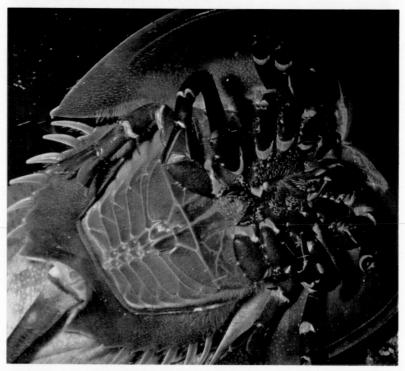

Close-up of the arms of the crinoid *Himerometra robustipinna* from the Indo-Pacific. Notice the thick, regular pinnules and the distinct discs (G.R. Allen photo).

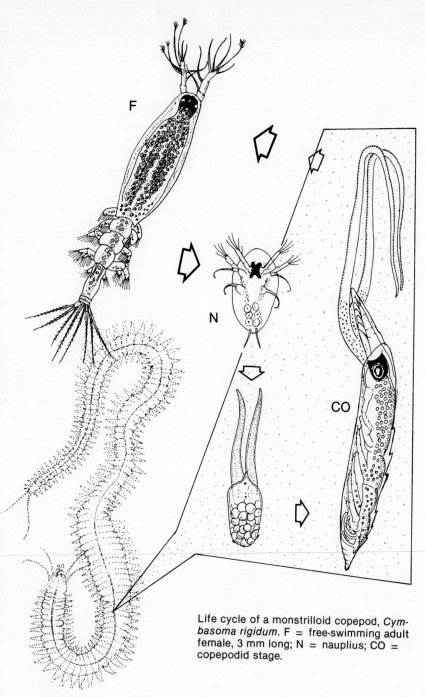

Life cycle of a monstrilloid copepod, *Cymbasoma rigidum*. F = free-swimming adult female, 3 mm long; N = nauplius; CO = copepodid stage.

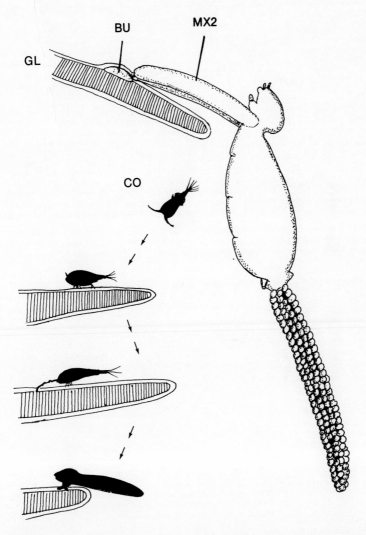

Stages in the life cycle of the copepod *Salminicola salmonae*, a lernaeopodid parasitic in salmon. GL = salmon gill; BU = bulla; MX = maxilla; **CO** = copepodid stage;

Ventral views of two feather stars. In the species above, the cirri are strongly developed and numerous, while in the species below they are small or absent (both Indo-Pacific; W. Deas photo above, Y. Takemura and K. Suzuki photo below).

Two Caribbean feather stars. At right, the swimming *Analcidometra caribbea;* below, the nocturnal *Comactinia echinoptera* (P. Colin photos).

vertebrates. *Scambicornus lobulatus*, for example, is found on the body surface of the sea cucumber *Bohadschia graeffei*.

The majority of harpacticoid copepods are free-living, less than 2 mm long, and have tapering bodies, although in a few species the body is flattened. Female harpacticoids, such as *Mesochra lilljeborgi*, may be seen often with one or a pair of egg sacs attached to the ventral surface of the first abdominal segment. As swimming is not particularly well developed, only a few species occur in the plankton. The majority crawl on the surface of sand and mud or live under the surface. Such forms often have greatly elongated bodies, such as *Stenocaris arenicola*, or are strongly flattened dorsally, as in *Porcellidium echinophilum*. Some harpacticoids live associated with other animals. *Balaenophilus unisetus* lives on the baleen plates of the sulphurbottom whale, often forming yellowish patches due to their vast numbers. The maxillipeds and thoracic legs are well adapted for clasping the baleen of the whale, for rarely do they become dislodged by the torrents of water that flood over these plates when the whale feeds. The copepods appear to feed upon the green algae that grow profusely on the baleen plates. Harpacticoids, similar to members of the other orders, have exploited many unusual habitats. *Cancricola plumipes* and *Anthillesia cardisomae* live in the moist gill chambers of the land crab *Cardisoma guanhumi*. Only when the crab returns to the sea for a short period of ten days or so to breed can these copepods reproduce and re-infect other crabs.

The adults of monstrilloid copepods occur in marine plankton, but immature stages, as far as is known, parasitize bristle worms, marine snails, and brittlestars. The female of *Cymbasoma rigidum* is rare in plankton samples and probably feeds upon stored food as the gut is not developed. From the eggs produced hatch free-swimming larvae (nauplii). If one succeeds in penetrating the skin of a bristle worm it enters the body cavity, passes into the dorsal blood vessel of its host, loses its limbs, and develops two anterior processes; food at this stage is absorbed through the body surface. These processes elongate and the parasite develops into the next larval (copepodid) stage within the skin envelope that surrounds it. When maturity is reached it escapes from the host by breaking through the blood vessel and then the skin. Then begins a short free-swimming stage during which time more eggs and free-swimming nauplius larvae are produced to repeat the life cycle.

Notodelphoid copepods are chiefly associated with sea squirts. The bodies of some species are very swollen, as seen in *Demoixys chattoni*, or a brood pouch may be developed for carrying eggs, as in *Notodelphys reducta*. The larval stages and males of some species are free-swimming, but it seems that the females always remain with their host.

Caligoid copepods are external parasites of marine and freshwater fish, aquatic mammals, and rarely squids; they sometimes swim freely in the plankton. Commonly associated with fish, they have been named fish lice. The mouthparts are reduced or modified for grasping and sucking, an adaption for a secure attachment to their host. The female of *Lepeophtheirus pectoralis* measures some 5.6 mm in length when matured and is a common parasite of plaice (a type of flatfish), occurring on the body surface. The nauplius larva swims freely in the plankton and in the later (copepodid) stage attaches to a fish by a frontal filament where it continues to molt through to the adult form. The female of the largest caligoid, *Pennella balaenoptera*, grows to 300 mm in length, excluding the egg strings. It parasitizes the common rorqual or finner whale, the head buried deep in the whale's blubber. Little is known about its development, but the larval stages may be parasitic on squid. Other species of *Pennella* are found on swordfish, oceanic sunfish, and flying fish. Because of the striking branched and feather-like processes on the abdomen, these parasites have been termed 'feathers' of flying fish, and the name *Pennella* is derived from the Latin *penna* or feather.

Lernaeopodid copepods are parasitic upon marine and freshwater fish and are so modified at some stages that they barely resemble copepods. *Salminicola salmonea* is a common parasite of the Atlantic salmon. The female grows to 8

Two Caribbean species of the large feather star genus *Nemaster*. Above is *Nemaster discoidea* of deep reefs; below is *Nemaster rubiginosa* of shallower reefs. Notice the difference in density of pinnules on the arms. (P. Colin photos).

Nemaster rubiginosa increases its plankton-filtering efficiency with fine projections ("hairs") on each pinnule (Caribbean; P. Colin photo).

Coenolia trichoptera, a small feather star from shallow waters of the Tasman Sea (T.E. Thompson photo).

mm in length and attaches to gills of these fish by her large second maxillae. Permanent and secure attachment is achieved by the enlarged ' bulla ' buried beneath the skin of the fish's gills. The male is much smaller than the female. While the fish is in fresh water the female produces some 900 eggs contained in two elongated egg sacs. These hatch as advanced (copepodid) larvae and attach to the host's gills or infect other salmon. Secure attachment is achieved by a frontal filament that grows into the hole made by the head appendages. Hatching and attachment occur while the salmon is migrating to spawning areas in the upper reaches of rivers. Following the second larval stage, males appear, mature rapidly, and, although still attached to their host, move about freely. Mating then occurs and the males die. The females produce eggs before the fish migrates back to the sea. During this sea-dwelling period the parasite grows but awaits the return of the salmon to fresh water before producing new larvae that will again reinfect the original host or find other hosts.

SUBCLASS BRANCHIURA

Branchiurans or fish lice are external parasites upon marine and freshwater fish, although one species at least, *Argulus arcassonenis*, has been taken from the skin of a dead cuttlefish. *Argulus laticauda* occurs on the eel, skate, summer flounder and other fish. The body is strongly flattened dorsally and the wing-like expansions of the carapace conceal the thoracic limbs. The first pair of antennae bear strong recurved hooks; the mandibles are reduced and hidden behind a posteriorly directed proboscis above which there is a forward-pointing spine. The basal segments of each first maxilla are modified as a tube-like sucker, while those of the second bear numerous prehensile spines and processes. In contrast to these head appendages, modified for clinging, piercing, and sucking, the four pairs of forked setose thoracic limbs are adapted for swimming. The small bilobed abdomen is without limbs.

Although *Argulus* can cling firmly to its host, it can also scuttle rapidly over the fish's skin with a walking motion of the first maxillae. The parasite usually takes up a position with its head pointing in the same direction as that of its host. The various spines on its ventral surface and limbs point downward or backward and bury deeper into the skin as the parasite is pressed harder against the body by water currents flowing over the fish when it swims. *Argulus* can easily detach itself and swim freely with a gliding motion. This frequently occurs during the breeding season when the male settles upon the female's back and mates with her.

The female *Argulus* deposits her eggs, often in rows of 10-12, upon a suitable substrate. *Argulus megalops* breeds in October off the western coast of America. The eggs take about 60-80 days to develop, and the larval stages (nauplius and metanauplius) are passed within the eggs. The juvenile stage that hatches has a full complement of swimming legs. After several molts it acquires fully developed suctorial mouthparts and is then sexually mature.

SUBCLASS CIRRIPEDIA

To this subclass belong the familiar barnacles and lesser known forms that are parasites or live within or on other crustaceans, corals, etc. The name of this subclass is derived from the typical feather- or cirrus-like feet characteristic of the free-living cirripeds, the barnacles.

The largest cirriped is *Lepas anatifera*, a stalked barnacle that grows to a length of 800 mm. The adult is permanently attached to pieces of driftwood or other objects by the head, which is developed as a long stalk. The head carries a minute pair of first antennae containing adhesive glands whose secretions cement the stalk to its substrate. Each half of the carapace is covered with large calcareous plates that protect and enclose the limbs and body organs. These four to six plates are formed during development and become strengthened by the deposition of mineral salts; one or two are keel-shaped, but the others are paired and flattened. The transversely placed adductor muscle, attached to each inner wall of the carapace, serves to draw each half of the

Two relatively primitive starfishes. Above, *Astropecten scoparius* (Japan; Y. Takemura and K. Suzuki photo); below, *Luidia alternata,* one of the most common West Atlantic starfishes (Caribbean; C. Arneson photo).

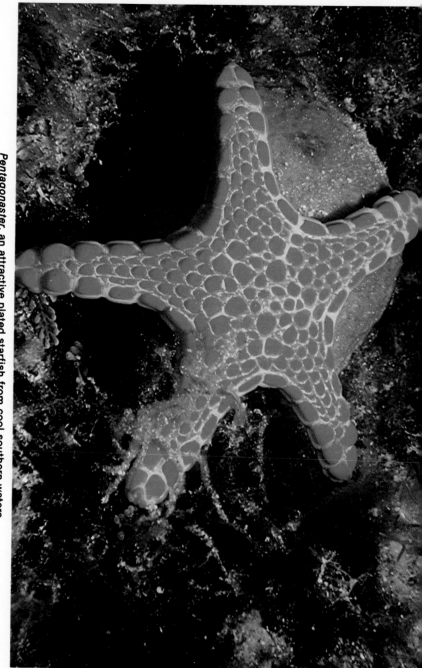

Pentagonaster, an attractive plated starfish from cool southern waters.

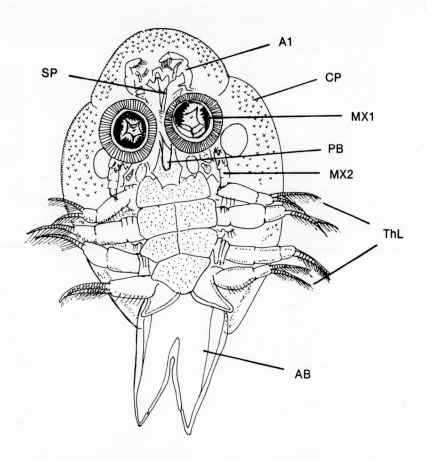

Above: The fish louse (branchiuran) *Argulus matuii*; female, 7 mm long. A1 = first antenna; SP = spines; CP = carapace; PB = proboscis; MX = maxilla; ThL = thoracic limbs; AB = abdomen.

Facing page: Stages in the life cycle of the stalked barnacle *Lepas*. 1) Metanauplius, 5 mm long; 2) Cypris larva, 4 mm long; 3) Adult, 80 mm long. SK = suckers; A1 = first antenna; ThL = thoracic limbs; ADG = adhesive gland; ST = stalk; FL = filamentary organ; M = mouth; ADM = adductor muscle; P = penis.

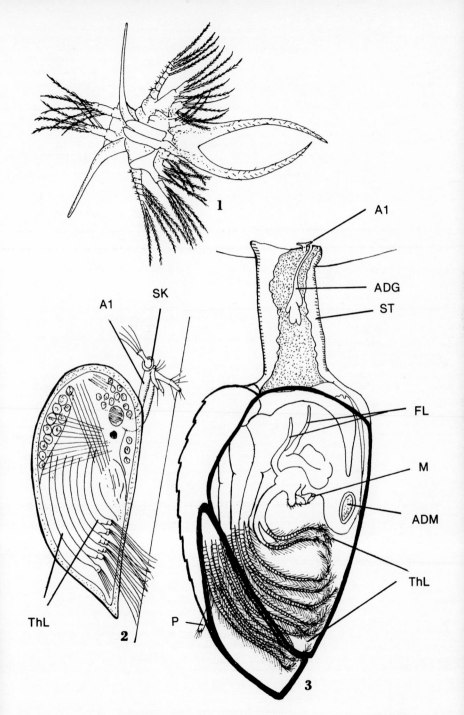

The plated starfish *Iconaster longimanus* has a unique shape and color pattern (West Pacific; A. Power photo).

The dried skeletons of *Goniaster tessellatus* are often sold for decoration. Although common in the Caribbean, it is also found in the East Atlantic and Indo-Pacific (N. Herwig photo).

Amphiaster insignis, an extremely spiny plated starfish from the East Pacific (A. Kerstitch photo).

shell together. The thoracic limbs move with a regular rhythm during which their long setae strain small particles of food from the surrounding water. The abdomen has lost all recognizable segmentation and has become somewhat folded so that the mouth and anus lay at the same level. *Lepas* breathes through its carapace, but the filamentory organ may also be used for this purpose.

Similar to most other barnacles, *Lepas* is a hermaphrodite. The eggs of *Lepas* are fertilized within the body cavity after being liberated. The larvae (nauplii) hatch and are then released into the surrounding water. Each nauplius molts to a metanauplius, a short-lived stage, and at the next larval stage (cypris) a bivalved shell is developed that gives the young barnacle the appearance of a minute cyprid ostracod. At this stage all limbs are present. This cypris does not feed but seeks an area on which to settle. It explores suitable substrates, walking with its first antennae as if on stilts. These antennae bear terminal suckers, and the little cyprid larva may make several attachments before finally secreting cement from special glands. The cement will permanently fix it to the chosen surface. It then becomes transformed into a sessile stage when the final barnacle-like form is developed and revealed after a further molt.

The body shape of thoracicid barnacles ranges from the striking goose barnacles, such as *Lepas anatifera* and *Pollicipes polymerus*, in which the head region is stalked, to the sessile forms in which this region is flattened into a broad disc for attachment as seen in *Verruca gibbosa* and *Balanus balanoides*; this last species has many additional calcareous plates. Some stalked barnacles (limestone burrowers such as *Lithotrya valentiana* and the deepwater *Megalasma striatum*) have very rudimentary stalks. By comparison, in conditions of crowding the heads of some sessile species (*Balanus balanoides*) may become elongated to several times their normal length.

Although stalked barnacles are often attached to driftwood, many species form clusters on a variety of floating objects including *Velella* and the marine purple snails, *Jan-*

thina. The cement gland secretions form a conspicuous yellow mass that serves as a bubble float should the barnacle fail to attach to an object. Some species of stalked barnacles have their calcareous plates very reduced or lost as seen in *Conchoderma virgatum*, which is often attached to sea turtles and swimming crabs, and in *Octolasmis california*, found in gill chambers of some spiny lobsters.

Many sessile barnacles are intertidal. The acorn barnacle *Chthamalus stellatus* lives high on the shore and can withstand long periods of exposure to air. Sessile species usually occur in dense populations and are often the most abundant crustaceans of rocky shores. The release of larvae from each individual is timed to occur almost simultaneously and when food is abundant in the plankton. This massive release of larvae results in a dense settlement. Growing in such close proximity to each other ensures efficient cross-fertilization within the crowded colonies. Such dense settlements of cyprid larvae often occur where the rock surfaces are already covered with living or dead barnacles as the shells of such contain proteins that appear to stimulate such settlement as well as revealing the suitability of the area for such a purpose.

Some ascothoracic cirripeds are parasitic within the body cavities of echinoderms, below the horny layer of some corals and sea fans, or attached to the body of such hosts and sucking their body fluids. The mouthparts of such cirripeds are modified for piercing or sucking or may be reduced. In some species the sexes are separate. The small (1 mm) male of *Ascothorax ophioctenis* lives attached to the female (3-4 mm) that is endoparasitic in the brittlestar *Ophiocten sericeus*. Each half of the carapace of the female is greatly expanded, often obscuring the thorax and abdomen. The cyprid larval stage develops from the nauplius within the female's brood chamber. After leaving the female the small cyprid larvae search out a new host in which to develop into the adult form.

Acrothoracic cirripeds bore into corals, other barnacles, plates of sea urchins, bivalves, and marine snail shells. All

Above, *Mediaster aequalis* (East Pacific; D. Gotshall photo). Below, *Nidorellia armata* (East Pacific; A. Kerstitch photo).

Two common and colorful species of the shallow-water starfish genus *Oreaster*. Above, the Caribbean *Oreaster reticulatus* (P. Colin photo); below, the East Pacific *Oreaster occidentalis* (A. Kerstitch photo).

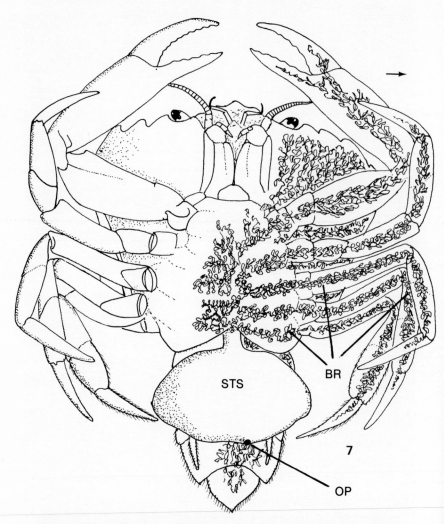

Above and facing page: Stages is the life cycle of the parasitic rhizocephalan barnacle *Sacculina carcini*. 1) Nauplius larva, 0.23 mm long; 2) Cypris larva, 0.2 mm long; 3) Cypris attached to seta on host crab (SE); 4) Cypris at later stage (OS = outer shell of cypris); 5) Later stage (SC = sac containing parasite); 6) Young parasite (SP = hollow spine); 7) Adult stage in host crab (STS = stalked sac, OP = opening of sac, BR = root branches in crab's body).

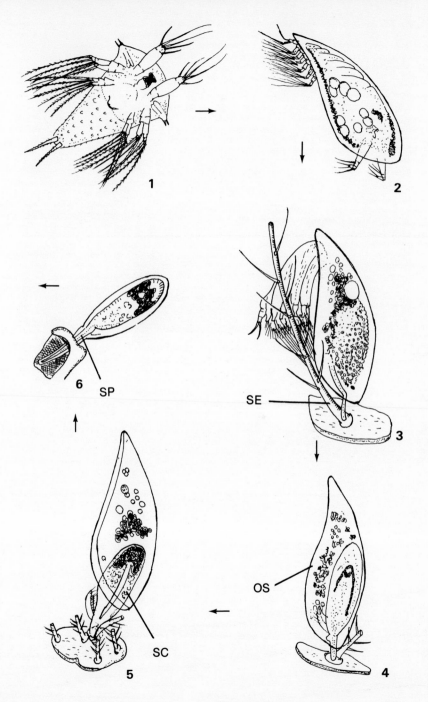

SP

SE

OS

SC

1

2

3

4

5

6

Culcita novaeguineae, the pincushion starfish of the Indo-Pacific. This species is able to change its shape from rather normally five-armed to basketball-shaped by inflating with water (G. Marcuse photo).

Ventral view of the East Pacific *Oreaster occidentalis* (A. Kerstitch photo).

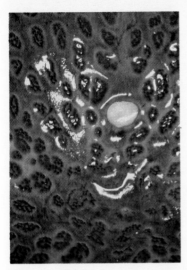

Full-body view and close-up of *Dermasterias imbricata,* a common smooth starfish of the East Pacific (K. Lucas photo at Steinhart Aquarium on left, D. Gotshall photo on right).

The somewhat similar Indo-Pacific smooth starfish *Choriaster granulatus* (H.R. Axelrod photo).

have separate sexes. The female of *Trypetesa lampas* grows to 2 mm in length but the male, attached to the female, is very small, only 0.4-1.2 mm long. This species lives in snail shells inhabited by hermit crabs.

The rhizocephalans are the most specialized of all the parasitic cirripeds. The larval nauplii of *Sacculina carcini* hatch from eggs in the brood pouch of the female, swim freely in the sea for about eight days, and there continue to develop through further naupliar stages while feeding upon stored yolk. The cypris stage that follows swims freely for a further four or five days, after which time, during darkness, it attaches by its first antennae to a young shore crab, *Carcinus maenas*, about 12 mm long. The point of attachment is at the base of a seta on the crab's body or legs. The inner shell of this cypris larva then draws away from the original outer shell, forming a sac that contains a group of specialized cells. The outer skin of this sac hardens and the remaining parts of the old cypris larva are cast off. A hollow spine forms that is pushed into the base of the seta and through which the group of cells pass, entering the crab's body. The adult form of *Sacculina* then begins to develop from this group of cells that become attached to the midgut of the crab; branches are soon formed that invade spaces between the other organs.

A part of the parasite now breaks through the base of the crab's abdomen and grows out as a stalked sac. The end of this has a small opening through which a male cyprid larva can pass and whose sperm will fertilize the large numbers of eggs within the sac. The enormous numbers of nauplius larvae produced from these fertilized eggs pass out through the opening into the sea, where they continue their development to the cypris stage and settle upon another crab to again repeat the life history. About 6% of the infected male crabs do not develop testes, and the abdomen becomes broadened and similar in shape to that of the female.

Rhizocephalans occur in many crustaceans. *Thompsonia* parasitizes hermit crabs, *Mycetomorpha* shrimp, and *Duplorbis* isopods. The rarest known cirriped is the apodan *Proteolepas bivincta*, known from only a single preserved specimen taken from within the West Indian stalked barnacle *Heteralepas*.

SUBCLASS MALACOSTRACA

Malacostracous means ' soft shelled '; it is perhaps unfortunate that this subclass should have been so named, as it contains some of the most strongly calcified crustaceans (lobsters and crabs). However, the name was used originally by early naturalists to distinguish the so-called ' soft shelled ' crustaceans from the ' hard shelled ' oysters and clams. These last mentioned animals now of course belong to the phylum Mollusca.

Malacostracans are, on the whole, larger than the crustaceans so far considered and show a greater diversity in habitat and behavior. Within this subclass are placed the isopods and their relatives, mantis shrimp, opossum shrimp, krill shrimp, true shrimp, lobsters, and crabs, whose body segmentation (with one exception) has become stabilized to nineteen segments, although this number is not always obvious at first sight.

SUPERORDER LEPTOSTRACA (= PHYLLOCARIDA) ORDER NEBALIACEA

The large suboval carapace of *Nebalia bipes* is flattened from side to side, each half fused to the other along the dorsal midline and able to be drawn together by the strong adductor muscle; the hinged rostrum can close the anterior gap of the carapace. The long antennae have many sensory setae and the compound eyes are set on long stalks. *Nebalia* breathes through the outer branches and false gills on the thoracic legs and through the inner lining of the carapace. These surfaces are kept free from debris by the brush-like action of the palps on the first maxillae. The abdomen has seven instead of the six segments typical of all other malacostrocans.

Many nebalids inhabit shallow water. *Nebalia bipes* lives beneath stones and rocks of the lower shore, feeding upon

Two extremely colorful Indo-Pacific starfishes. Above, *Fromia* sp.; below, *Neoferdina ocellata* (A. Norman photo above, A. Power photo below).

The bright blue *Linckia laevigata* is one of the most familiar Indo-Pacific starfishes, but it is difficult to keep in the aquarium.

Late postlarval stage of an unidentified stomatopod or mantis shrimp. (C. Arneson photo taken in the Caribbean.)

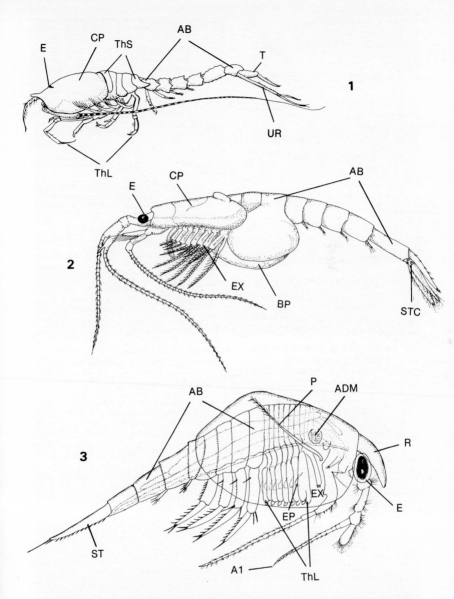

1) *Diastylis cornuta*, a cumacean; male, 12 mm long. 2) A female opossum shrimp or mysidacean, *Gastrosaccus sanctus*; 13 mm long. 3) *Nebalia bipes*, a nebaliacean; 12 mm long. E = eye; CP = carapace; AB = abdomen; EX = exopod; BP = brood punch; STC = statocyst; ThS = thoracic segments; T = telson; UR = uropod; ThL = thoracic limbs; EP = epipod; A1 = first antenna; ST = stylet; P = mandibular palp; ADM = adductor muscle.

Unlike many other groups of animals, starfishes do not become more drably colored in cooler waters—or at least there are enough colorful exceptions to catch the collector's eye. Above is *Uniophora granifera* (W. Deas photo); below is *Plectaster decanus* (T.E. Thompson photo) (both Tasman Sea).

Multiarmed starfishes seem to be common in the East Pacific. To the right is *Crossaster papillosus* (D. Gotshall photo); below is *Solaster stimpsoni* (T.E. Thompson photo).

detritus and decaying flesh. The female lays eggs in a brood pouch formed from the long setae of the outer branches of the thoracic limbs; such females do not feed. The young hatch then molt twice; when their yolk store is used up they leave the brood pouch, at which time they measure about 1.5 mm in length and in most respects resemble the adults.

SUPERORDER HOPLOCARIDA
ORDER STOMATOPODA

Stomatopods or ' squillas ' have their second pair of thoracic legs (maxillipeds) very enlarged and developed as ' jack-knife ' claws. In this respect they superficially resemble the common garden mantis (an insect) and have been appropriately called ' mantis shrimp. ' Many mantis shrimp live in shallow coastal waters, but quite a few occur in deeper water. All apparently make burrows in sand or mud or live in burrows discarded by other marine animals; some live in tubes of dead coral or in rock interstices.

Squilla mantis occurs in the Mediterranean, where it burrows into mud or muddy sand. It is dorsally flattened; the carapace is thin and covers only the anterior thorax. The prominent kidney-shaped eyes can perceive movement in the semi-darkness in which the animal often lives. The whip-like flagella of the first antennae sense and feel prey; if these organs are amputated, *Squilla* may never again feed. The first thoracic limbs (maxillipeds) are used for grooming; the second and largest have long recurved spines on the distal segment that can close with a quick ' jack-knife ' movement and, assisted by the third to fifth pair of limbs, are very efficient at capturing prey. The cylindrical legs of segments six to eight assist the animal in walking over the sea bed. When walking, *Squilla* uses the middle segments of the second maxillipeds as a supporting elbow. The long and flexible abdomen has many distally spined dorsal carinae. Each swimmeret is coupled to its partner by hooks and each bears a tubular gill on its outer branch. The uropods of the sixth abdominal segment, together with the telson, form a tail fan that along with the large flattened antennal scale can be used to guide *Squilla* when it swims.

The female *Squilla mantis* after laying eggs in her burrow cements them together by secretions from glands on the posterior thoracic segments. Using the three posterior pairs of maxillipeds, the egg mass is formed into an elliptical ball that is turned frequently for about ten to eleven weeks in order to aerate them. The larvae that hatch from these eggs only vaguely resemble the adult. The abdomen is well developed, the carapace has three spines, and the raptorial second maxilliped and stalked eyes are very prominent. The larvae, which measure 20-25 mm long, spend their first two stages near the sea bed and then rise to within about 100 m of the surface, where they complete their development. They are carnivorous and, similar to the adults, make full use of their raptorial claws for catching and holding prey; the spines on the second maxillae also assist in this task.

Some species of stomatopods occur in moderate depths in oceanic waters. *Bathysquilla microps*, for example, occurs in the Gulf of Mexico down to 952 m, while *Indosquilla manihinei* is known only from a single specimen taken from the mouth of a fish in the Indian Ocean at 420 m.

In contrast to these species, *Gonodactylus glabrous* lives in crevices and vacated burrows of other marine borers and in dead coral; small specimens live among eel grass (*Zostera*). Females are frequently found guarding their mass of greenish yellow eggs located at the bottom of their burrows. This species makes a loud snapping sound effected by the second maxilliped (raptorial claw) and by the sudden release of one segment (the propodus) from a groove in the preceding one (the merus). Other species are known to produce sounds in a similar manner.

An efficient burrower is *Lysiosquilla excavatrix*, which makes deep, almost vertical burrows in muddy sand. The animal rests at the entrance, remaining almost immobile and hidden except for the occasional movement of the eyes and flagella of the first antennae. Prey is captured with one swift move from the burrow and with the vicious ' jackknife' action of the second maxillipeds assisted by the others; food is often stored in the bottoms of burrows.

Flattened starfishes with shallowly notched or webbed arms are commonly called bat stars by aquarists. Shown here is the Japanese *Asterina pectinifera*, with
a close-up of its partially extended stomach (Y. Takemura and K. Suzuki photos).

The common East Pacific bat star, *Patiria miniata,* is often kept in aquaria (H.R. Axelrod photo).

The Tasman Sea bat star *Patiriella calcar* is unusual in that it typically has eight arms (T.E. Thompson photo).

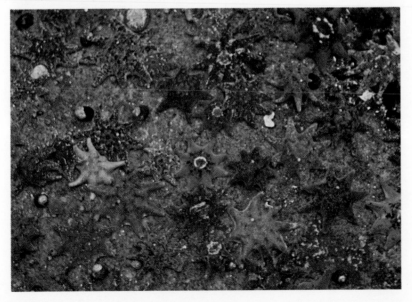

SUPERORDER PERACARIDA

To this superorder belong the opossum shrimp (Mysidacea), scuds and beachhoppers (Amphipoda), the aquatic woodlice and relatives (Isopoda), and the less familiar forms belonging to the Tanaidacea and Cumacea. There are over 10,200 species of peracarids. Females can be recognized by having brood plates arising from the basal segments of some or all of the thoracic limbs. These plates are directed inward and usually cross each other; with the fine setae arising from their margins they form a ' brood pouch ' or marsupium into which the eggs are laid. The young, hatched in the marsupium, usually closely resemble the adults in body shape.

ORDER MYSIDACEA

Opossum shrimp have a well developed carapace fused to some of the thoracic segments, a thin integument, stalked eyes, and large outer branches to their thoracic limbs. In one family (the Mysidae) a statocyst is present in the base of each outer branch of the uropod, a feature by which members of this family can be distinguished from eucaridan shrimp. Mysids feed either by browsing upon large masses of food picked up by the thoracic limbs or by filtering organisms from the plankton. In common with other crustaceans, many species of mysids migrate vertically in the water. *Eucopia unguiculata* and *Boreomysis microps* make vertical migrations through a range of some 200-600 m.

Mysids mate at night. Males are attracted to mature females only after the latter have liberated their young, have molted, and have ripe eggs ready for fertilization. The developing embryos are closely packed together in the brood pouch, where water is circulated over them by continuous pulsation of the brood plates. In cold Arctic waters females breed only once a year, in February and March, after which the female molts to a non-breeding form with the brood plates reduced in size. Temperate water species breed continuously. *Praunus flexuosus* begins reproducing at the end of April and produces five successive broods before dying in July or August. The young produced at the end of the season overwinter and form the first breeding colony of the following year.

Shallow water mysids tend to assume the tone and color of their surroundings. *Praunus neglectus* commonly lives among green algae, where it is bright green in color with brownish markings. If placed on a black background, its body takes on a dark olive color in an attempt to harmonize with its new background. Such changes are brought about by the redistribution of pigments in specialized cells (chromatophores) whose processes form a complex network in the body tissue. The number and position of these cells are established in the juvenile stage of each species and remain constant.

The largest known mysid is *Gnathophausia ingens*, which grows to 350 mm in length and occurs in depths down to and beyond 2000 m. This species produces a luminous secretion from glands in the first pair of maxillae. Occurring in even deeper waters, from 1000-6500 m, is *Eucopia unguiculata*. The first four pairs of thoracic limbs are subchelate; this species is a predator of smaller invertebrates. In contrast, *Gastrosaccus sanctus* lives in coastal waters 1-3 m deep, buried in sand by day and swimming at the surface by night. When disturbed it jumps some distance out of the water with considerable agility.

Some species of *Heteromysis* are commensal with other invertebrates. *H. actiniae*, for example, lives among the tentacles of the anemone *Bartholomea annulata*, and *H. harpax* lives in marine snail shells already occupied by hermit crabs.

ORDER CUMACEA

Cumaceans are small, 0.5-2.0 cm, sand-dwelling peracarids recognized by their often enormously developed carapace and thorax in contrast to the small abdomen with slender rod-like uropods. They spend most of their time buried in sand, although some species swarm at the water surface during the night at certain times of the year. Cumaceans burrow by using their posterior three pairs of thoracic

Pteraster tesselatus (East Pacific; D. Gotshall photo).

Henricia leviuscula, a common East Pacific starfish often called the blood star because of its color (D. Gotshall photo).

The crown of thorns starfish, *Acanthaster plancii,* is of course notorious for its reef-destroying habits. There is considerable variation in color and development of spines in individuals of the species (above, Australia, K. Gillett photo; below, Sri Lanka, R. Jonklass photo).

limbs to shovel aside the sand. The animal sinks into a hole further enlarged by the backward and forward movements of the abdomen. The sand inhabited by cumaceans is usually rich in detritus, and some species 'graze' off the sand grains, turning each grain with the maxillipeds while the setae on the second maxillae and mandibles brush off algae and detritus. Other species sieve mud to obtain microorganisms and organic detritus. The common species *Diastylis rathkei* measures 13-18 mm long and occurs in 5-800 m in northern European waters. Several similar species inhabit the northeastern United States.

ORDER TANAIDACEA

Tanaids are small peracaridans. Most of the species have subcylindrical bodies, and the second thoracic legs are modified as fully developed or partially developed (subchelate) claws. Many species burrow into mud or sand, where they make tubes. The large claws are used for burrowing. Females of *Heterotanais oerstedi* take only one minute to bury themselves in mud; males (which have larger claws) can burrow more rapidly. This species consolidates its burrow with a secretion from glands in the body segments. The beating swimmerets set up water currents through the tube, drawing in microorganisms that stick to the tube walls. Such algae (diatoms, etc.) provide food for newly hatched young. The adult female feeds upon clumps of detritus, breaking up large pieces with her claws. Males of *Heterotanais oerstedi* only appear for three to four months of the year and wander out of their tubes only during the daytime. The mouthparts degenerate as the animal matures; they do not feed and probably die soon after mating.

ORDER ISOPODA

The most familiar members of this order are the terrestrial pillbugs. A marine relative of these is the sea slater, *Ligia oceanica*, which occurs in crevices just above tide level. Marine isopods occur from the intertidal region down to some 10,000 m. Others are, at some stages of their lives, semiparasites or permanent parasites of fish and crustaceans. Typically the body of isopods is dorsally flattened, the limbs are without well developed claws, and the last pair of abdominal limbs are modified into large or small stylets or flat blades (uropods). Most isopods are nocturnal; *Ligia oceanica* remains concealed during the day in rock crevices above tide level and emerges at night when large numbers descend to the shore to browse upon seaweed.

Of different habits are adults of the small *Gnathia dentata*. These measure 4 mm in length and live in burrows dug by the males into firm mud. The males can be recognized by their large forceps-like mandibles; the mouthparts of the females are reduced. Some species live in tubes of worms and sponges. The larval stages of *Gnathia* are ectoparasitic upon fish. Their mouthparts are modified for piercing and suck the host's blood. These small larvae become so gorged with blood that the soft membranes between the body segments become distended. The larva leaves its host and sinks to the sea bed to molt into an adult that can never again feed as the mouthparts are now unsuited for the purpose.

The small isopod *Microcerberus pauliani* grows to only 0.7 mm in length. It lives between sand grains in underground or cave waters of Mediterranean beaches. Similar to many other invertebrates inhabiting this specialized habitat, the body is very elongated and highly flexible.

The gribble, *Limnoria lignorum*, of northern Atlantic and Pacific waters feeds upon submerged timber, to which it causes considerable damage. The gribble bores 10-20 mm below the wood surface and often makes winding burrows with smaller secondary ones made by the young. These burrows have series of holes spaced at intervals along their lengths to admit respiratory water currents. These burrows considerably weaken the timber surface, which is then broken away by wave action, forcing the gribble to burrow deeper. This repeated erosive action causes the eventual collapse of the structure. The gribble population is often dense, with some 300-400 individuals in just over 20 cubic millimeters of wood. Pier pilings can be destroyed in a little over two years.

Acanthaster ellisii, a small and less spiny East Pacific relative of the crown of thorns (A. Kerstitch photos).

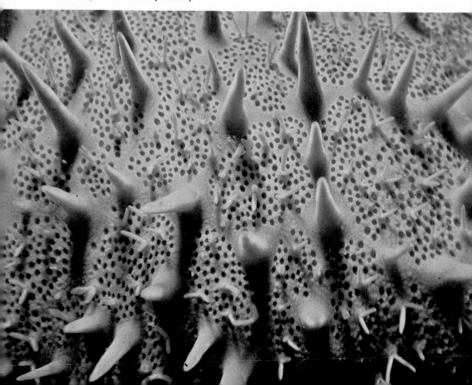

Among the large starfish fauna of the East Pacific are several species of *Pisaster*. Shown here is the short-spined starfish, *Pisaster brevispinus,* an attractive species that can reach a very large size (T.E. Thompson photo above; K. Lucas photo at Steinhart Aquarium below).

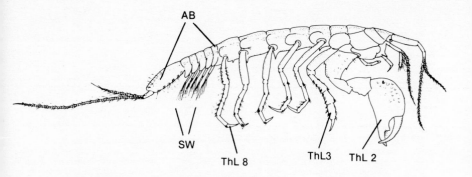

Top: Male tanaidacean, *Apseudes spinosa*; 12 mm long. **Bottom**: A sea slater (isopod), *Ligia oceanica*; 18 mm long. AB = abdomen; SW = swimmerets; ThL = thoracic limb; ST = stylet.

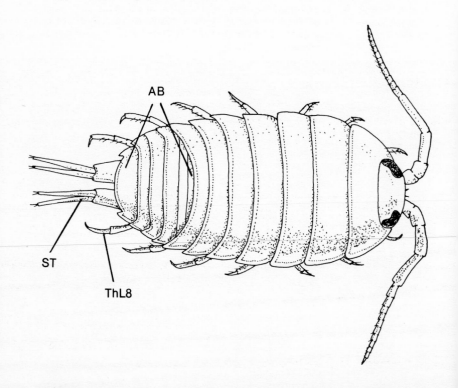

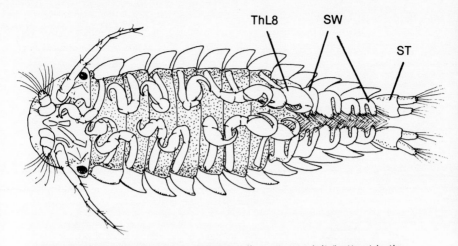

The epicarid isopod *Cancricepon elegans* lives as an adult (bottom) in the gill chambers of the crab *Pilumnus*. Its free-swimming cryptoniscan larval stage (top) is quite different from the adult. ThL = thoracic limb; SW = swimmerets; ST = stylet; BP = brood pouch; M = dwarf male.

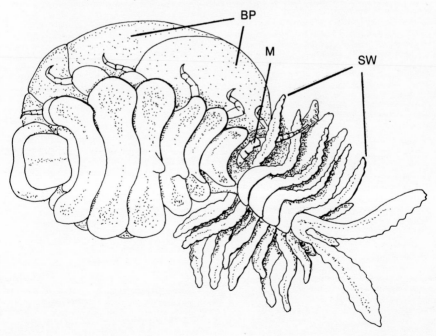

The common purple starfish of the East Pacific, *Pisaster ochraceus*, spawns like most other starfishes: the sexes emit clouds of eggs or sperm. This sperm cloud is easily seen in the male above. Below is the bipinnaria larva of the same species (T.E. Thompson photos).

The small *Leptasterias aequalis* is unusual in that it carries its young near the mouth; there does not appear to be a free-swimming larval stage (East Pacific; K. Lucas photo at Steinhart Aquarium).

Although appearing spiny, the integument of *Astrometis sertulifera* is actually quite soft (East Pacific; D. Gothsall photo).

Left: The parasitic isopod *Ourozeuktes owenii* in its host, the filefish *Meuschenia hippocrepis*. (U. E. Friese photo.)

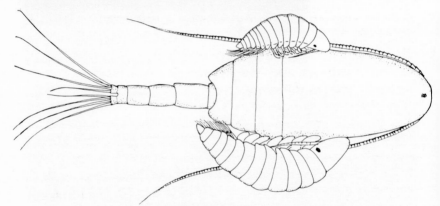

Above: Intermediate microniscus stage of an epicarid isopod attached to the copepod *Calanus*. **Below:** A scud or amphipod, *Marinogammarus marinus*; 15 mm long. ThL = thoracic limb; ST = stylet.

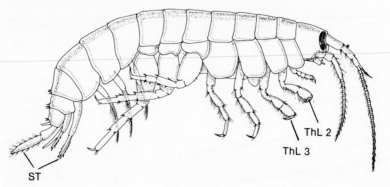

The largest known isopod is *Bathynomus giganteus*, which grows to 350 mm in length. It was first discovered in the West Indies and is rather common in the Gulf of Mexico. Another large species, *Glyptonotus antarcticus*, reaches 300 mm in length. It inhabits the Antarctic coasts from the shore down to 600 m and feeds chiefly upon brittlestars and marine snails.

Some isopods are ectoparasites. *Aega psora* lives upon the skin of many species of fish, feeding upon the host's blood. The mouthparts are modified for piercing and sucking, and the thoracic legs have strongly hooked claws that dig into the skin and hold the parasite firmly attached. The gut of *Aega* becomes distended with blood that partly solidifies. When extracted and dried it is known as ' Peter's Stone ' in Icelandic folklore and was once considered to have medicinal and magical properties.

The adult parasitic isopod *Bopyrus fourgerouxi* lives in the gill cavity of the shrimp *Palaemon serratus*. The area of the carapace covering the parasite is usually swollen, revealing its presence. *Bopyrus* is still recognizable as an isopod, although mature females are asymmetrical. The brood plates are very large, and between the abdominal appendages is found the diminutive male that closely resembles the last free-living larval stage of the parasite. This last stage usually invades a very juvenile shrimp and grows progressively with its host. The epicarid isopods, to which *Bopyrus* belongs, develop through three larval stages before invading their final host where they take up a fully parasitic life. The first larva resembles juveniles of other isopods such as the sea slater and is free-swimming in the plankton. In the second stage the larva attaches to a copepod such as *Calanus* and while on this host molts into a third stage that then seeks out the final host, another crustacean.

ORDER AMPHIPODA

The familiar examples belonging to this order are the ' shore hoppers ' or ' beach fleas ' of the upper shore above the waterline. The majority of amphipods, however, occur

The Japanese *Coscinasterias acutispina* in the process of reproducing by fission (Y. Takemura and K. Suzuki photo).

The common European *Marthasterias glacialis* ranges as far south as South Africa. In the photo of a regenerating specimen below, notice the small pedicellariae around the bases of the large spines (S. Frank photo above; T.E. Thompson photo below).

in the intertidal waters, in plankton, in the mid-water region, or as bottom-dwellers down to the abyssal depths of 10,000 m. A few are external parasites of marine mammals. The body of a typical amphipod is compressed sideways, although in a few it is flattened dorsally. The second and third pairs of thoracic limbs are often modified as large claws, and the last three pairs of abdominal limbs are developed as stylets used by some species to obtain a grip upon sand or rock surfaces.

An actively swimming amphipod such as *Marinogammarus marinus* shows lateral compression of the body and has well developed claw-like anterior thoracic limbs. In contrast, the tube dwelling *Corophium volutator* is flattened dorsally and has the second pair of antennae greatly enlarged in the male.

The wood-boring *Chelura terebrans* invades submerged timber previously bored by the gribble isopod. The burrows made by *Chelura* are shallow and usually run parallel to the wood grain.

The body of the skeleton shrimp, *Caprella*, is thin and almost tubular; the fourth and fifth thoracic legs are absent or reduced, the second and third pairs are modified for catching prey, and the sixth to eighth are adapted for gripping onto stems of colonial hydroids or branching seaweeds. *Caprella* can remain motionless in such a situation, mimicking the color of its surroundings among hydroids or seaweeds of a similar shape. One species, *Caprella grahami*, lives on the starfish *Asterias forbesi*, moving over its host with a looping-caterpillar motion and feeding upon detritus on the starfish's body.

Related to the skeleton shrimp are the whale lice. *Cyamus boopis*, 10 mm long, lives upon the skin of the humpback whale, often in great numbers. Young whale lice, after leaving the brood pouch of their mother, quickly dig themselves into the host's skin using the sharp curved claws of the thoracic limbs. These parasites are probably transferred from one whale to another by body contact, particularly when the mother humpback is suckling a calf.

Hyperid amphipods have greatly enlarged eyes and a correspondingly enlarged head. They are sometimes abundant in plankton communities. *Hyperia galba* grows to 20 mm in length and occurs from the New England coast through the Arctic to northern European waters. It often swims free but can attach to the subumbrella or tentacles of jellyfish, where it penetrates into the body tissue.

SUPERORDER EUCARIDA

This superorder includes the krill shrimp (order Euphausiacea), the true shrimp, lobsters, squat lobsters, mud shrimp, hermit crabs, and true crabs (order Decapoda). Eucarideans have a conspicuous carapace that in most forms is fused dorsally with all the thoracic segments. The often conspicuous eyes are set on stalks.

ORDER EUPHAUSIACEA

The krill shrimp at first sight are easily mistaken for opossum shrimp (mysidaceans) or for true shrimp. Their carapace, unlike that of opossum shrimp, is always dorsally fused to all thoracic segments and the sides are short, leaving the gills exposed. Nearly all have unspecialized and uniformly shaped thoracic limbs that separate them from the true shrimp.

The integument of the euphausid *Meganyctiphanes norvegica* is poorly calcified. There are eight pairs of unspecialized thoracic limbs, and the seventh and eighth pairs are the smallest. Branched tubular gills arise from the basal parts of the limbs. The outer branch of each thoracic limb is developed for swimming, giving these appendages a ' split-footed ' appearance, a feature that was once used to group the euphausids along with the mysids into the old order ' Schizopoda.' The abdominal plates are well developed, the telson elongated, and the uropods flattened.

Although many euphausids shed their eggs, some (the Californian *Nyctiphanes simplex*, for example) carry the eggs in two pear-shaped egg sacs attached to the outer branch of the sixth to eighth thoracic limbs. These sacs can

The long tubefeet of *Asterias* and related genera are efficient in applying slow but constant pressure to bivalve shells. Under this pressure even the strong muscles of oysters relax and open the shell enough for the extendible starfish stomach to find a way in (Y. Takemura and K. Suzuki photo).

Asterias forbesii, the common shallow-water starfish of the northwestern Atlantic. Notice the bright orange madreporite that helps separate it from its closest relatives (H.R. Axelrod photo).

The sunflower starfish, *Pycnopodia helianthoides,* is probably one of the largest or even the largest known starfish, reaching widths of over a meter (East Pacific; K. Lucas photo at Steinhart Aquariun).

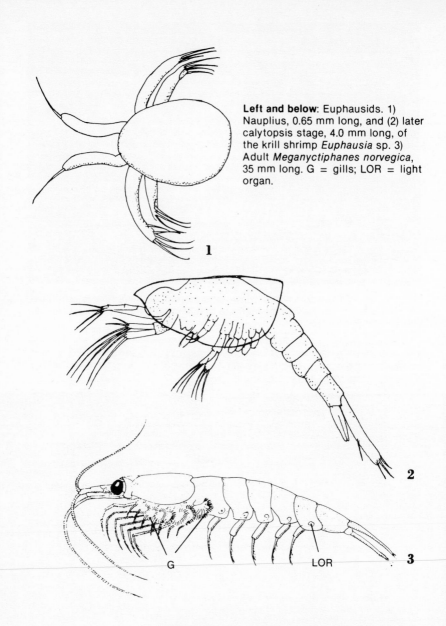

Left and below: Euphausids. 1) Nauplius, 0.65 mm long, and (2) later calytopsis stage, 4.0 mm long, of the krill shrimp *Euphausia* sp. 3) Adult *Meganyctiphanes norvegica*, 35 mm long. G = gills; LOR = light organ.

1

2

G   LOR   **3**

Facing page: Stages in the life cycle of the shrimp *Penaeus kerathurus*. 1) First nauplius, 0.42 mm long. 2) First protozoea. 3) Last protozoea, 2.3 mm long. 4) Last mysis, 4.3 mm long.

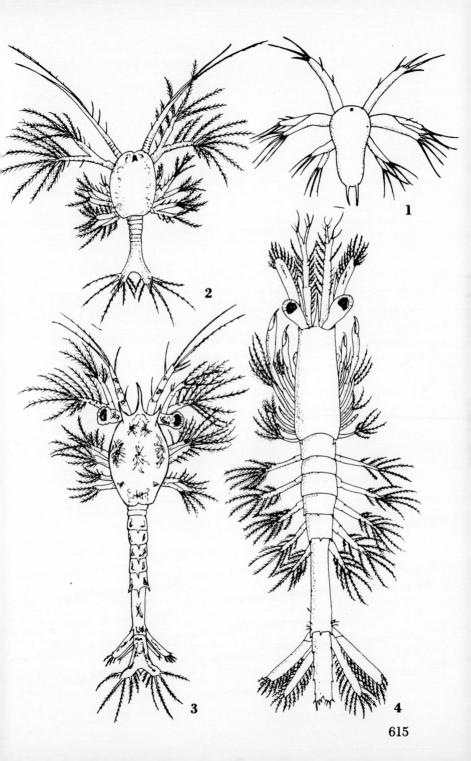

Brittle stars are the most flexible echinoderms and often occur in intertwined clusters on the sea floor (above) or wrapped around gorgonians or other reef animals (below)(East Pacific, K. Lucas photo at Steinhart Aquarium above; Australia, L.P. Zann photo of *Astrobrachion adhaerens* below).

Ventral view of *Ophiocreas* sp., a serpent brittle star. These deeper water brittle stars are much less spiny than typical and have the plates only barely visible (Indo-Pacific).

contain as many as thirty eggs. The first larval stage (nauplius) is passed within the egg sac and the larvae emerge as minute metanauplii. This stage is followed by three so-called ' calytopis' stages at which the abdominal segments and carapace develop. Then follow many ' furcilial ' stages, the last of which closely resembles the adult. *Meganyctiphanes norvegica* matures in one year and mates between January and February. Females spawn in March and April in northern European waters. The larval stages take some two to three months to reach juvenile form, mature through the winter, and live for 1½-2 years.

Most euphausids live in the open sea and swim continuously, never resting on the bottom. A number of species occur in enormous swarms, and fish such as herring and sardines, in addition to whalebone whales, feed almost entirely on this so-called ' krill. ' Some 28 % of the stomach contents of Barents Sea herring were found to be euphausid material, and a 26-meter blue whale had about 5 million *Euphausia superba* in its stomach. Whales usually feed only from the dense upper layers of euphausids in the plankton. High catches of baleen whales are often closely related to the abundance of euphausids in Atlantic and Antarctic waters. In daylight late furcilial larval stages of some species are often found in the surface waters, while older animals descend to some 200 m. *Thysanoessa spinigera* occurs in large numbers in surface waters off La Jolla, California, and quantities are sometimes swept ashore and stranded upon beaches, providing abundant food for scavenging sea birds. The Australian *Nyctiphanes australis* often occurs as dense patches of an acre or more at the surface, where they form a superabundance of food for mutton birds.

Nearly all species of euphausids luminesce. Both sexes bear light-producing organs on the eyestalks, on the coxal segments of the second to seventh thoracic limbs, and on the fourth abdominal segment. The structure of these light-producing organs is similar on all appendages except that those on the eyestalk do not have lenses. A concave reflector throws back light produced by the rod mass, and this light is concentrated through the biconcave lens. *Meganyctiphanes norvegica* starts to luminesce when the reproductive organs become active. This light production may be a signal for the swarming that takes place just before mating, for afterward most animals cease to luminesce continuously, doing so only if disturbed.

Most euphausids are microplankton feeders. The thoracic limbs and setae of *Euphausia superba* form a 'filter basket' that can strain from the plankton plant cells (usually diatoms) as small as 0.04 mm in diameter. Such food can be quickly transformed into vitamin A. Only exceptionally are the limbs of euphausids modified for other than plankton feeding. *Stylocheiron suhmi* is probably a predator on other small invertebrates, catching its prey with the modified thoracic limbs.

ORDER DECAPODA

This order contains some of the largest and most highly socially developed crustaceans. The carapace of decapods is well developed, extending downward to the bases of the legs and enclosing and protecting the often elaborate gills. The first three pairs of thoracic limbs are always modified as 'foot jaws' (maxillipeds) for holding food, and in most decapods one or more of the three following pairs of limbs are modified as claws for handling prey or for defense. The most active and abundant decapods are the shrimp.

The terms shrimp and prawn have no exact meaning. *Penaeus aztecus*, for example, is called the brown shrimp and the Gulf prawn, while a closely related species, *Penaeus indicus*, is known as the Indian prawn. Generally prawn is applied to edible large species. We will use shrimp for all. Many species of shrimp are fished to supply man with food, particularly the penaeid shrimp. Some penaeids reach a length of 300 mm, and the meat from the 'tails' (abdomens) is a highly esteemed luxury in many countries and commands a high price. Exports of such a commodity materially contribute to the economics of nations fishing such shrimp.

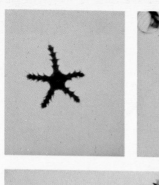

Growth series of aquarium-raised *Ophiarachna incrassata* (J. Mougin photos courtesy Nancy Aquarium, France).

Close-up view of the oral (feeding or ventral) side of the disc of the green brittle star, *Ophiarachna incrassata* (Indo-Pacific; J. Mougin photo courtesy Nancy Aquarium, France).

In the eastern Gulf of Mexico the most important economic species is the pink shrimp, *Penaeus* (*Melicertus*) *duorarum*. In the Tortugas-Sanibel region of the Gulf yearly catches of 8,000,000 kg have been reported. The color of the pink shrimp varies from pink to grayish brown, and young specimens are sometimes almost colorless. The pink shrimp ranges from the lower Chesapeake Bay down to the Gulf of Mexico and off Central America, with highest concentrations occurring between 2 and 37 m.

Mating probably occurs throughout the year, when hard-shelled males mate with soft-shelled females; a peak in mating activity is reached about May to June. The male deposits a spermatophore into the female's sperm pouch. Females spawn in open oceanic waters in depths from 8-47 m, probably early in the morning, and the fertilized eggs are shed into the water.

The last of the first five larval stages (naupliar stages) molts into a form in which the abdomen is clearly segmented (protozoeal stage). There are three such stages followed by three further (mysis) stages in which the larva begins to resemble the adult. Truly shrimp-like features are attained at the next stage, the juvenile. The minimum time taken to accomplish this whole process in the laboratory is 15 days. In the sea these juvenile shrimp now seek out inshore waters. After migrating to these 'nursery grounds' they grow rapidly, moving back toward the sea as they approach maturity.

The pink shrimp feeds at night upon algae, sand, mud, organic matter, and a variety of invertebrates and fish larvae that occur near the sea bed. It lives upon hard bottom substrates that contain shell, silt, and mud. The adults become active as night falls and are often fished at that time. During daylight the shrimp usually remain buried in the sea bed. Light seems to be the most important factor controlling their activity; fishermen know that offshore catches will be low if fishing is done during a full moon. Many other species of the genus *Penaeus* are of commercial importance in other areas.

A somewhat aberrant but easily recognized penaeid shrimp is *Lucifer faxoni*, occurring in the Atlantic waters of North and South America and off West Africa to the Azores. Populations of this mostly coastal species often occur planktonically in the open sea, but these are thought to be nonbreeding communities that have been carried from coastal waters by the Gulf Stream. Female lucifers carry their eggs attached to the thoracic legs and swim near the surface just before the eggs hatch.

Of the caridean shrimp, *Pandalus montagui* is one of the most important commercial species occurring in European north Atlantic waters; it prefers waters of low temperatures and high salinity. This commercial shrimp, also called pink pandalid and Aesop shrimp, is known to change sex during its life span. Up to 50 % of the population begin their lives as males and then change to females.

Female pink pandalids spawn at the onset of November and continue until February. The fertilized eggs are carried attached to the swimmerets. The majority of egg-carrying females are found in deeper water of 55-73 m off the southeastern coast of England. The eggs are bright green when newly spawned but change to greenish violet as they develop. The shrimp-like larvae (zoeae) that hatch measure 2.4-3.4 mm in length and, in the laboratory, are known to pass through eleven stages, although those living in the sea may have fewer stages. These larvae superficially resemble the adult even in their first stage; in the last the rostrum has acquired both its dorsal and ventral teeth.

The pink pandalid lives for 3-4 years. Populations tend to frequent shallow waters in early spring, moving to deeper waters in the autumn when they prepare for breeding. The species is extensively fished in European waters.

A number of species of shrimp have developed highly specialized social relationships with other marine animals. The advanced behavior of the cleaner shrimp allows them to enter mouths and gill cavities of various coral-dwelling fish in order to remove parasites and food particles and to clean injured tissues off their hosts. These species of shrimp

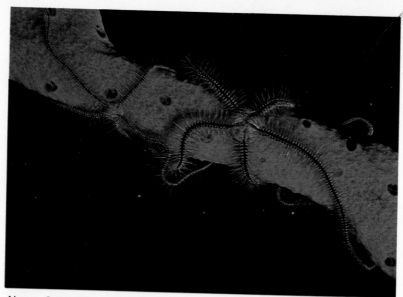

Above, *Ophiothrix suensoni,* a common Caribbean brittle star often found on sponges (P. Colin photo). Below, *Ophiocoma* sp., a rather typical brittle star in both shape and pattern (Indo-Pacific; A. Power photo).

Astrocaneum spinosum, an East Pacific basket star; the branching of the arms is clearly visible in this clumped specimen (A. Kerstitch photo).

An unidentified Caribbean brittle star that apparently feeds on plankton filtered from the water (P. Colin photo).

are possibly immune from being eaten by their host, and such is their importance in the ecosystem of many reef communities that when they are removed many fish disappear, while the incidence of fish with ulcerated sores becomes high. In well developed relationships of this type fish swim to special cleaning stations and will line-up waiting their turn to be cleaned. Some cleaner shrimp signal their presence to the fish by being brightly colored, displaying and moving their conspicuous antennae and perching upon coral prominences on which they sway back and forth. The fish, for their part, change color, move toward the shrimp, and indicate their willingness to be cleaned by extending their gill covers and opening their mouths.

Pederson's cleaner shrimp, *Periclimenes pedersoni*, 40 mm in length, occurs in depths of 17 m or more on reefs in the West Indies. Pairs of shrimp are often found hanging onto the anemone *Bartholomea annulata*, and specimens have been known to spend three weeks or more upon the one anemone. Numerous species of reef-dwelling fish are attracted to these shrimp. Only when the fish is resting will the shrimp climb onto it and move rapidly over its body, pulling off parasites and thoroughly cleaning tissues or wounds. The fish raises its gill covers as the shrimp approach and allows them to enter this very delicate area. The mouth is also opened to allow the removal of parasites that are often located in that region. On completing their tasks the shrimp return to the anemone.

The banded cleaning shrimp, *Stenopus hispidus*, is the largest cleaner, growing to 75 mm. It is widely distributed throughout the Indo-Pacific from the Red Sea to southern Africa and to the Hawaiian Islands; in the western Atlantic it occurs from Bermuda and southern Florida to the South American coast. *Stenopus* tends to occur in quiet shallow waters and often on the ceilings of caves or grotto entrances. It has been observed cleaning a moray eel by placing its longest pair of thoracic legs around the eel's body and using the smaller pair for cleaning.

Some species of shrimp live in permanent association with

other invertebrates or plants. *Gnathophylloides mineri* from Barbados lives on the test and spines of the sea urchin *Tripneustes esculentus*, picking pieces of detritus from the urchin's spines; the shrimp also feeds upon plankton. The small *Hippolyte acuminata*, 15 mm long, is often perfectly camouflaged on the pelagic seaweed *Sargassum*. The colored bands on the body seem to divide the shrimp into two parts, each half resembling a ' berry ' of the seaweed.

Species of pistol or snapping shrimp have the fourth thoracic leg on one side developed as an elaborate robust claw. *Alpheus californiensis*, which grows to 50 mm, lives in pairs within burrows and ' shoots off ' its claw with a loud pop when disturbed. The distal segment is provided with a small plug that fits into a pocket of the preceding segment. How this noise is produced is not fully understood. It may occur when the fingers are opened rapidly and plug suddenly withdrawn or when the fingers are suddenly closed. However, the shock waves produced by this report are sufficient to stun fishes that come too close to the burrow opening. The semiconscious fish is dragged into the burrow and torn apart to be shared with the other occupant.

Juveniles and adults of lobsters, spiny lobsters, and their relatives are bottom-living crustaceans; their highly calcified, usually thick integument makes them much heavier than shrimp. The chief features of the common European lobster have been described earlier. Its practically identical relative, the American or Canadian lobster, *Homarus americanus*, is extensively fished off the Canadian and American Atlantic coasts.

A social hierarchy exists in lobster communities. Animals will fight to win a place in which to shelter. The loser lowers its antennae, waving them feebly while the victor holds its antennae high, beating the water vigorously. Hard-shelled lobsters usually fight when placed together, but the approach of a male to a newly molted female in the breeding season is quite different. This is now slow and cautious; the male walks around the female and strokes her with his second antennae, which will appease a hostile female. Such behavior often leads to mating and is induced

Gorgonocephalus caryi, a basket star (East Pacific; D. Gotshall photo).

Astrophyton muricatum, the common basket star of the Caribbean in a fully extended state (P. Colin photo).

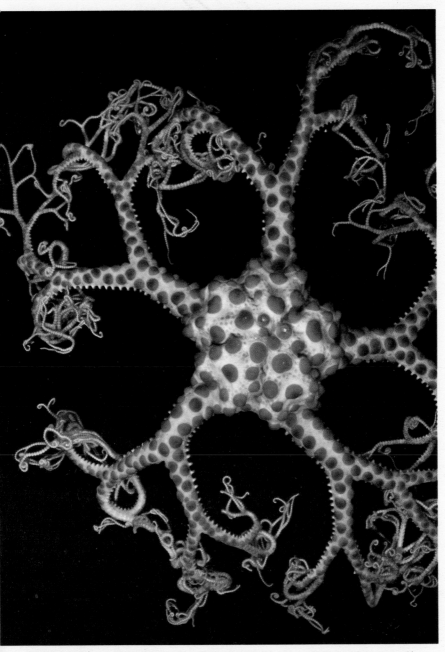

The unusually ornamented Tasman Sea basket star, *Conocladus amblyconus* (K. Gillett photo).

when a chemical that suppresses aggression is released by the female. The American lobster mates up to twelve or so days following a molt by the female, and several males can mate with the same soft-shelled female. Males with large claws generally have an advantage over those with smaller claws during inter-male fighting before mating. The development stages following hatching have been described. The greatest change occurs from the third to fourth stage, when the claws have become longer, the abdomen wider, and the tail (telson) shorter and broader. This stage marks the change from a swimming planktonic life to a walking one that lives on the sea bed.

A species of more slender build is the Norway lobster or Dublin Bay prawn, *Nephrops norvegicus*. This species is fished off Iceland, in the North Sea, off the Scottish and French coasts, and south to the Mediterranean and Adriatic. The tails (abdomens) are sold as ' scampi. ' The Norwegian lobster occurs chiefly upon muddy bottoms where it excavates burrows in soft mud where it lives for much of the time, leaving only at dusk and dawn to forage for food. Its diet is very variable, consisting of bristle worms, other crustaceans, molluscs, and starfish.

In European waters the female spawns each year and carries her eggs for nine months. These are green at first and gradually turn pale pinkish brown before hatching. The first larval stage (zoea) measures 5-7 mm in length, and the third and last (the megalops) that appears some two to three weeks later is between 10-12 mm. The first postlarval stage is very similar in shape to the adult and lives on the sea bed.

There are some 130 species of spiny or rock lobsters. The name crawfish is often applied to them but is perhaps more properly associated with the freshwater lobster-like decapods. Many species are fished extensively off the American, South African, and Australian coasts. The thoracic limbs are usually subcylindrical and often have half-formed claws (subchela), while many species have the carapace surface invested with numerous spines. This group is easily distinguished from the true lobsters by lacking heavy claws and having the antennae greatly enlarged.

Over three and a half million kilograms of the South African rock lobster, *Jasus lalandii*, are fished every year and exported. This species is abundant in the inshore waters down to about 128 m; off the South-West African coast it lives on rocky bottoms, often in the kelp seaweed zone where it feeds upon species of black mussels. Large males are often found isolated in holes although lobsters of all sizes will congregate in crevices and caves. They often hang upside down on the ceilings of caves or cling to vertical walls with their heads pointing toward the cave opening.

In South African waters females become sexually mature at a carapace length of about 70 mm and carry eggs from June to September. These eggs hatch during October and November. The small larva (called a naupliosoma) that emerges is quite unlike the adult and molts, often within hours, into a ' glass shrimp ' stage (phyllosoma). This is a thin transparent planktonic larva often difficult to see when in the water. This glass shrimp probably feeds upon smaller invertebrates, but some Australian forms are known to eat jellyfishes. After several molts the larva becomes transformed into a small lobster-like form (puerulus) that soon molts into a juvenile spiny lobster.

The American Atlantic spiny lobster, *Panulirus argus*, is known to form single-file queues of many thousands of individuals. In order to queue, an individual lobster approaches the rear of another, touching the abdomen of its partner with its second antennae. It then may hold onto its partner's abdomen using its thoracic legs. The reason for this single-file migration behavior is unknown, but it may provide some form of protection when a population is moving across open areas of the sea bed.

Related to the spiny lobsters are the slipper lobsters. *Scyllarus arctus* lives on muddy bottoms and is fished and eaten in the Mediterranean area. The female breeds throughout the year, but egg-carrying females are more common during the summer months. The small glass shrimp larvae have a brief planktonic stage, sink to the sea bed, and molt into lobster-like larvae, similar in shape to those of the spiny

Diadema antillarum, the long-spined banded urchin (Caribbean; P. Colin photo).

An East Pacific long-spined urchin, *Centrostephanus coronatus* (D. Gotshall photo).

Echinometra lucunter, a sea urchin that is able to burrow into rock with its Aristotle's lantern (Caribbean; P. Colin photo).

Because of their spines, urchins are favorite habitats for schools of small fishes; this urchin is *Astropyga* sp. (Caribbean; C. Arneson photo).

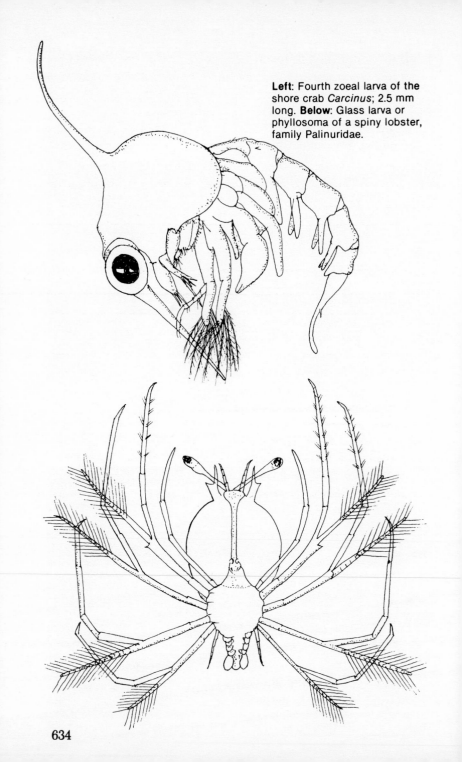

Left: Fourth zoeal larva of the shore crab *Carcinus*; 2.5 mm long. **Below**: Glass larva or phyllosoma of a spiny lobster, family Palinuridae.

lobsters, and then to juveniles. Several small species of slipper lobsters occur in the Atlantic and Indo-Pacific, many of which make good additions to the aquarium.

Lobster-like forms with soft abdomens are the mud shrimp or ghost shrimp. The greatest number of species belong to the genus *Callianassa* and dwell in mud or sand where they build tubes. Mud shrimp are generally whitish yellow, sometimes almost translucent, hence their name ghost shrimp. The eggs, which can often be seen through the transparent integument, are usually bright yellow or pale orange. The most characteristic feature of mud shrimp is that the right or left fourth thoracic leg is enlarged into a claw. This claw is believed to have a protective function as in some species it can be folded back over the thin carapace. Similar to most lobster-like forms, mud shrimp swim with reluctance; while doing so the large claw is held out straight. The thoracic legs assist the animal in burrowing into mud. The claws of the fourth and fifth thoracic legs loosen the mud and scoop it backward; the sixth leg levers the body forward into the depression formed in the mud. The distal segments of the third leg loosen mud and scoop it toward the mouth, forming it into a ball that is carried out of the hole as the shrimp backs out and is then dumped. By repeating this sequence many times a short cavernous burrow is formed in which the mud shrimp can turn with a quick somersault. Further tunnelling is made in various directions, and some of these open at the surface. At least two openings are necessary; respiratory water is pumped through the burrows by the beating of the abdominal limbs.

Callianassa californiensis occurs on the Pacific coast of America from Alaska to California and can deposit as much as 50 cc of sand around its burrow entrance in 24 hours. Burrows are made to a depth of only 0.5 m as most of the organic-rich material on which the animal feeds is found in the surface layers of the sand. These burrows also provide shelter for other animals; up to ten different species have been found living as commensals with this mud shrimp. For example, the small goby *Clevelandia ios*, the pea crabs

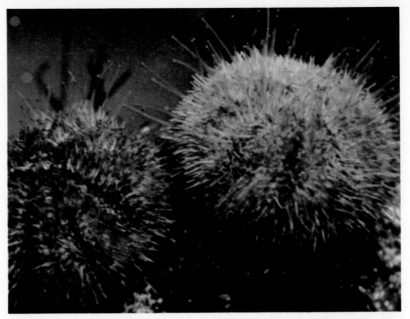

As strange as it may seem to some tastes, sea urchin eggs (roe) are a gourmet item in several parts of the world. *Hemicentrotus pulcherrimus* (above) is used for food in Japan, while *Tripneustes esculentus* (below) is a Caribbean favorite (Y. Takemura and K. Suzuki photo above; P. Colin photo below).

Strongylocentrotus is a common East Pacific genus of heavily spined sea urchins. Above is *Strongylocentrotus franciscanus;* below is *Strongylocentrotus purpuratus* (T.E. Thompson photo above; D. Gotshall photo below).

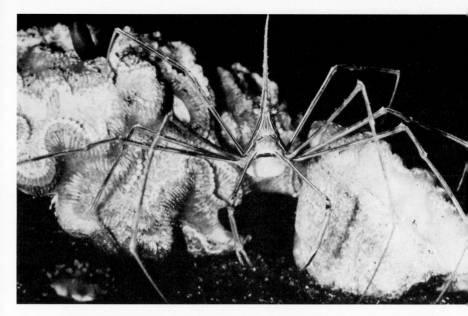

Above: *Stenorhynchus seticornis*, the arrow crab commonly kept in marine aquaria (H. Hansen photo). **Below**: *Petrolisthes galathinus*, a porcelain crab (C. Arneson photo).

638

Scleroplax granulata and *Pinnixa franciscana*, and the bristle worm *Hesperonoe complanata* are all found with *Callianassa californiensis*.

The largest mud shrimp is *Thalassina anomala*. This species grows to 300 mm in length and is a reddish color. Its burrowing activities can be recognized by the nearly one-meter-high mounds frequently seen in large numbers on mangrove flats. This mud shrimp can burrow down to almost one meter, and such activity often undermines levees and mud dams, so the species is regarded as a serious pest in some areas.

With few exceptions the hermit crabs are shell-dwellers. The soft abdomen is vulnerable and is coiled into the spiral of the shell of a marine snail. As the crab grows a new larger empty shell is selected into which the hermit quickly changes. The large claws, usually unequal in size, can effectively close the entrance to the shell when the crab withdraws if danger threatens. Most hermits communicate chiefly by visual displays of their antennae and claws.

Clibanarius tricolor and *C. corallinus*, and many other species as well, are gregarious when they are not feeding and concentrate in damp localities as they follow the tide down on its recession; they are frequently found in hundreds beneath rocks and boulders.

The mating behavior of the soldier hermit, *Pagurus bernhardus*, is quite elaborate. The male carries the female around, holding the rim of her shell with his smaller claw and facing her away from him. The female may be carried in this manner for several days but is eventually turned to face her mate. Then follows a ritual tapping by the male against the female's claws combined or interspersed with jerky pulls of her toward him. The female then strokes the male's claws. Shortly afterward the pair partly emerge from their shells to mate. The soldier hermit produces successive batches of eggs. The female often mates while she is bearing eggs from previous matings that are almost ready to hatch.

A Mediterranean hermit, *Dardanus arrosor*, is known to detach anemones (*Calliactis parasitica*) from rock surfaces

Lytechinus williamsi is an uncommon Caribbean urchin usually found on coral reefs. The very large purple pedicellariae (below) are conspicuous even from a distance (P. Colin photo above; T.E. Thompson photo below).

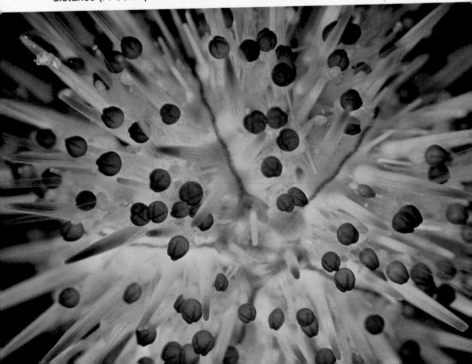

The massive spines of *Heterocentrotus mamillatus* were once used as chalk, hence the name slate pencil urchin (Indo-Pacific; K. Gillett photo).

and place them upon its shell. This anemone transfer behavior has been studied carefully in the Hawaiian hermit *Dardanus gemmatus*, on whose shell lives the anemone *Calliactis polypus*. The hermit, after finding an anemone, begins to tap and scratch at the anemone's body, applying pressure around the middle and upper parts. Such behavior is continued until the anemone becomes relaxed. Then the hermit inserts the tips of its walking legs and claws beneath the margin of the foot and detaches it from the rock surface, holding the anemone against its shell until the tentacles or base adhere and the anemone becomes fixed.

Other invertebrates also attach to the shell of hermit crabs. The shell of *Paguristes bakeri* is often covered with a dense growth of sessile barnacles that so weigh down the crab that it finds difficulty in moving around. Sharing the shell with this species is the bristle worm *Halosydra brevisetosa*. The shell of the Californian *Pylopagurus varians* is covered with bryozoans, while that of *Pagurus cuanensis* is enveloped in a sponge that dissolves away the shell as the sponge grows.

In some hermit species the abdomen is only slightly curled or almost straight. *Pylocheles miersii* of the Indian Ocean conceals its abdomen within small lengths of bamboo; similarly, *Pylopagurus minimus* of the American West Coast lives within the straight shells of the marine mollusc *Dentalium*.

A few hermits have adapted to a life on land although they need to return to the sea to breed. The most successful and largest of these is the coconut or robber crab, *Birgus latro*. The young inhabit shells only for the first two years of their lives, after which they forsake this protection and spend the rest of their existence with the fleshy abdomen exposed. The species gets its name from its habit of climbing palm trees; it is believed to cut down coconuts for food. The robber crab can probably break open damaged coconuts and is an efficient and agile climber. It can also rapidly ascend the sides of buildings and the apparently smooth surfaces of door and window frames. The pointed claws of the walking legs enable it to easily find foot-holds.

In areas of New Guinea the natives are said to fasten a ring of grass part way up a tree but below the robber crab. If the crab descends backward down the trunk, on touching the grass the crab mistakes it for the ground, releases its hold, and comes crashing down stunned or dead. In the Celebes the natives place stones around the tree base to provide a hard area onto which the crab will fall. The coconut crab, however, often defeats this trap by descending head-first. Hunting the crab is a more efficient method for capturing it and is a recognized sport in the South Pacific islands. Hunting is usually restricted to the hours of darkness, at which time the crabs will venture out of hiding to feed. Lighted torches and flashlights seek out the crabs that are momentarily stunned by the light and are then grasped by the carapace and held firmly until secured. The powerful claws can quickly amputate a finger, and once an object is grasped the vice-like grip can be maintained for hours.

The robber crab feeds upon a diet of coconuts although it will eat dead carrion on the beach. Its abdomen is rich in oil, yielding over a liter of this commodity. The flesh is esteemed by the Chinese, and crabs are frequently kept in captivity (minus their claws) and fattened for the table. Although a land-dweller frequently found at altitudes of several hundred feet, females must return to the sea to hatch their eggs. The minute larvae that emerge are similar to those of the true marine hermits.

Crab-like forms related to the hermits are the porcelain crabs. These have a well calcified and flattened carapace, broad flattened claws, and the last pair of walking legs reduced in size and dorsally placed. *Porcellana sayana* occurs on the western Atlantic coast from Cape Hatteras to Venezuela and Surinam. Groups of crabs are often found in dead oyster shells or as commensals of hermit crabs. Some porcelain crabs are commensals with other invertebrates. *Megalobrachium soriatum* lives in the canals of sponges, while *Polyonyx gibbesi* occurs in the tubes occupied by the bristle worm *Chaetopteris variopedatus*.

In *Colobocentrotus atratus* a peripheral row of spines is flattened and used to help the animal adhere to rocks in heavy surf; this unusual urchin is found in several Indo-Pacific localities (Hawaii; S. Johnson photo).

Toxopneustes roseus, a venomous urchin from the East Pacific. Notice the debris being held to the test by the large pedicellariae (A. Kerstitch photo).

Detail of the pedicellariae and spines (the pedicellariae are like three-petaled flowers) of the venomous Indo-Pacific urchin *Toxopneustes pileolus* (W. Deas photo).

Lobster-like crustaceans having some affinities to hermit and porcelain crabs are the squat lobsters (galatheids). In many species the last pair of walking legs are small and hidden beneath the carapace. The 'lobster krill' of the southern Atlantic and Pacific Oceans is represented by two species of squat lobsters, *Munida subrugosa* and *M. gregaria*. 'Krill' is the Norwegian word for 'whale food,' and these crustaceans form an important part of the diet of some species of whales. Lobster krill frequently occur in enormous shoals. Both adult and larval stages of *Munida gregaria* are found together just below the sea surface, where their prolific numbers color the water bright red. Small shoals are often circular and are familiar sights to whalers. It is not uncommon for vast quantities to be stranded upon beaches following high tides and onshore winds.

The small red or pelagic crab *Pleuroncodes planipes* of the Gulf of California, despite its name, is another squat lobster that swarms in vast numbers. It is believed that such swarms occur in regions where algal blooms occur upon which the red crab larvae 'graze.'

Lithodid crabs are often called 'king crabs'—a vernacular name also used for the marine merostomates or 'horseshoe crabs,' to which they are not at all related. Lithodids are probably descended from hermit crab ancestors. They can be distinguished from the true crabs, which they superficially resemble, by the position of their second antennae that are placed outside the eyes instead of between them as in the true crabs. In addition, the last pair of thoracic legs of lithodids are very small and concealed beneath the carapace.

Paralithodes camtschatica, the North Pacific king crab, ranges from the Gulf of Alaska to the Bering Sea and is fished intensively, chiefly by Japanese and Russian fleets. Adult crabs congregate on sandy bottoms in deep water for most of the time but move to shallow areas once a year during the breeding season in April and May. Individuals have been known to travel 110 miles, but most populations seem to migrate not further than about 35 miles.

The larvae after hatching take about three and a half months to complete their development. The early stages are passed near the sea surface and later stages move to deeper water before developing into young crabs. These reach maturity in about eight to nine years, but it has been estimated that adults can live at least seventeen years. The king crab gives a high meat yield, the outstretched legs of large specimens spanning some one and a half meters from tip to tip; a live four-and-a-half-kilogram crab will contain about nine hundred and eighty grams of meat. The king crab has been fished by Japanese fishermen for many years. A canning industry was started as far back as 1892, followed by rapid development of floating processing plants. In Alaska the species has been fished since about 1909 and improved techniques of freezing have enabled the fresh flavor of the meat to be retained. This and intensive advertising have considerably improved sales of king crab meat in the USA.

The mole crabs are highly adapted for burrowing into and living beneath the sand surface. *Emerita analoga* occurs in many locations in great numbers. Their presence is hinted only by a small V-shaped depression at the sand surface that indicates the position of the tips of the first pair of antennae. This species of mole crab is particularly plentiful in the area of the beach covered by wave wash. The ever-shifting sand of that region and the sudden dangers from shock waves and predators have made *Emerita* a very nimble burrower. The thoracic legs and the last pair of abdominal ones are modified for this purpose. The crab enters the sand backward with its head pointing toward the sea and burrows obliquely downward. When at rest in the sand the tips of the first antennae just protrude above the surface and the ' bailers ' of the second maxillae vibrate rapidly, drawing water from the sand surface down the channel formed by the opposing halves of these antennal flagella and into the gill cavity. This water then passes over the gills and leaves beneath the posterior margin of the carapace.

Sponge crabs derive their name from having a piece of

Irregular urchins are elongated rather than rounded and have bilateral symmetry rather than radial, at least superficially. They burrow in the silt or sand by day and are active by night. Above is the large Caribbean *Plagiobrissus grandis* (C. Arneson photo); below the more colorful Indo-Pacific *Brissus* sp. (S. Johnson photo).

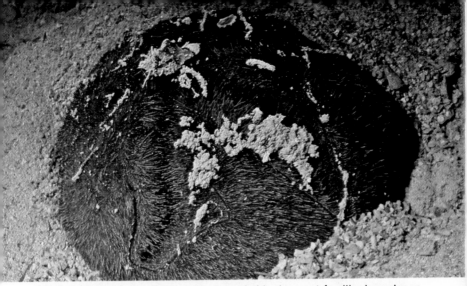

Meoma ventricosa, the heart urchin, is probably the most familiar irregular urchin in the Caribbean. It is active only at night but can be very common in some localities (P. Colin photo).

Sand dollars, such as this common West Atlantic *Mellita quinquesperforata*, are merely irregular sea urchins that have evolved an extremely flattened shape with greatly reduced spines (N. Herwig photo).

sponge covering and concealing the dorsal surface of their carapace. This protective covering is supported by the last two pairs of thoracic limbs, which are dorsally placed and reflexed. Various species of sponge are carried, but sea squirts, algae, and even pieces of paper or rag are sometimes held. *Dromia personata*, the sleepy sponge crab, cuts out pieces of encrusting sponges from rock surfaces or sometimes tears a sponge off the surface of a mussel shell, thus obtaining a piece of the correct contour to fit its highly vaulted carapace.

A similar type of behavior is seen in dorippid crabs that carry dead bivalve shells upon their back. The Japanese name for *Dorippe dorsipes* is ' Kimemgani ' or ' crab with a demon face' because the carapace contours are so shaped as to give the appearance of a sinister human face. In parts of the Orient *Dorippe* is regarded as sacred because the ' face ' is thought to be that of a deceased relative whose soul has passed into the crab.

The box crabs or calappids have their mouth frame prolonged forward and narrowed anteriorly, but their most interesting feature is the specialized use of one of the claws, in some species, for crushing shells of marine snails to gain access to the soft parts for food. This modified claw, usually the right, has an oval projection at the base of the last segment and two corresponding processes on the preceding one against which the projection fits. *Calappa hepatica* hunts for marine snails by sweeping the surface of the sand with its walking legs. The prey is held against the body by the claw while the following pair of legs turn the shell until its opening is facing upward. With the shell thus supported the specialized part of the claw is now brought into use to crack away the shell a piece at a time.

The masked crab, *Corystes cassivelaunus*, lives for most of its time buried beneath the sand surface in offshore waters of the European coast. Occasionally dead specimens by the hundreds are cast upon the shore after heavy gales. The second pair of antennae are very long; each flagellum is fringed with setae, and when each is apposed one to the other a tube is formed by the interlocking setae. The crab burrows into fine sand, leaving only the tips of the antennae protruding. The water current takes a path similar to that described for the mole crab, but *Corystes* will often reverse this current, drawing water in beneath the carapace and ejecting it through the antennal tube.

The Atlantic and Pacific coasts of the USA can claim a number of edible crab species. The Dungeness crab, *Cancer magister*, of the Pacific coast and the rock crab, *Cancer irroratus*, of New England, along with the blue crab, *Callinectes sapidus*, of Atlantic waters, all form the bases of important fisheries. Other species forming smaller fisheries are the Jonah crab, *Cancer borealis*, of New England and the stone crab, *Menippe mercenaria*, of the southern Atlantic and Gulf coasts. A recent addition to this list is the red crab, *Geryon quinquidens*, living in depths from 40-2,155 m off the American Atlantic coast.

Females of the Dungeness crab, *Cancer magister*, mature at about 100 mm in breadth and males probably mature at a slightly larger size. The female, similar to most crab species, has a broad abdomen while that of the male is narrow and triangular. Mating occurs from April to September, at which time the male often carries the female until she molts, after which event he mates with her. The pair often bury themselves in sand at this time. In some parts of its range the female carries eggs from October to June, sometimes into August, and hatching occurs from December until as late as June of the following year. The early larval stages (zoeae) swim in the plankton in large numbers near the surface, and the last of these (megalops) appear in July and August. These small megalops larvae can also swim, and one was observed to hold its own against a current of some 2 mm per minute. By the end of November the third crab stage has appeared. The Dungeness crab molts buried in the sand; this process apparently takes only a few minutes, but the new shell takes two weeks or more to completely harden.

The name *Callinectes* means 'beautiful swimmer' and

Molpadia arenicola, a sea cucumber without tubefeet (East Pacific; D. Gotshall photo).

Perhaps the most easily identified cucumber, *Bohadschia argus* is widely distributed in the Indo-Pacific (A. Power photo).

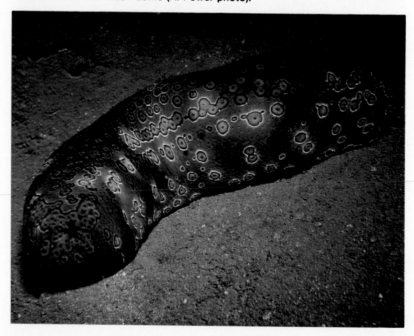

At right, *Holothuria leuco-spilota;* below, an unidentified but rather strikingly patterned cucumber (both Indo-Pacific; K. Gillett photo at right).

sapidus ' savory. ' The blue or Chesapeake Bay crab, *Callinectes sapidus*, forms a large and important fishery from Cape Cod to Mexico. A conspicuous feature, besides its color, is the flattening of the distal segments of the last pair of thoracic legs into swimming paddles. The male carries the soft-shelled female during mating, and the female can store sperm that remain viable for one year. The female spawns from two to nine months after this mating and carries some 7,000,000 to 20,000,000 eggs attached to her abdominal limbs. Blue crabs commence a northward migration in the fall that stops at the onset of cold weather when the crab settles onto the sea bed and ceases to feed or grow. When spring arrives migration again continues and the crabs reach Maryland waters, where most of them mate. The impregnated females then seek out waters of higher salinity in lower Chesapeake Bay, but most males remain in the breeding area and overwinter in deep holes or in river estuaries.

The dense mangrove swamps of the Florida south coast provide suitable shelter for many species of grapsid crabs. The most abundant of these is the small mangrove or tree climbing *Aratus pisonii* that occurs on both the Atlantic and Pacific coasts from Florida to Brazil and from Nicaragua to Peru as well as on all the major Caribbean islands. The mangrove crab feeds upon the terminal leaves of mangrove trees and will climb, often very swiftly, to considerable heights to obtain such food. Small specimens of less than 5 mm carapace width live just below the water surface, clinging to tree roots on which they feed by picking away at parts of the outer covering. Larger specimens readily leave the water to search out mangrove trees but need to return frequently to avoid desiccation. The sharp-tipped walking legs enable the crab to obtain a secure grip on the tree when climbing, and the sharp spoon-shaped tips of the fourth pair of legs (chelipeds) are well adapted for tearing small pieces of leaf and passing this food to the mouthparts. This habit of repeatedly leaving and returning to the water makes the mangrove crab more vulnerable to predators than other crab species, particularly during periods of molting. This crab however, molts very quickly, the new skin being already quite hard before the old shell is shed.

The brightly colored livid shore crab, *Pachygrapsus crassipes*, occurs from the Oregon coast down to California and along the coast of Japan and Korea. An active species similar to most grapsids, it seems equally at home both in and out of water and can survive a fair amount of desiccation. This species feeds on algae, and numbers can be seen frequently scraping away at the green algae that line shore pools and damp crevices. The hoof-shaped tips of the claws on the fourth pair of thoracic legs are well adapted for this task. The majority of females spawn between April and September although a few are found carrying eggs throughout the winter months. Females will spawn twice during the breeding season. The eggs take about one month to develop and hatch.

The presence of numerous small sand pellets arranged in a defined pattern on the sand surface is the result of industrious activity by sand bubbler crabs. One such species, *Scopimera inflata*, occurs on Queensland beaches and lives in burrows in the region just below high tide. During the day, when the tide is low, these crabs emerge to feed. After clearing its burrow the crab forms any remaining sand into pellets and pushes each to the surface out of the burrow opening. Each pellet, carried by the claws, is deposited 50-100 mm away from the burrow. After cleaning the burrow to its satisfaction the crab commences to feed upon the sand surface, scooping sand into its mouthparts and so forming a shallow trench as it moves. This imbibed sand is again made into pellets that are passed beneath the body and line one side of the trench. It is along this trench that the crab runs to make a rapid retreat into its burrow if danger threatens. Each trench is made parallel to the preceding one, and a radial pattern is soon formed as the crab moves around its burrow while feeding. At the end of this foraging period the crab returns to its burrow, places a ring of pellets around the entrance, and finally seals itself

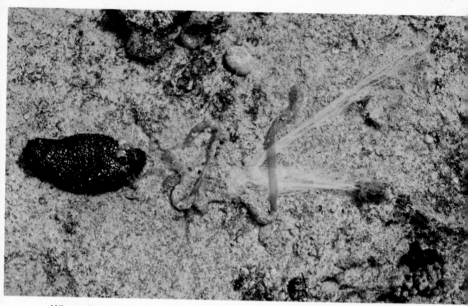

When disturbed, some sea cucumbers can eject the gut and sticky Cuvierian tubules into the face of the enemy and regenerate these internal organs at a later date (T.E. Thompson photo of *Holothuria leucospilota*).

Parastichopus californicus. Species such as this leave no doubt how the name sea cucumber originated (East Pacific; T.E. Thompson photo).

Filter-feeding sea cucumbers. Above, the brilliantly colored *Pseudocolochirus axiologus* (Pacific; R. Steene photo). Below, a more typical though still attractive cucumber, *Cucumaria insolens* (South Africa; T.E. Thompson photo).

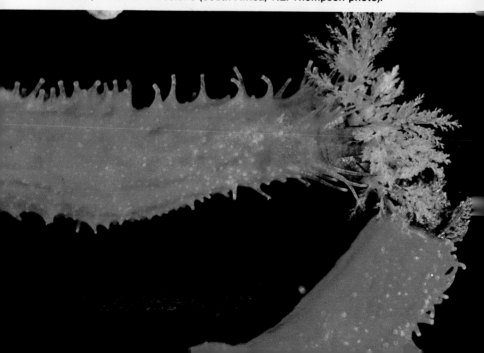

within the burrow by plugging the entrance.

Of the numerous species of xanthid crabs, the spectacular spotted rock crab, *Carpilius maculatus*, inhabits shallow waters of the Indo-Pacific region. In some regions it is fished by the local communities, particularly at night, at which time it comes out of hiding from the coral crevices that it normally inhabits. The large claws contain much meat. The legend of how this crab acquired its red spots is delightfully related (*Aust. Mus. Mag.*, 1937, 6:215) by Mr. Melbourne Ward as follows:

> "On reefs of ancient Hawaii the sea god was wont to search for delicacies to assuage his divine hunger, and one fine morning he espied a fine big crab, fat of body, smooth, shining, and of a uniform yellowish-pink shade, very beautiful to behold. Seized with a desire to add this delectable morsel to his repast, the god grasped the crab. This sudden attack surprised the crab, and, seizing the god by the fingers, it drew blood. The god, in surprise and pain, dropped the crab, leaving a row of red finger-marks on its back. Quickly overcoming the shock of the first encounter, the god seized the crab again, only to relinquish his hold of the powerful creature for the second time, leaving a second set of finger prints. For the third and last time the god caught the crab, which had no doubt been greatly weakened by this contest with divinity, and killed it. The descendants of this beautiful tropical species, Carpilius maculatus, all display the red imprints of the god's fingers."

Temperate and tropical intertidal beaches are the homes of fiddler crabs. The males have one claw of the fourth thoracic leg greatly enlarged and often much longer than the body. The repeated waving or beckoning with this often brightly colored claw is the most conspicuous activity of fiddlers and perhaps warns off other males from the crab's territory or attracts females. This display varies according to the species. The males of the Indo-Pacific *Uca manii* and *U.*

rathbunae, which inhabit protected estuaries and tidal streams, make lateral motions with this claw and after pursuing the female mate with her on the mud surface. By contrast, males of *Uca tetragonum* and *Uca zamboangansis* live upon exposed sandy beaches and beckon with a more or less circular motion of the claw. The mating approach to the female is more elaborate as the male attempts to attract his mate down the burrow that he enters first.

The most familiar of the frog crabs is *Ranina ranina*, the red frog crab, which digs into sand in offshore Indo-Pacific waters. In some areas it is fished by native communities, and in some parts of the Philippines it occurs in large numbers and has some commercial value as the meat is considered a delicacy.

The females of pea crabs (pinnotherids) spend most of their lives within bivalve molluscs, starfish, and sea squirts. A few species are known to damage their host and are therefore considered to be parasites. The males are much smaller than the females. On the eastern coast of America the pea crab *Pinnotheres oestreum* lives within the American oyster, *Crassostrea virginica*, where it feeds upon food strings picked from the oyster's mantle.

A truly spectacular representative of the spider crabs is *Macrocheira kampferi*, the giant Japanese spider crab and the largest living arthropod known to man. An adult male can span 3 m between the outstretched tips of its claws. The body, however, is a little more than 300 mm wide. This crab occurs on sandy or muddy bottoms between 50-300 m deep and is found only on the southeastern coast of Japan. The crab was first reported by a Dr. Engelbert Kaempfer in his account of the history of Japan. He called it ' Simagani ' (striped crab). The modern Japanese name is ' Takaashigani ' meaning ' tall-legged crab. ' Kaempfer compared the size of one of the large claw segments to that of a human leg bone. His specimens came from a cook shop in Suruga; the crab is eaten today but not in large quantity.

Coral or hapalocarcinid crabs are small and highly modified for living most of their lives in the confines of cavities in

Occasionally small, colorful sea cucumbers appear in pet stores, but they are almost impossible to identify without study of the small plates buried in the skin (A. Norman photo).

Psolus chitonoides, a sea cucumber with a distinct foot or creeping sole, heavy armor, and distinct head and tail ends (East Pacific; T.E. Thompson photo).

Side view of *Psolus* sp. (East Pacific; A. Kerstitch photo).

Opheodesma, a greatly elongated burrowing sea cucumber from the Indo-Pacific, and a close-up of its "head" (R. Steene photos).

coral colonies. As a juvenile the female of *Hapalocarcinus marsupialis* settles between two adjacent branches of coral and in some way alters the growth of the coral, causing the branches to form a gall that engulfs the crab. This gall or chamber in which the female is now imprisoned has a number of holes through which water can flow in addition to providing an entrance for very small males. After mating the female produces large broods of larvae. The modified mouthparts presumably filter plankton drawn into the gall cavity by the respiratory current.

PYCNOGONIDA

The Pycnogonida are a wholly marine group of arthropods often referred to as sea spiders because of their superficial similarity to the true spiders. There is virtually no fossil record and therefore their relationship to other groups is somewhat vague. Their body is greatly reduced, the head and thorax being fused to form the so-called prosoma. The abdomen (or opisthosoma) is merely a small unsegmented protuberance with the anus at its tip. They are unusual among arthropods in having the mouth at the end of a pro-

Lateral and dorsal views of pycnogonids, the anterior end at the top.

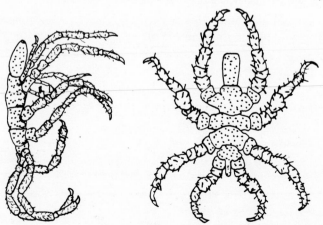

boscis which may attain half the length of the body in some species. The legs are attached to the prosoma and are usually eight in number, although it can be ten or even twelve. In addition to the walking legs there may be a pair of palps which are sensory and sometimes assist in feeding, a pair of chelifores used for feeding, and a pair of ovigerous legs which are frequently only present in males and have a reproductive function.

Sea spiders have a wide geographic range, being represented in both polar and tropical seas, and occur at a wide range of depths from the intertidal zone to depths of 6000 meters or more. Generally they are found only in small numbers, but where their main food supply is abundant they occur in much larger numbers. Most live on the sea bed throughout their lives, crawling slowly over the substrate. Some, however, particularly those with attenuated bodies and long thin legs such as *Anoplodactylus*, *Pallene*, and *Nymphon*, are capable of swimming freely. They propel themselves by means of the walking legs which beat mainly in a vertical plane in relation to the axis of the body, thus moving them dorsal side first. They also move passively in the ocean currents while attached to seaweeds, hydroids, and bryozoans.

Pycnogonids in general are either white in color or have a color which matches the background (which is usually their main food). As there is no evidence to suggest that they can change color at will, this would indicate a long term adaptation to a particular habitat which would confer some protective coloration or camouflage. *Anoplodactylus lentus* occurs on *Eudrendrium* (a hydroid) colonies which have a pale cream color, whereas *Anoplodactylus angulatus* resembles the seaweed *Ascophyllum*. Deepsea species, in common with other creatures, usually have an orange-red coloration.

Mating has been observed in a few species, although the actual fertilization process has not. *Anoplodactylus lentus* breeds toward the end of August on the coast of the USA. The male crawls over the back of the female and onto her

An acorn worm, *Ptychodera flava,* removed from its burrow. Below is a close-up of the anterior end showing the proboscis and collar (Hawaii; S. Johnson photos).

A small group of salps (right) (East Pacific; D. Gotshall photo). Tubular pelagic colony of the thaliacid tunicate *Pyrosoma* (below); such colonies may reach gigantic proportions (Caribbean; C. Arneson photo).

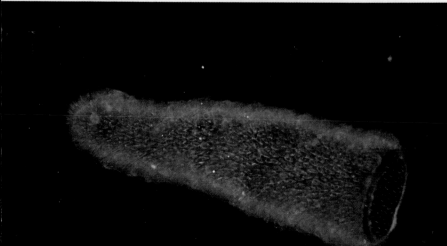

ventral surface, leaving their ventral surfaces touching and their genital apertures opposed. The hooked ovigerous legs of the male then remove the extruding egg masses and form them into a ball. In *Pycnogonum littorale* the male rests on the back of the female with the genital orifices touching. In this species the eggs form a single large ball with both ovigerous legs embedded in it. In most other species a number of balls are made which attach to the ovigerous legs. Shoreline species such as *Nymphon gracile* have a seasonal release of eggs, but offshore species often release eggs at all times of year. When the eggs hatch they form a larva called a protonymphon which is not dissimilar to the crustacean nauplius larva. Larvae which form from the type of egg which has little yolk soon leave the ovigers of the male. However, those which form from eggs with a large amount of yolk generally remain attached to the ovigers for a longer period of time. In some species they remain attached until metamorphosis is complete. Metamorphosis takes place by means of a series of molts, after each of which a new pair of appendages form behind the pre-existing ones and the larval appendages gradually regress.

The feeding behavior and food preferences of pycnogonids are very poorly known. Most of the evidence is based on inference from observations of pycnogonids in association with other animals, usually bryozoans or hydroids. The type of association with bryozoans falls into three main categories: a) the bryozoans provide a substrate on which they can safely move about, such as *Pycnogonum littorale* on the bryozoan *Flustra foliacea*; b) the pycnogonids feed on epibionts or debris on the surface of the bryozoan; c) the pycnogonids actually feed on bryozoans, for example *Achelia echinata* on *Flustra foliacea*. There are many instances of pycnogonids feeding on hydroids. The actual method of feeding varies depending upon the armature around the mouth, shape of the proboscis and whether the species has chelifores or palps.

XIPHOSURA

The Xiphosura are better known under their common names of king crabs or horseshoe crabs. In spite of the name they are in fact only distantly related to the true crabs and are close to the Arachnida (the spiders and scorpions). Looked at from above, as the name suggests, they present a horseshoe-like outline, from the posterior end of which projects a long spine. The whole body is covered by a smooth chitinous sheath varying from sage-green to black in color. From the rounded dorsal surface project a number of spines arranged in a median and two lateral rows. The anterior median spine overhangs the median eyes, and the anterior lateral spine on each side overshadows the large lateral eyes. The lateral eyes are compound eyes consisting of a number of ommatidia lying beneath a single lens, very akin to the situation in flies. The median eyes are simple ocelli which although incapable of image formation are highly sensitive to ultraviolet light and are analogous to the pineal eye of vertebrates.

The numerous jointed limbs are not visible from above, being completely encased by the carapace. There are a pair of chelicerae (as found in spiders) which are characteristically in front of and associated with the mouth, followed by five pairs of similar appendages. All of these bear pincer-like structures (chelae) at the ends, apart from the fifth which bears terminal spines used in digging. These are the walking legs. The first pair of walking legs in males is modified to perform a clasping function during mating. The mouth lies between the first pair of legs. King crabs respire by means of external gills found on the appendages of the ninth to thirteenth segments. Each gill consists of a series of leaves like the leaves of a book, some 150-200 in number. They are in fact sometimes called gill-books. A constant stream of fresh water is passed over them by the movements of the legs.

Horseshoe crabs are generally sexually mature after about three years. When mating the male grabs the hind end of the female's carapace with his specialized first pair of

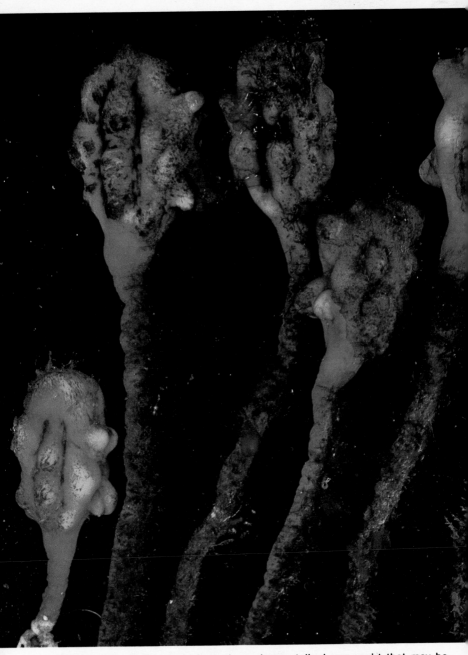
The sea-tulip, *Pyura pachydermatina,* a large stalked sea squirt that may be brightly colored (Australia; K. Gillett photo).

Above, *Clavelinella* sp., a transparent solitary sea squirt (Atlantic). Below, colony of *Botryllus scholsseri,* a common East Atlantic encrusting tunicate (T.E. Thompson photo).

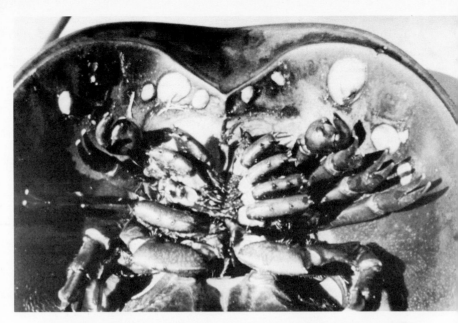

Ventral views of the anterior body of *Limulus*, showing sexual dimorphism. In the male (top) the distal segments of the first walking legs are inflated and claw-like, while in the female (bottom) the first legs end in pincers much like those of the other legs (N. Herwig photos).

walking legs. Thus fastened together the pair then move into shallow water, the female finally digging burrows in the intertidal zone into which are deposited several hundred eggs. These are fertilized externally. Several months later the eggs hatch as pelagic larvae which grow by a series of molts. It is not until the final molt that they are sexually mature and capable of mating since it is only then that the first pair of legs of the male become modified.

Although mainly marine, king crabs occasionally occur in brackish estuarine waters. They burrow and forage for bivalves and worms, particularly nereids, in mud and sand with a head-on shoveling action of the carapace while the sixth pair of appendages push the disturbed sand out behind. There are only three genera: *Tachypleus* and *Carcinospinus*, which occur off the coast of Southeast Asia, and *Limulus*, which occurs off the coast of North America.

Halocynthia, the sea peach, a distinctive solitary sea squirt of the East Atlantic.

Phylum Echinodermata

The Echinodermata (from the Greek words *echinos*—spiny, and *derma*—skin) are a group of closely related animals (a phylum) which includes the starfish (Asteroidea), brittlestars or serpent stars (Ophiuroidea), sea urchins and sand dollars (Echinoidea), sea lilies and feather stars (Crinoidea), and sea cucumbers (Holothuroidea).

Echinoderms are one of the groups thought to be most closely related to the phylum Chordata, to which man and the other backboned animals belong. In fact, they may be looked upon as a transitional group between the lower animals and the vertebrates. Although most echinoderms are radially symmetrical as adults, this is really a secondary trait; their larval stages are always bilaterally symmetrical. Like the majority of the lower invertebrates, echinoderms have unsegmented bodies, although they tend toward a serial repetition of certain parts, which points toward segmentation. They possess a spacious body cavity, lacking in more primitive phyla. Their digestive system has two openings, one for ingestion of food and one for excretion, a characteristic of higher, more motile animals, while lower, more sedentary animals usually have but one opening serving both purposes. The echinoderm circulatory system is intermediate between the simple diffusion system of lower invertebrates and the more sophisticated vascular system of higher animals. Finally, there are no colonial echinoderms, just as there are no colonial higher animals; each animal exists as an individual.

Echinoderms are readily distinguished from all other phyla by the uniqueness of several major characteristics, the principal one being pentamerous symmetry; that is, in most species their bodies are made up of five equal and similar parts. Another important characteristic is that each hard part—spine, plate, spicule, or tooth—is essentially a single crystal of calcium carbonate. A third structural difference common to all echinoderms is a unique organ system called

The common Caribbean starfish *Luidia clathrata* is a fairly typical starfish —it has five long arms that are broad at the base, a relatively large disc, visible plates and spines on the arms, and is active when disturbed. (C. Arneson photo).

the water vascular system, which is mainly a filtering mechanism and fluid circulating system.

The echinoderms are exclusively marine—neither terrestrial nor freshwater forms are known. There are nearly six thousand known species of living echinoderms, and many thousands more are known from the fossil record. These numbers are being increased every day by new discoveries.

The most obvious and striking feature that distinguishes the echinoderms from other animals is their *pentaradiate* (five-part) *symmetry*. A body divided into five equal and similar parts is rare in the animal kingdom; higher animals are all divisible into two equal and similar parts and are said to possess bilateral symmetry. Many lower animals either lack symmetry altogether (sponges) or have bodies divisible into four, six, or eight parts (radial symmetry). Echinoderms, in fact, hover uneasily on the borderline between pentaradiate and bilateral symmetry; all are at some stage in their life cycle bilateral and at another stage pentaradiate. Some species of starfish, for instance, may have many arms (not necessarily in multiples of five) as adults, but they start their adult life at metamorphosis with five arms. All larval echinoderms are bilaterally symmetrical, and it is thought that the ancestors of echinoderms were probably also bilateral. Just when, how, or why echinoderms became pentaradiate is not known.

As possession of a backbone is a major and unique distinction of the phylum Chordata, so the *water vascular system* is the major characteristic of the phylum Echinodermata. There is no organic system comparable to the echinoderm water vascular system in any other animal group; at the same time, all echinoderms possess this system. In this one fluid system are combined the functions that in other animals are divided among several fluid systems. The development of this hydrostatic system is a specialization which equips echinoderms to cope with life in the sea, but which also limits them to a saline environment, since it is not adapted to regulating salt concentrations in the body but is open to the free passage of sea water.

Let us examine this system in some detail. It is a multipurpose tubular system, open at one end to the free passage of sea water through the sieve plate (madreporite) or hydropore. It furnishes hydraulic power for locomotion by operating, through a complex of valves, reservoirs, and contractile muscles, the *tubefeet* on which the echinoderm moves. Central to the water vascular complex is a tubular ring around the esophagus, with tubes running out from it along the ambulacra. An ambulacrum is that part of the body which bears the rows of tubefeet (branches or extensions of the tube); there are usually five ambulacra, so there are five double rows of tubefeet. The areas bearing the tubefeet alternate with an equal number of areas of the body wall which are without tubefeet and are called interambulacra or interradii.

In one area of the water ring around the esophagus, between two of the radial water vessels, a tube arises in the direction away from the mouth (toward the anus), and this tube, strengthened with spicules or small plates of calcite, is called the stone canal. It is capped by a stony, porous plate, something like the perforated head on the spout of a watering can. The radial water vessels which extend along the ambulacra have numerous short side branches which terminate in the tubefeet or tentacles.

The arrangement of the tubefeet is not random but very regular, either branching off from the radial vessel in pairs, one on each side, or alternately from side to side. A combination of muscles and cilia serves to draw water into the system through the stony plate (called a *madreporite*) and to expel it through the same plate via the stone canal. Bulbous reservoirs called ampullae, one just under the madreporite, one to several around the ring canal, and frequently one above each tubefoot, serve to store fluid. The water vascular system operates the hydraulic tubefeet by maintaining the proper hydrostatic pressure to enable them to act as locomotory organs, which, after food gathering, is their prime and proper function. Tubefeet also have become adapted to a number of secondary functions such as burrow-building, sensory activities, and respiration. In the

676

echinoderms, as in many other invertebrates, ' feet ' serve a number of functions, not being restricted as are our own feet to locomotion.

The spiny, warty, or granular covering of the echinoderm body differs from the hard carapace of crustaceans, insects, molluscs, and other shelled animals in being composed of a large number of (usually) polygonal plates, closely or loosely jointed and covered with a thin layer of ectoderm, with spines or other hard protuberances attached to the outer surface of the plate.

Each skeletal part of an echinoderm is composed of calcium carbonate in the form of the mineral calcite. Each piece is, for all practical purposes, a single crystal of calcite, whether the piece involved is a large spine, a plate of the body wall, a section of the mouth frame, or the spiky skeletal rods supporting the tiny body of a larva. Since hard skeletal material is practically indestructible, this peculiar crystalline structure is found not only in living animals, but also in fossils, thus making recognition of isolated fragments of echinoderms in the fossil record easier and simplifying the task of the paleontologist.

Two classes of echinoderms, the asteroids and the echinoids, possess tiny peculiar organs called pedicellariae. These are small pincer-like external appendages used to remove debris from the body, to capture small prey, and to pass bits of food along toward the mouth. These were once believed, by such prominent zoologists as Lamarck and Cuvier, to be some sort of parasite living on the seastars and sea urchins. Johannes Muller, a prominent echinologist of the mid-nineteenth century, described them as separate parasitic animals and even gave them names.

Respiration in echinoderms is not confined to any one organ or body system. Principal respiratory mechanisms are the projections of the water vascular system, the tubefeet or tentacles. Oxygen is absorbed almost entirely via the water vascular system in sea lilies, where the tentacles of the arms and pinnules offer the only respiratory surfaces. In starfish the thin walls of the tubefeet ' inhale ' oxygen from the surrounding seawater, and tiny thin evaginations of the body wall in the shape of sacs or little glove-like projections (papulae) of the body wall stick up between the plates of the dorsal surface to perform the main part of the respiratory function.

The sea urchins go a step further and have ten bushy gills in the membrane around the mouth, in addition to the oxygen-absorbing surface of the tubefeet. In the brittlestars, the respiration of the tubefeet is supplemented by a respiratory membrane lining five pouches in the body wall, called bursae, through which sea water is pumped in and out by a rhythmic motion of the body.

The sea cucumbers go still further in the development of special respiratory mechanisms, having, in addition to normal tubefeet, bushy or digitate tentacles (modified tubefeet) around the mouth which, although primarily food-gathering organs, also serve for respiration. Some sea cucumbers have also developed special internal organs, the respiratory trees, which arise inside the body from the cloaca and through which water is circulated by being pumped in and out through the anus.

The ability of many echinoderms to throw off parts and then regrow them is well known. Indeed, it was from this habit that the brittlestars got their name. A brittlestar can shed its arms and sometimes its disc with the greatest of ease when roughly handled, when attacked, or just when, through a shortage of food or extreme changes in temperature or salinity, its surroundings do not suit it—a sort of echinoderm temper tantrum. Sea cucumbers are even more prone to self-mutilation in defense of their lives, but as they have little to lose in the way of external appendages, they shed their internal organs instead. This defense mechanism is intended either to distract the predator or to entangle him in the sticky mass thus formed. Numerous starfish, particularly those with many arms or with a fragile skeleton, can throw off their arms in self-defense, while the best sea urchins can do is throw off their spines when conditions are unfavorable.

As you might guess from the fact that they so light-heartedly part with parts, echinoderms have unusual powers of regeneration, and with a skeleton of many breakable pieces and soft tissues of incredible thinness, it is a good thing they do. An injury, even quite a major one, need not necessarily be a tragedy, for spare parts can be grown and mutilated tissues repaired in a very short time. The rate of regeneration in echinoderms varies from species to species, but usually lost parts can be replaced in just a few weeks. In fact, Japanese fishermen taking sea cucumbers for the market frequently reseed the fishery by throwing back pieces of the animals in the hope that they will grow into whole animals and, indeed, some of them do. No attempt will be made here to discuss how each and every species differs from or modifies the systems and structures heretofore discussed in this chapter, but a brief description of each class is necessary to the understanding of diversity among the echinoderms.

The basic structural plan of a **crinoid** is a cup-like body, or calyx, of many closely joined plates, with a membranous or solid ' lid ' called the tegmen and five arms arising from the rim of the cup. The cup, or calyx, contains the central parts of the main organ systems. The mouth is in or near the center of the lid, and the anal opening is usually at the tip of a tube or cone projecting slightly off center from the tegmen.

Some crinoids are stalked, others (the comatulids) are unstalked. In stalked forms the calyx sits on top of a slender jointed column of circular or pentagonal plates through the center of which is a space containing extensions of the body cavity and nervous system. In the comatulids the stalk is reduced to a single plate called the centrodorsal. The arms in both forms are usually branched, in some cases many times, and bear rows of thin, regular, pointed, twig-like projections along each side, alternating from side to side. These are called pinnules. Among the stalked forms, the number of arm branches is usually less than forty, while the comatulids may have more than forty branches. Crinoids from deeper, colder waters generally have fewer arm bran-

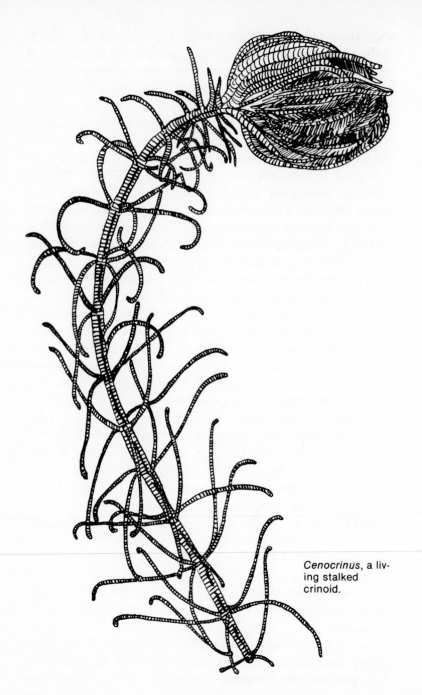

Cenocrinus, a living stalked crinoid.

ches; those with more than forty are characteristic of warm shallow seas. The length of the arms also seems to be a temperature-determined character; short-armed forms are from cold water and longer arms occur in warm temperatures.

The arms and pinnules are the food-gathering system of the crinoids. A groove on the oral side of each pinnule leads to a groove in the arm, like tributaries joining a river, and these food grooves continue across the surface of the tegmen to the mouth. The arms and pinnules are made up of a series of jointed plates, and at each joint a cluster of three tubefeet arises. A series of cover plates, or lappets, borders each groove and can be closed over it for protection. The tubefeet produce strands of mucus in which plankton is caught and conveyed along the ciliated grooves to the mouth.

The crinoid mouth is surrounded by five or more specialized tubefeet whose function is mainly sensory. The mouth leads to a short esophagus and it in turn leads to an intestine and a short rectum which projects above the surface of the tegmen, terminating in an anal opening. The projecting tube has been observed to contract rhythmically and is thought to serve a respiratory as well as an excretory function.

The **Asteroidea** are free-living echinoderms, pentagonal or stellate in form, with a flattened, more or less flexible disc continuous with the arms, which are typically five in number. Species regularly having six or seven arms are not uncommon, and there are some with as many as fifty. Each arm contains gonads and digestive ceca, as well as extensions of the water vascular system. The arms have open ambulacral grooves on the under side provided with two or four rows of tubefeet. The mouth is located at the center of the under side of the disc, and the anus is at or near the center of the upper side. The skeleton is made up of separate calcareous pieces bound together by connective tissue and usually ornamented with projecting spines, knobs, or tubercles.

The skeletal plates and their ornamentation are important to the classification of starfish species. The endoskeleton of calcareous ossicles consists of the main supporting skeleton and its more superficial parts—spines, tubercles, warts, granules, etc. The whole is clothed with a thin layer of skin which may rub off the more projecting parts. The innermost part of the skeleton consists of paired rows of arched plates flooring the ambulacral grooves in the center of the oral surface of each arm. The ossicles of the body wall may be rounded, polygonal, squarish, rod-like, or crescentic. They are bound together with connective tissue and may overlap slightly at their corners, forming a reticulate network, may fit closely together like a mosaic pavement, or may overlap like slate roofing.

Some starfish have a hard texture and a fairly rigid body due to numerous close-fitting plates, while at the other extreme are some species which have a very reduced skeleton of small plates connected by long muscle strands and a fleshy upper surface supported by long slender spines. Some seastars have a conspicuous margin of two rows of large, block-like plates (marginals), and many have a bordering fringe of spines around the periphery. An inconspicuous anus is located in the center of the dorsal surface, and between the center and the margin, in an interradial position on the upper side, is the stone plate or madreporite. In some starfish there is a median row of plates, the carina, down the aboral surface of each ray.

The paired ambulacral plates, consisting of two rows of elongate, laterally flattened ossicles, form an arched channel through which the radial water vessel runs, with pores between the plates for the extrusion of the tubefeet. They never bear spines. The edges of the ambulacral grooves are formed by a series of plates, the adambulacrals, which correspond one to one with the ambulacral plates. These plates always bear strong movable spines which can mesh together to protect the ambulacral groove or even close it completely. Spines and tubercles are usually present on other plates as well; some species are quite smooth, but spininess reaches perhaps its greatest development in the Pacific reef-dwelling starfish *Acanthaster*, where spines up to an inch in length are mounted on tall columns and thickly clothe the entire dorsal surface.

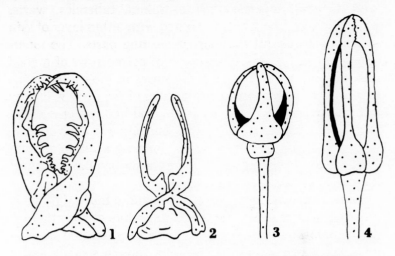

Pedicellaria from starfish (1 and 2) and irregular echinoids (3 and 4).

Pedicellariae are present in all orders of asteroids except the Spinulosida. They are more varied in seastars than in echinoids (the only two classes in which they occur). In the order Forcipulatida the pedicellariae somewhat resemble echinoid pedicellariae, being stalked, but in the other orders sessile pedicellariae, formed simply of two or more movable spines attached directly to the skeletal ossicles, or alveolar pedicellariae, similar to sessile ones but partly sunken into a depression in the supporting plate, are found.

The open ambulacral system is the feature that most distinguishes the Asteroidea from other echinoderms. The radial parts of the water vascular system lie wholly outside the skeleton, lacking even the cover plates, or lappets, of the crinoids and protected only by the spines of the adambulacral plates.

On the oral, or under, surface the mouth is central. The ambulacral grooves radiate out along the middle of the oral surface of each arm to the tip. Each contains two or four rows of tubefeet, either pointed or with flat terminal discs or suckers. At the tip of the arm is an unpaired terminal tubefoot with a cluster of light-sensitive cells. The ampulla above each tubefoot is a reservoir from which fluid is forced into the tubefoot to extend it, or into which water is drawn when the tubefoot is retracted.

The madreporite is a circular, grooved plate, always on the aboral surface in an interradial position, which is usually larger and more conspicuous than the other dorsal plates but may be quite small and concealed by the other plates. It is usually single, but some species may have as many as sixteen madreporites.

The asteroid mouth leads to a vertical esophagus, above which the stomach is divided into two parts, the large oral cardiac portion below and the smaller aboral pyloric portion above. Two hepatic ceca are given off from this portion into each arm. Lastly, there is a short rectum, which in some species ends blind, while in others it terminates in a dorsally located anus.

The basic **ophiuroid** body plan is simple and unvaried, except for the relatively few species which have branching arms: a small, flattened disc and five (rarely six or seven) long slender arms of somewhat solid construction which are not extensions of the disc but are inserted into it or under it and are without external ambulacral grooves. The arms and disc may be smooth, spiny, or granular. Ophiuroids have neither intestine nor anus and, with rare exceptions, no organic system extends into the arms except the radial branches of the water vascular system and its connected nerves. The tubefeet (in this group they are called tentacles) do not serve primarily for locomotion or for feeding, as they do in the other echinoderms, but have an almost purely respiratory function.

In the Ophiuroidea the disc may be round, pentagonal, or scalloped, and the disc covering varies widely from species to species. All sorts of arrangements exist, from thick naked skin (common in the group with branching arms) to heavy plating covering the entire disc. In many species the disc is ornamented with spines, tubercles, or granules which sometimes completely conceal the disc plates. At the base of each arm on the aboral surface of the disc are a pair of plates called the radial shields. These may be concealed by

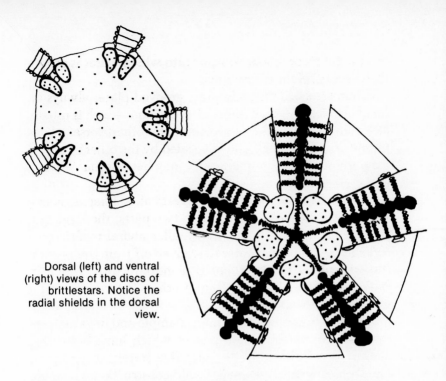

Dorsal (left) and ventral (right) views of the discs of brittlestars. Notice the radial shields in the dorsal view.

the disc covering or ornamentation, but they are almost always present. They are frequently very large and are usually the most conspicuous of the aboral plates.

Ophiuroid arms are cylindrical in cross-section and from five to ten times as long as the diameter of the disc in most cases. The Ophiurida (true brittlestars) cannot coil their arms as the Euryalida (basketstars) do because the vertebral ossicles are articulated to one another in a different manner in the two groups. Although five is the normal number of arms, species with six or seven arms are known, and a species from the Antarctic has eleven arms.

You will have noted that most of the ossicles of the ophiuroids are called shields. This should indicate to you that, while these plates are of varying thicknesses, they are mainly two-dimensional. The plates called shields are chiefly external and protective. The internal skeletal elements of the arms are called vertebral ossicles because of their resemblance to a vertebrate spinal column. The 'vertebrae' are discoid, thick in the middle and thinner at the edge, and in most ophiuroids one side of the vertebra has a depression in the center and the other side a projection, so the ossicles fit into one another. This sort of arrangement limits the movement of the arm to a horizontal plane, so the Euryalida (which, you remember, can coil their arms up) have a different arrangement; their vertebral ossicles have an hourglass-shaped projection on each side, horizontal on one side and vertical on the other, so the arms can move in any direction. Each vertebra represents an arm joint and each has its own set of arm shields. The vertebral ossicles probably represent a fused pair of ambulacral plates.

The arms continue under the disc to the mouth frame. On the oral side of the disc, on either side of the arm, a slit in the body wall called the bursal slit leads to bursal pouches within the disc. In most species the slit is long and narrow. The bursae, unique to the Ophiuroidea, occupy most of the space within the disc. Normally, there is a bursa corresponding to each slit, but in certain species the bursae open into each other, forming a single large tubular space surrounding the esophagus. The internal surface of the bursa is heavily ciliated and a water current is constantly circulated through the bursae, which serve as the main respiratory mechanism in the ophiuroids, as well as outlets for the reproductive cells and as brood chambers in brooding species. The water currents which circulate through the bursae are probably not simply a passive respiratory device but seem to be aided by pumping movements of the disc itself.

Where spines are present in the ophiuroids, these are borne on the sides of the arms. The lateral arm shields bear a vertical row of from two to fifteen spines. They may be attached to a ridge near the center of the arm shield and stand out at right angles to the arm, or they may be on the distal edge of the arm shield and pressed against the arm. The spines vary in size and shape from species to species and are rarely smooth, being either minutely roughened or very thorny. In some species they are modified into glassy hooks. Some species have glandular spines which may be poisonous.

686

These spines are covered with a thick fleshy epidermis containing many gland cells.

There are no pedicellariae and no respiratory papulae in the Ophiuroidea. Spines, present on the lateral arm shields and sometimes on the disc, are never present on the dorsal or ventral arm plates. Ophiuroid spines are movable, being mounted on a tubercle as in the echinoids and not fixed directly and immovably to a plate as in most asteroids.

The Ophiuroidea have no external ambulacral system. The only visible indications of the water vascular system in living animals are the tentacles, a pair to each arm joint, which issue from holes between the lateral arm shields and the oral arm shields. The tentacle pores are frequently protected by one or more flattened, rounded spines called tentacle scales. Strictly speaking, the ophiuroids have no madreporite. In most species a single hydropore pierces one of the buccal shields. A few species have more than one hydropore (up to 12), and in one, *Trichaster elegans*, the interradial hydropore is not associated with any skeletal element at all. The buccal shield (one of five large flat plates around the mouth) containing the hydropore is sometimes called the madreporite, but in the Euryalida it is not uncommon for all the buccal shields to have numerous hydropores.

A radial canal runs from the water ring in each radius out to the tip of the arm through a ' tunnel ' in the vertebral ossicles, terminating in a single sensory tentacle at the end of the arm. In each arm joint the radial canal gives off a pair of podial canals leading to the tubefeet, or tentacles, and there are no ampullae. Although the tentacles are not employed in locomotion in the ordinary way, in some species they do produce a sticky secretion which aids the brittlestar in climbing over smooth surfaces. The tentacles are also used by burrowing species in digging a burrow in which the whole animal except the arm tips lies buried.

The mouth of ophiuroids is in the center of the oral surface of the disc and is surrounded by a series of plates, the mouth frame, which, when viewed from above (with the

A basketstar, probably one of the most impressive echinoderms (Tierney & Killingsworth photo).

disc removed) looks much like a compressed Aristotle's lantern. This rather complex structure of many pieces is usually ornamented with teeth and modified spines called papillae. The true mouth actually lies above the jaw apparatus, within the disc. It leads through a very short esophagus to a sac-like stomach which fills most of the disc not occupied by the bursae, and there is no intestine and no anus. Material not digested is simply passed out again through the mouth. The Ophiuroidea get their common name of brittlestars from their habit of casting off one or more arms when handled or disturbed. These are readily regenerated. Certain species can also cast off most of the disc, including stomach and gonads, and later regenerate them.

The **Echinoidea** are globose, egg-shaped, or discoid echinoderms with a shell, or test, of closely fitting plates arranged in rows running from top center to bottom center and covered with spines. The mouth is on the under side, and the tubefeet pass through pores in the plates. The continuous skeleton of semi-fused calcareous plates which leaves only the peristome and periproct free constitutes the main feature of the echinoids. The plates of the hard, rigid test are arranged in twenty rows, ten pairs of ambulacra alternating with ten pairs of interambulacra. The ambulacral plates are pierced with holes for the tubefeet. The interambulacral plates are usually broader than the ambulacral plates and bear the larger spines.

The test is immovable except in the family Echinothuriidae, where the test is leathery and flexible and the plates are embedded in it. The shape of a regular echinoid is generally spherical, somewhat flattened at the poles, or sometimes slightly pentagonal or oval. Irregular echinoids are egg-shaped, flask-shaped, or flattened like a cookie.

The test, spines, pedicellariae, and other external appendages are all covered with a thin skin which may be worn away at the tips of the spines. Muscles of the body wall are confined to those which operate the movable appendages and to the muscles of the Aristotle's lantern.

Four test modifications in sand dollars. 1) *Dendraster excentricus*; California. 2) *Mellita longifissa*; Baja California. 3) *Encope grandis*; Baja California. 4) *Rotula arbeculus*; West Africa. (All K. Lucas photos at Steinhart Aquarium.)

Close-up of the Aristotle's lantern of the sea urchin *Sphaerechinus* (G. Marcuse photo).

As echinoids are oriented mouth downward, the oral surface is flattened or concave. The spines, which cover the body thickly, may be nearly uniform in size, but usually those on the oral surface are shorter than those on the aboral surface, and short and long spines are frequently closely mixed. Irregular urchins generally have a thick coat of short spines, while the spines on some regular urchins may be as much as a foot long.

As would be expected, the spiniest echinoderms are also the group with the greatest variety of spines. Spine development has certainly gone to every conceivable extreme in the echinoids. There are tiny hair-like spines, flat paddle-shaped spines, flat-topped spines that look like small heavy thumb tacks, spines like miniature spears or arrows, thorny spines, smooth spines, hollow spines, ribbed spines, spines round, triangular, or polygonal in cross-section, fragile delicate spines, and gross heavy, nearly unbreakable spines.

Echinoid spines are mounted on tubercles on the test and articulated by a ball-and-socket joint. The spines are movable by a set of muscles around the base of each spine, attaching it to the test in a ring around the tubercle. Pedicellariae are present in all species of echinoids and are usually very abundant. They are mounted on a stalk and generally have three valves, or jaws. The stalk is movable on the test, and the head of the pedicellaria is movable on the stalk. They are used both defensively and offensively, to catch prey or ward off attackers, as well as to keep the test clean.

Echinoids are mostly well equipped with numerous tube-feet which protrude through the pores in the ambulacra and are capable of great extension. They are mainly locomotory in function and are furnished with a terminal disc or sucker, supported internally by a delicate rosette of calcite. In locomotion, the tubefeet are aided by the spines.

At the upper pole of the test, opposite the mouth, a membranous area called the periproct and containing one to many embedded plates, also contains the anus. It is surrounded by five large and five slightly smaller plates; this plate series is called the apical system. The large plates are the genital plates, each being pierced by a pore for the passage of eggs or sperm. The other five plates are called the oculars and are light-sensitive. One genital plate is usually larger than the others and has numerous tiny pores; it is the madreporite. During growth new plates are added at the outer edge of the ten plates of the apical system. As the test grows, the apical system remains roughly the same size.

In irregular echinoids the anus has retreated along the posterior radius and is thus removed from the apical system, which maintains the central aboral position. The arrangement of plates of the apical system varies widely in irregular echinoids and is important to their classification.

The jaw apparatus of echinoids, the Aristotle's lantern, is one of the most intriguing mechanisms in the animal kingdom. Present in all regular echinoids but reduced or lacking except in young stages of most irregular echinoids, this complex structure of forty calcareous pieces and their attendant muscles is used for grasping and chewing food. The lantern supports and operates the five long, strong teeth in such a way that they can move up and down and toward and away from each other to rasp food from the substrate and chew it. The principal large cavity inside the echinoid test contains the gonads and intestine. A separate cavity, completely enclosed in membrane, houses the Aristotle's lantern. These cavities are filled with a fluid containing wandering blood cells.

In **holothurians** the number and shape of the tentacles, presence or absence of a ring of calcareous teeth around the anus, location of the tubefeet, body shape, and ornamentation are all important in classification, but the single most important character in distinguishing one species from another is the type of spicules in the body wall. There is a truly amazing variety of forms. Perhaps the most peculiar are the well-known ' wheel and anchor ' plates typical of the families Chiridotidae and Synaptidae. Other types are smooth or warty rods, various fenestrated plates, buttons, tables, rosettes, and baskets. Those embedded in the surface of the skin provide a rough texture which prevents the sea

693

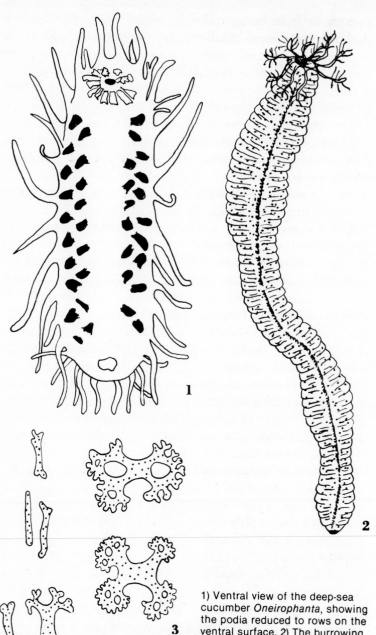

1) Ventral view of the deep-sea cucumber *Oneirophanta*, showing the podia reduced to rows on the ventral surface. 2) The burrowing cucumber *Leptosynapta*. 3) Wall ossicles of the sea cucumber *Actinopyga*.

cucumber from being rolled helplessly along the bottom. Although composed chiefly of calcium carbonate, like the skeletal elements of other echinoderms, certain of the Molpadiidae have a deposit of iron phosphate which is formed on the spicules as the animals age and which gradually replaces the calcium carbonate.

Holothurians have no spines and no pedicellariae. Except in the Psolidae, the body wall lacks large calcareous plates. The water vascular system consists of a ring canal which encircles the pharynx and from which two kinds of structures arise, the polian vesicles and the stone canal. Polian vesicles are sac-like organs which function as expansion chambers for the water vascular system, like the ampullae in starfishes. Usually they are single, but in some species there may be up to fifty or more. The stone canal terminates in a calcified madreporite. Again, this is single in some holothurians, multiple in others. In most holothurians the madreporite hangs free inside the body cavity, thus making the sea cucumbers the only echinoderms with a completely internal water vascular system. In some species, however, the madreporite opens to the exterior. The five radial water vessels, simple or branched, connect the ring canal with the tubefeet and with the tentacles (modified tubefeet).

Some species of holothurians have numerous white, pink, or red very sticky tubes called cuvierian tubules. These are used as a defense against predators and are emitted from the anus in slender threads to entangle the attacker. In England these species are called cotton-spinners, and fishermen hate to find them in their crab and lobster traps, for they can make an unbelievably sticky, hard-to-remove mess. Certain holothurians can discard all of their insides in order to distract predators, but they quickly grow a new set. A toxin called holothurin, deadly to fish, is present in many sea cucumbers.

The echinoderms might be called middle-sized animals; they are all macroscopic as adults, but the majority of them are smaller than most of the vertebrates. Even within these limits, however, their size range is very great. The smallest known adult echinoderm is probably *Nannophiura lagani*, a pinhead-sized brittlestar that lives among the spines of certain sand dollars. There are a couple of echinoids, *Echinocyamus* and *Fibularia*, which are smaller than a baby pea when fully grown. Very small starfishes are unknown as adults, but little juvenile asteroids, like living snowflakes, are frequently found in clumps of seaweed. The smallest known crinoid is probably *Holopus*, a funny little beast like a tiny clenched fist, about one and a half inches high.

The longest echinoderm is undoubtedly *Synapta maculata*, an Indian Ocean sea cucumber which reaches an expanded length of six to eight feet. However, it is far from being the bulkiest echinoderm, for it can contract to a much smaller size and if removed from the water becomes a mere handful of thin tissue. The bulkiest living echinoderm yet described is probably *Pycnopodia*, a starfish with many arms which grows large enough to fill a washtub. But by far the largest echinoderm that ever lived (and possibly the largest non-colonial invertebrate that ever lived) was a Devonian sea lily with a stalk over seventy feet long.

Echinoderm sexes are generally separate, but some individuals combine both sexes, and in several species this is the usual condition. A few echinoderms are known to undergo an alternation of sex, being male when younger but becoming female as they reach a larger size, or vice versa. Quite a number of starfish reproduce themselves by fission; that is, they split in two and each half grows into a complete new star. The genus *Linckia* can reproduce whole stars from single cast-off arms. Reproduction by fission is also known for certain sea cucumbers and brittlestars.

A sort of copulation occurs in several species of starfish; during the breeding period, large numbers of pairs congregate in shallow water, the male resting on top of the female with his arms alternating with hers. Among brittlestars, several species have dwarf males, tiny animals living attached to the body of the females and with the sole function of fertilizing the eggs. Although nothing resembling copulation is known for the other three classes, many pairs of sea cucumbers have been observed in shallow waters around Puerto Rico lying together with the bodies closely adhering to one another through about half their length. They cannot be separated without tearing the flesh. Whether this is sexual behavior is unknown, but it would be worth investigating. And then there is an Antarctic brittle-star which is thought to reproduce without benefit of males at all. It is fairly common, and large numbers of specimens have been collected, but no male has ever been found.

Echinoderm larvae are mostly microscopic creatures, the largest known being but half an inch long; most are very much smaller. They are crystalline little beauties, drifting hither and thither at the mercy of the currents, together with countless billions of other plankters.

Nearly a hundred echinoderms which brood their young are known at present, and more are constantly being discovered. Many of these have special chambers for protecting the young and quite a few brood them in a sunken area around the mouth. Brooding echinoderms are particularly common in Arctic and Antarctic waters. Only a few such echinoderms are known from tropical areas. Egg-brooding is particularly common among seastars of the southern hemisphere. Usually the large yolky eggs are held under the disc in a compact mass against the parental peristome; they are frequently attached by a strand, somewhat like an umbilicus, to the adult. In those brooding stars with short broad rays and less flexible bodies, the young are sometimes brooded on the aboral surface among the disc plates. Some starfish with many arms brood their young between the bases of the rays, and a special brood chamber, between the outer dorsal surface and the actual upper body wall, has been reported in several species of Pterasteridae.

At least one species of ophiuroid attaches its yolky eggs to rocks or seaweed rather than shedding them into the sea, and many species brood their young inside the bursae. Although, as with other echinoderms, brooding species are especially abundant in cold polar waters, many brooding species are also known from the tropics, a thing unusual in the other groups. Only one species of brooding basketstar is known.

Echinoids that brood their young are fairly common, particularly in the Antarctic. Spatangoids and cidarids have many brooding species. The brooding spatangoids have deeply sunken petaloids which serve as brood pouches, while the young are brooded around the mouth or anus in brooding cidarids. In both, spines crossed over the brooding area help hold the young in place.

In south polar waters, more than thirty species of brood-protecting holothurians are known. Some species brood the young on the ventral surface (or sole), some on the dorsal surface, and some in special brood pockets. These vary from simple depressions in the surface of the parental body to pockets at the bases of the tentacles, and even to deep internal incubatory sacs.

As with many marine creatures, little is known about the growth rates of echinoderms. It apparently varies from species to species and is probably slower in colder northern waters.

Echinoderms have been collected at all depths, in all the seas of the world. They form a conspicuous part of the fauna in tropical waters, and in the Antarctic they are one of the most abundant groups of animals. They are found in very shallow waters or even washed up on the beaches, and they have been collected from the deepest trenches in the ocean.

There are species of echinoderms adapted to almost any type of sea floor, but species adapted to rocky or hard substrates cannot live on muddy or oozy bottoms, and vice versa. A fragile echinoderm with a thin and flimsy skeleton would be pounded to pieces on a rocky shore, while one with a thick, weighty skeleton and blunt, heavy spines would sink and be smothered in soft mud.

Featherstars are generally found on coral reefs or on rocky bottoms, seldom in mud or sand. They are mostly shallow-water animals, reaching their lower limits at about fifteen hundred meters. Sea lilies, on the other hand, seldom occur at less than a hundred meters and are most abundant between two hundred and five thousand meters; they seem to prefer muddy bottoms. Crinoids tend to occur in large aggregates, and an account of dredging operations by the U.S. Fish Commission vessel 'Blake' off Havana, Cuba in 1878 tells of the dredge coming up heavily laden with nearly a ton of broken crowns and stalks of one species, having apparently passed through a veritable forest of sea lilies.

Most asteroids are bottom-dwellers, and they frequently bury themselves in sand or mud. Many species live in crevices on rocky coasts, and others spend their lives crawling around on coral reefs. It is thought that one deepsea family, the Benthopectinidae, may be at least partly pelagic, as they are equipped with paired dorsal muscle bands on the arms which could possibly be used to thrash the arms up and down in a swimming motion, as in the comatulids.

Because of their habit of living concealed under rocks, seaweeds, shells, and sponges, few people notice brittlestars at the seashore or have any idea of their abundance. However, they inhabit all sorts of marine environments and are present in large numbers in shallow water close to shore as well as at moderate depths. Many habitually live buried in the substrate, and some spend their lives attached to other animals. Some species of ophiuroids aggregate in large numbers, and in some parts of the world large areas of the sea floor are literally covered with dense mats of brittlestars. Some observers have reported at least two species which swim by actual rowing motions of the arms.

All echinoids live on the ocean bottom or on rocks and reefs, with the mouth directed downward against the substrate. Regular echinoids usually inhabit rocky or shelly bottoms and coral reefs, while irregular echinoids normally live on or in sand and muddy bottoms. The rock-dwelling species are quite adept at climbing around, using their many tubefeet like tiny suction cups to keep from falling. Although large aggregates of echinoids comparable to brittlestar and crinoid aggregates are unknown, echinoids on sandy bottoms do tend to 'herd' together in small groups of a dozen or less, and these groups 'gallop' briskly around on the sand together, moving as a unit, particularly at night when they are feeding.

Most sea cucumbers are sluggish beasts, some living under rocks or in crannies in coral and others among clumps of seaweed or other branching organisms. Many of the larger, more conspicuous species simply lie around, fully exposed, on the sandy bottom. A good many live partly or wholly buried in mud, with the posterior end sticking out for respiration. The class Holothuroidea contains the only known truly pelagic adult echinoderms, the bathypelagic elasipods and the surface-swimming *Pelagothuria* and its relatives.

All echinoderms are, of course, marine; they live in every ocean and at every depth. Not only are they found in the deepest trenches of the ocean, but some starfish, stranded by the receding tide, can even survive long periods well above the tide line, entirely out of water, by extruding their stomach pouches and pushing them down into the damp sand. Echinoderms have also been found frozen, but still alive, in chunks of ice in the Antarctic.

Some echinoderms are vegetarians, feeding mainly on marine plants. Most of them are detritus feeders, eating bits of organic matter from the ocean floor. Many are scavengers and live on the dead bodies of fish and other marine animals. A few are carnivorous, and among these are some of the starfish which do extensive damage to commercial oyster beds, as well as the crown-of-thorns starfish which is eating the coral reefs of the Pacific.

In most echinoderms the tubefeet and tentacles play an important role in food gathering. The tubefeet (tentacles around the mouth of sea cucumbers are modified tubefeet, as are the tentacles of brittlestars and sea lilies) secrete a mucus which traps small bits of organic matter or micro-organisms, and these are passed along the ambulacra to the mouth. The strands of food-bearing mucus are pushed along toward the mouth by fine hair-like projections of tissue called cilia, which beat constantly with a whip-like motion in the direction of the mouth. Many echinoderms seem to be omnivorous, eating whatever comes their way, and, as with respiration, they have evolved more than one method of collecting food. Sea lilies stick mainly to the simple method outlined above, but the other groups have additional, more elaborate methods. Some starfish (but by no means all) evert their stomach through the mouth, and digestion of their prey takes place outside the body. This enables them to take advantage of food sources which would be too large to be taken into the body cavity. Also, starfish lack the elaborate grinding and crushing mouth frame present in brittlestars and many sea urchins, and by enveloping and digesting their food externally, these asteroids simply absorb what they need and eliminate the problem of elimination.

A sea cucumber can hardly be considered a gourmet, as its idea of a good meal is whatever is in front of its mouth. It simply shovels great gobs of sand or mud into its mouth with its tentacles and apparently finds sufficient organic nutrients in it to sustain itself. Since crinoids are ciliary-mucus feeders, their food consists of small organisms like diatoms, unicellular green algae, radiolarians, foraminifera, other protozoans, small crustaceans, and various planktonic larvae, as well as other particulate organic matter. These are swept into the mouth by the combined action of arms, pinnules, and tentacles.

Some starfish are ciliary feeders, but by far the most interesting (though by no means the most numerous) are those species which feed on bivalves. The extraordinary pull the starfish exerts with its tubefeet on the shells of the mollusc to pull it open (up to 10,000 grams or 20 pounds) and its rather revolting habit of extruding its stomach and digesting the bivalve outside its own body are interesting to us but a nuisance to oystermen. One species, *Asterias vulgaris*, preying on oyster beds in Delaware Bay, accounts for a loss in revenue running into thousands of dollars annually. For years, oystermen cleaning starfish from the beds would break the animal in two and throw it overboard. Biologists finally pointed out that, because of the seastar's phenomenal powers of regeneration, they were increasing the population rather than eliminating it. Now starfish removed by means of huge 'mops' dragged over the bottom are either thrown out on shore or scalded in boiling water.

Ophiuroids generally avoid direct light and actively seek the dark. Most of them therefore probably feed at night. They seem to eat principally bottom material, both animal and plant, but a few are strictly vegetarian. Some are cannibals and eat each other.

Echinoid spines and tubefeet serve not only for locomotion and defense, but also to move food along the test to the mouth, where it is macerated by the lantern. Sea urchins will apparently eat almost anything, and most are general scavengers, but some tend to carnivorous habits, while some live almost exclusively on plant material. Among the most interesting of the latter are some very deep water species which seem to live entirely on wood pulp obtained from trees which are washed out into the sea and sink, waterlogged, to the bottom.

In sea cucumbers the mouth is surrounded by a circle of from eight to thirty food-gathering tentacles which are actually tubefeet specialized for obtaining food. In some holothurians the mouth is directed down when feeding, and as they creep across the bottom ingesting detritus, they look like some type of primitive vacuum cleaner. In others, especially the Dendrochirotida, the mouth is directed up. These holothurians are plankton-feeders, waving their sticky branching tentacles about to capture minute organisms and bits of drifting detritus. All other holothurians are mud-swallowers, simply shovelling the substrate into their mouths with their tentacles and voiding it at the other end after the organic nutrients have been digested. It has been estimated that a coral reef with a population density of two thousand cucumbers per acre can pass more than sixty tons of sand, mud, and other material through their digestive tracts each year.

Echinoderms are voracious beasts and for the most part are not too fussy about what they eat. Starfish in particular do not hesitate to bite off more than they can chew and then swallow it whole. Asteroids in aquaria have been observed to catch and swallow fish as big as they are (sluggish ones, to be sure), and in the collection at the U.S. National Museum are a number of specimens of *Luidia clathrata* which have somehow managed to ingest sand dollars much larger than their own discs, stretching and rupturing the disc. However, with their extraordinary powers of regeneration, such ruptured starfish are probably only temporarily inconvenienced. Mud-gulping holothurians, too, swallow more than their body weight in bottom sediments, and one frequently finds, on collecting these beasts, that one has in hand much more mud than sea cucumber.

Echinoderms are sociable animals, if sometimes unwillingly so. They are seldom found in isolated situations and, because most of them do not move around rapidly, may play host to numerous other animals. The sedentary crinoids are almost never without visitors, either accidental or obligate. A few protozoan parasites live either in the digestive tract or on the surface of many sea lilies, and hydroids frequently find the stalks of stemmed forms a good place to attach. Scale worms living on crinoids adapt their colors to harmonize with those of their host.

One entire group of very curious worms, called myzostomes, are found almost exclusively on crinoids, where they frequently form galls or cysts on the arms or creep about freely on the disc. These are true parasites and occur in great numbers, as many as three to four hundred on a single crinoid. Small parasitic snails also infest some crinoids, boring holes in the hard parts to get to the soft tissue. Certain species of shrimp, crabs, and other crustaceans are obligate commensals (that is, they live only on certain species) and match their hosts remarkably well in color and pattern. In the Red Sea a small clingfish lives exclusively on comatulids, slithering over the disc and among the arms and sharing meals with its host. Ophiuroids of the genus *Ophiomaza* live on the discs of crinoids and correspond in color pattern to their hosts. However, apart from the parasites which prey upon them, crinoids have few enemies.

Protozoans also inhabit sea cucumbers, living either in the respiratory tree or in the intestine. Several kinds of worms live either in or on holothurians, and a small rotifer lives in pits in the skin of some synaptids. A number of quite peculiar parasitic molluscs live in association with various holothurians. Several small bivalves live either on the exterior or in the intestine of some sea cucumbers, while many greatly modified gastropods parasitize other species of holothurians. Some of these are so degenerate that in the adult stage they are simply tubular sacs containing only the digestive and reproductive systems of the snail. Several pinnotherid crabs live in the cloacas of holothurians or among the tentacles around the mouth.

The most famous associates of sea cucumbers, however, are pearl fish, *Carapus* and allies, slender little fish which occupy the stem of the respiratory tree with the head protruding from the holothurian's cloaca. It enters the body through the cloaca, usually tail first. A fish may be as much as six inches long. This is not a true parasite, as the fish does no harm to its host but simply uses if for shelter.

Asteroids are less subject to protozoan parasites than are other echinoderms but are far more apt to play host to scale worms. These are most often found living in the ambulacral grooves of starfish and may actually poke their heads into the asteroid's stomach to obtain food. Parasitic snails also infest some asteroids, and one genus of mollusc, *Thyca*, is found exclusively on starfish of the genera *Linckia* and *Stellaster* in the tropics. Several copepods and amphipods parasitize starfish or their egg clusters, and the degenerate barnacles known as ascothoracids infest the body cavity of some asteroids. The cushionstar, *Culcita*, sometimes provides a home for the same kind of pearl fish that is found in tropical holothurians.

Despite their protective armor, echinoids are preyed upon by a number of animals. Sea gulls carry them up into the air, drop them on rocks to break them open, and devour the meat. The sea otter becomes a tool-using animal when it eats echinoids; it brings up a stone and several urchins from the bottom, turns over to float on its back, and, placing the stone on its chest, smashes the urchins vigorously on the stone to crack them open. Many fish eat sea urchins. The parrotfish, with their powerful beak-like teeth, carefully nibble off the spines before attacking the shell. Crabs, well protected by their own armor, just move in and break open the urchin with their heavy claws.

Many small animals such as fish and little crustaceans use sea urchins for protective hiding places, like the little protectively colored shrimp which cling to the spines of *Diadema*. Sessile organisms find the spines of some echinoids, particularly the cidarids, a suitable place to attach. Sponges, bryozoans, tube worms, small sea anemones, and barnacles commonly cover cidarid spines and help camouflage the echinoid from predators. Ciliates in large numbers infest the digestive tract of echinoids, many species being obligate commensals in certain types of sea urchins. Both flatworms and polychaetes are frequent associates of echinoids, usually on the spines or test, but sometimes living inside their host.

Small bivalves and snails live on or in echinoids and prey upon them, and parasitic copepods frequently infest them. Many of these copepods form large galls on the tests or spines of the urchin, distorting them grotesquely. Little crabs may force their way into the epiproct of an echinoid and, as they grow, become trapped there, only able to move their claws to catch food. Several of the smaller species of ophiuroids have formed an association with echinoids, living among the spines or around the mouth. Small molluscs of the genus *Montacuta* live among the spines around the periproct of spatangoid sea urchins, apparently feeding on echinoid feces.

Because of their small size and active habits, one would expect fewer parasites to prey on ophiuroids. This does not, however, seem to be the case. The usual echinoderm plagues of protozoans, polychaetes, molluscs, and crustaceans infest the brittlestars. In addition, they have a few associates uniquely their own. A unicellular green alga grows in the skeletal meshes of several brittlestars, gradually destroying the skeleton and killing the ophiuroid. In the Antarctic a sponge, *Iophon*, commonly grows thickly on the disc and upper arm plates of a species of *Ophiurolepis*, impeding the movements of the ophiuroid but otherwise apparently not harming its host. Although polychaetes are not commonly found associated with ophiurans, those curious parasites of crinoids, the myzostomes, are also occasionally found on brittlestars. Copepods are perhaps the most abundant ophiuroid parasites, and a luminescent copepod living in the bursal slits of *Amphipholis squamatus* led to false reports of luminescence in that species.

Because of their small size and clinging arms, ophiuroids are more frequently found in association with other animals to which they attach than are other echinoderms. Sponges and coelenterates (gorgonians, corals, sea pens, etc.) are favorite places of attachment; *Ophiocnemis marmorata* lives under the umbrella of a jellyfish, over eight hundred specimens having been found on a single medusa. Several species of brittlestars are only found clinging to crinoids, and the smallest known brittlestar (and smallest echinoderm), *Nannophiura lagani*, lives among the tiny spines of a sand dollar. There is also another small species, *Amphilycus androphorus*, which lives on sand dollars and has a dwarf male attached to the disc of the female. Dwarf males are known in several ophiuroid species but are not known in the other classes of echinoderms.

Although there are no truly parasitic echinoderms, some of the serpent stars living in a close "table-mate" association with featherstars might be said to approach this condition, but they cause no harm to their hosts. There are, too, no colonial echinoderms, but brittlestars again seem to approach this condition in certain species and under certain circumstances; dense mats of intertwining ophiuroids carpet the bottom in limited areas of the sea floor, notably in the North Atlantic. Ophiuroids which live on soft corals and sea pens have been known to nibble polyps, but their normal food is probably planktonic animals. Their small size and flexible arms fit many ophiuroids for a sort of clinging vine existence on other animals. Large sponges may be regular tenements for ophiurans, housing hundreds of individuals.

Not all molluscs are as helpless before the attacks of starfish as are oysters. At the first touch of a groping asteroid tubefoot, the scallop flaps madly off, clapping its shells together like castanets in hasty if erratic flight, a reaction which someone once compared to ' a galloping herd of agitated dentures. ' Most gastropods, too, put on remarkable bursts of speed at the approach of a hungry starfish. Surprisingly enough, this is also the reaction of a sea anemone, *Stomphia*, usually among the most sedentary of animals. Here, however, the reaction is apparently triggered not by touch but by sensitivity to a chemical secreted by the starfish. Even an unruly octopus can be easily subdued by having a starfish or two thrown on it.

Starfish have even been observed catching and devouring small fish, and they are not beyond preying on other echinoderms in a very cannibalistic way. Coral polyps are the favored item of diet with the Indo-Pacific starfish *Acanthaster*, which in recent years became a serious menace to the great coral reefs of northern Australia and the atolls of the South Pacific; for some unknown reason the *Acanthaster* population there exploded at an incredible rate, but seems now to have subsided.

Ophiuroids are eaten by a number of different animals and seem to be the principal food of hake and cod in the North Sea. In the Antarctic, a large marine bristle worm eats brittlestars almost exclusively, and they form the main food of *Glyptonotus*, a big Antarctic isopod. In the tropical Pacific, ophiurans are eaten by several species of starfish.

The ripe gonads of several species of echinoids are high in protein and rich in minerals and are eaten by people in the Mediterranean countries, the West Indies, and several other parts of the world. *Tripneustes ventricosus* is such a prized delicacy in the West Indies that restrictions have had to be placed on the commercial exploitation of this sea urchin. In the Bahamas the number of sea urchins that can be taken for the market is severely limited by law, even though they are used for local consumption only and none are exported. Japan, however, still cans and exports sea urchin eggs.

A few echinoderms, mostly in tropical waters, are dangerous to handle. None, of course, actively attack man, but divers have been injured by handling or brushing against the long needle-sharp spines of the large black tropical sea urchin *Diadema*. Fishermen in Japan have died from handling a fortunately rather rare deep-water sea urchin which sometimes is caught in their nets. A very spiny starfish in Australian waters, the *Acanthaster* mentioned before as a menace to coral reefs, has been reported to cause severe injury and toxic reaction if stepped on.

Some sea urchins have great destructive powers. Many are rock-borers, and the ability of almost any regular urchin to gnaw holes through the sides of wooden tanks is known to everyone working at a marine laboratory equipped with such tanks. There have even been reports of echinoids boring into underwater steel girders and pilings. They are now being blamed, too, for the disappearance of the kelp beds in California, where they apparently devour the kelp holdfasts. However, the urchins and the kelp have maintained a healthy balance for years, and the real culprit here is man, who by over-harvesting the kelp beds has upset the natural balance.

Although no definite proof exists of the toxicity of any holothurian to man, some tropical species are toxic to fish and are widely used in the islands of the Pacific to poison rock pools and lagoons. Japanese fishermen regularly use starfish to persuade octopuses from the traps they set for them. Merely touching a starfish arm to the octopus causes the many-armed mollusc to shoot out of the trap as if it had been jabbed with a pin.

Many echinoderms live in quite shallow water at the seashore and can be easily collected by hand. Once your eyes are accustomed to spotting them, you will be surprised at the numbers you can pick up this way. Turning over rocks or chunks of coral can be quite rewarding, and remember too that large sponges are virtually tenement houses for ophiuroids. Commercial fishing boats frequently bring up interesting specimens, and friendly fishermen can sometimes be persuaded to save a few specimens for you.

Most echinoderms may be preserved either wet or dry. Holothurians, of course, can only be preserved wet in jars of alcohol. Any echinoderm preserved wet must be kept in alcohol, never in formaldehyde. Formaldehyde is an acid and will quickly dissolve the delicate spicules, spines, plates, and other hard parts so necessary for identification. Other echinoderms may be dried. They should first be killed in formaldehyde and left in it not more than twenty-four hours (less for small specimens). They should then be rinsed off in fresh water and dried thoroughly in the sun or in a warm dry place. They can, of course, be killed in alcohol (in which case they should be allowed to soak for three or four days), but this will quickly remove all the color. Unfortunately, the color will fade eventually, anyway, no matter what the original method of preservation.

No collection is worth keeping unless proper records are kept. Each specimen should be labelled with the exact locality from which it came, the date it was collected, and the name of the collector. Any additional information, such as time of day, type of bottom, methods of collection, and observations on color and behavior, will add to the interest of your collection. Rough identifications can usually be made from a manual of seashore animals appropriate to your area, but as your collection and interest grow you may want to check your identifications with a specialist at some large museum or university.

Echinoderms often require more aquarium space than hobbyists are willing to allow them, but hardier types are easy to care for. The common shallow-water echinoids, like *Arbacia punctulata* and *Cidaris tribuloides*, eat mainly algae, and if you allow at least one wall of your tank to remain uncleaned, a healthy growth of green algae will keep your echinoids quite happy. Of course, they would be even happier with a dense ' soup ' of algal growth all over the tank, but whose happiness is more important, yours or theirs? Sand dollars are less satisfactory in an aquarium. They usually die quite quickly and pollute the tank. The reason for this is probably that they are difficult to collect without bruising or damaging their delicate outer tissue; injury results in fungal or bacterial infections.

A tank with a clean sandy bottom is perfect for such easy-to-find asteroids as *Astropecten* and *Luidia*. Small molluscs such as baby clams are their favorite food, but beware—they will eat almost any other animal in your tank that is not swift enough to get away. The ideal aquarium pet must surely be the tropical seastar *Linckia*, which requires only clean water and a sandy bottom with some crushed coral in it and is completely unaggressive. Best of all, what they eat is a mystery—despite frequent and prolonged observations, they have never been seen feeding. Thick and fleshy seastars do not make good aquarium pets; many of them exude a mass of mucus which kills other animals in the tank and makes an awful mess.

Crinoids need a large, well-aerated tank (preferably to themselves) with a fairly vigorous circulating system. They may be fed with small planktonic organisms like brine shrimp.

Among the holothurians, very few are attractive enough to tempt the aquarist. One kind that is attractive and seems to do well in aquariums is *Psolus*. Its bright red color and plume of branching tentacles are very decorative. But it is best to avoid most other holothurians, as many of them are poisonous to fish and to other invertebrates.

Brittlestars usually do well in aquariums, but they are difficult to feed; each species seems to have its own food preference, and you must just experiment until you find what is acceptable. One researcher working on brittlestar behavior despaired of finding anything his finicky ophiuroids would eat; frustrated, he threw a lump of peanut butter into the tank and was amazed to see his animals devour it happily. If you want to keep ophiuroids, remember that they need a place to hide, so supply some rocks or coral.

Invertebrate Chordates

This group of animals comprises a rather heterogenous assemblage which lacks a vertebral column but none-the-less possesses some vertebrate features including a perforated pharynx, dorsal tubular nervous system, and a notochord. It includes the phylum Hemichordata as well as the Tunicata and Cephalochordata, both of which are distinct subphyla of the phylum Chordata.

PHYLUM HEMICHORDATA

This phylum is divided into the classes Enteropneusta, the acorn worms, and Pterobranchia. The former group is by far the most common and includes the well known genera *Saccoglossus* and *Balanoglossus*. The pterobranchs are less well known and occur in cold southern as well as northern waters; specimens of *Rhabdopleura* have been found close to European coasts and near the low tide mark in Bermuda.

Acorn worms are found in all parts of the world, commonly burrowing in mud, although there are species that live in tubes of sand and mucus and others that are free-living. They have a cylindrical worm-shaped body comprising three distinct regions. Anteriorly there is the short conical proboscis followed by the collar zone, the leading edge of which overlaps the base of the proboscis (hence acorn worm), while its posterior margin overlaps the third and major region of the body, the trunk, which bears the terminal anus. Enteropneusts vary in size according to species and may be anything from 10 to 50 cm long overall. The Brazilian *Balanoglossus gigas* can reach 1.5 meters and lives in burrows 3 meters long.

There is a tripartite coelom, each body region containing a separate coelomic cavity, those of the collar and trunk being paired. Much of the proboscis coelom has become filled

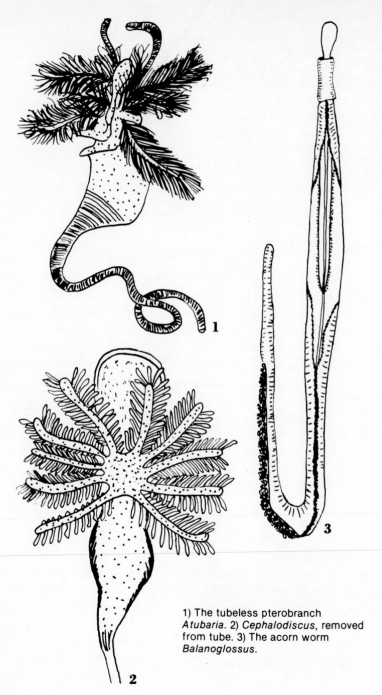

1) The tubeless pterobranch *Atubaria*. 2) *Cephalodiscus*, removed from tube. 3) The acorn worm *Balanoglossus*.

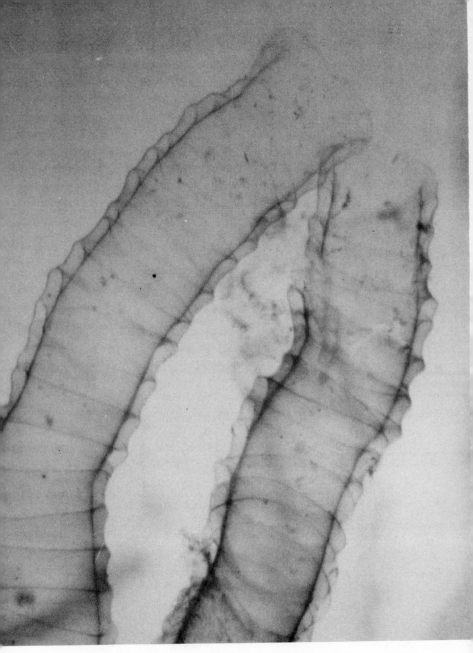

Close-up of two tubes of the pterobranch *Rhabdopleura* taken in shallow waters in Bermuda. The overlapping 'ring' structure is typical of tubes of this group (R. D. Barnes photo).

with muscle fibers and connective tissue, and it is this, together with the turgor pressure created when water is drawn into the remaining space, which gives the proboscis its impressive burrowing powers. The other two paired cavities have also become occluded with muscle fibers and connective tissue, and to a certain extent this replaces the typical body wall musculature. Peristaltic contractions of these muscles aid the major burrowing activities of the proboscis. Locomotion may also be influenced by the beating of the cilia of the epidermal cells which cover the whole body, together with a large number of mucous glands.

While many acorn worms feed by ingesting mud and organic matter from the substrate, suspension feeding brought about by the action of epidermal ciliary and mucous tracts play an important part in the lives of several species. Plankton and other organic material become trapped in ciliary tracts on the proboscis and are passed to the mouth, which lies ventrally in the anterior fold of the collar and is preceded by a special concentration of cilia, the preoral ciliated organ. Some particles are caught here and passed on to the mouth; others are rejected and carried out over the collar. Particles are carried in not only by cilia, but also with the assistance of a strong current of water flowing into the mouth and out through the pharyngeal gill slits which perforate the wall of the trunk. The number of slits can range from a few to over one hundred pairs, and new ones may be added posteriorly throughout life. Each slit comprises a branchial sac which opens internally as a U-shaped cleft and externally by way of a small pore into a dorsolateral groove on the trunk. The pharyngeal epithelium surrounding the slits is strongly ciliated, producing the water current which passes in through the mouth.

The digestive tract has a quite simple and straightforward tube-like design with certain demarcated regions such as esophagus and intestine. The buccal cavity which immediately follows the mouth bears a dorsal diverticulum extending into the proboscis; this diverticulum is known as the stomachord. For a number of years this was thought to be homologous with the notochord of true chordates, but it is now widely recognized as little more than a pre-oral extension of the gut.

Blood sinuses are present in the gill tissue and are thought to be involved in respiration. These vessels are part of an open blood-vascular system comprising a collection of sinuses and two main contractile channels. Blood passes anteriorly in a dorsal vessel into a central sinus at the base of the proboscis. Beneath this lies the heart vesicle, a closed contractile fluid-filled sac, pulsations of which aid blood flow. Beyond the central sinus is à specialized region of the blood system, the glomerulus, which is though to play some role in excretion. A ventral channel then carries blood posteriorly and connects by way of sinuses to the dorsal vessel once more.

The primitive subepidermal nerve plexi of enteropneusts is regionally thickened as nerve cords. These lie mid-dorsally in the proboscis, collar, and trunk as well as mid-ventrally in the proboscis and trunk only. The mid-dorsal collar cord is separated from the epidermis and in some worms may be hollow.

Acorn worms are dioecious (separate sexes), the serial gonads being located in the paired coelomic cavities of the trunk. In such genera as *Balanoglossus*, *Stereobalanus*, and *Ptychodera* the trunk bulges in these regions to form distinctive genital ' wings ' or ridges. The gonads each open by a single pore, frequently into the same longitudinal groove as the gill slits. Fertilization is external and early development in some genera (*Ptychodera*, etc.) is characterized by a ciliated planktonic tornaria larva strikingly like that found in some echinoderms. This eventually elongates and sinks, taking up the adult form. In other species development is direct. It is noteworthy that in some juveniles of *Saccoglossus horsti* there is a post-anal tail said to be homologous with the stalk region of pterobranchs.

Pterobranch hemichordates are represented by two tube-living genera, *Rhabdopleura* and *Cephalodiscus*, while the remaining *Atubaria* lacks a secreted tube. They share many of the features exhibited by enteropneusts but differ in some anatomical respects. Most species are found in deep water although *Rhabdopleura* has been found in near-tidal water of Bermuda. They have the tripartite body seen in acorn worms, but here the proboscis is shield-shaped. The most obvious departure from the enteropneust pattern is the occurrence of tentacled arms arising from the dorsal side of the collar; *Cephalodiscus* has up to nine pairs of arms while *Rhabdopleura* bears only a single pair. The trunk is short and in the colonial *Rhabdopleura* gives rise to the stolon connecting with other individuals. The epidermal layer is ciliated and mucous-secreting, trapped particles being passed to the mouth from the tentacles and other exposed regions of the body. The particles there enter the U-shaped digestive canal, at the end of which the anus opens anteriorly and dorsally on the collar. There is no assisting water current in *Rhabdopleura*, where the gill slits are absent. In *Cephalodiscus* a single pair of gill openings produces a gentle flow of water.

PHYLUM CHORDATA

Tunicates, or urochordates as they are sometimes known, are a familiar group of animals (subphylum Tunicata) comprising three major classes. There are the well known sedentary Ascidiacea or sea squirts of the sea shore; the pelagic and planktonic Thaliacea or salps; and thirdly the Larvacea or appendicularians. The ascidians are by far the most common members of this group and are found throughout the world. They are found attached by means of stalks or fine filaments to rocks, shells, reefs, piers, etc., as well as to ship hulls where they represent a considerable fouling hazard. The free unattached end has an opening, the buccal or inhalent siphon, while the atrial or exhalent siphon is usually a little distance away along the side of the animal. Size may range from a few millimeters up to 8 cm or more in such forms as the large sea peach, *Halocynthia pyriformis*.

The body of tunicates is entirely covered by a usually tough and resilient tunic or test. In many animals this has a dark warty appearance (*Amaroucium stellatum*), although in some species it may be light, delicate and translucent (*Ciona intestinalis*). The tunic has a fibrous cellulose matrix (called tunicine) within which amoeboid cells, blood cells, and, in *Ciona*, even blood vessels may be found. This curious feature is due to the migration of mesenchymal tissue into the substance of the tunic. Extensions of the tunic provide for the anchoring mechanism to the substrate.

The inhalent current of water passes into the pharynx through the anterior buccal siphon, guarded internally by a ring of oral tentacles. The pharyngeal chamber occupies much of the internal space, its walls being perforated by a large number of slits or stigmata. Along the floor of the pharynx runs a deep and strongly ciliated groove, the endostyle. Dorsally and running more or less parallel to the endostyle is the dorsal lamina or hyperpharyngeal band. From this, in *Ciona* and some other species, hangs a row of processes, the languets. The beating of the ciliated pharyngeal epithelium produces a strong current of water which passes out through the slits into the atrial cavity which surrounds the pharynx. Planktonic organisms are filtered from this flow by an intricate mucous net produced by the endostyle and passed up each side of the pharynx. Filtration is very efficient; particles down to a few microns can be retained and up to 200 liters of water an hour may be filtered by large species.

The mucous net is rolled into a 'rope' in the dorsal lamina and drawn into the alimentary canal by esophageal cilia. The anus eventually opens adjacent to the pharynx and just below the exhalent siphon next to the gonoduct.

The circulatory system is a modest open affair with a pericardial fold which serves as a heart; one vessel runs anteriorly from the heart beneath the endostyle and another posteriorly to the digestive and reproductive region. The subendostylar vessel leads to a network of respiratory vessels investing the pharyngeal basket. These unite dorsally as a hyperpharyngeal sinus which runs posteriorly to supply the digestive loop and other viscera. This open system is remarkable in that it is capable of periodic reversal of flow. The blood itself is similarly noteworthy, as it has a number of amoeboid phagocytic cells together with, in some species, certain cells capable of concentrating such unusual elements as vanadium and niobium to a very high degree.

The nervous system is a simple network concentrated between the two siphons as a cerebral ganglion. Beneath this lies a glandular organ, the subneural gland, which opens to the pharynx by way of a ciliated funnel. A number of functions have been attributed to this structure, stretching to homology with the vertebrate pituitary gland.

Larger ascidians such as *Ciona* and *Halocynthia* are solitary, but a number of smaller species are colonial or compound. In some, such as the common *Botryllus*, each individual shares a common tunic with several others. This forms star-shaped flattened colonies encrusting rocks, each individual with its own inhalent opening but sharing a communal exhalent aperture. In these colonial forms asexual budding and regeneration are frequently encountered. The bud, or blastozooid, may arise from any number of positions on the ' parent' according to species. In *Botryllus* the star-shaped colonies form by budding from a solitary newly settled larva. In hermaphroditic sexual reproduction the eggs produce the remarkable and enigmatic tadpole larva. This free-swimming planktonic larva has a distinct tail with notochord and dorsal neural tube. In some genera, for instance *Dendrodoa*, the larvae are released directly from eggs developing within the adult and escape via the exhalent siphon.

The thaliacians or salps are rather specialized pelagic tunicates. They have, in principle, a similar organization to ascidians, although here inhalent and exhalent siphons are at opposite ends of the body, which is barrel-shaped. The water current serves as a propulsive force as well as a feeding and respiratory current. There are three main types of salp: the Pyrosomida, Doliolida, and Salpida.

The pyrosomids bud to form a colonial cylinder which is closed at one end. The inhalent apertures open on the outer surface and the exhalent siphons open into the lumen of the cylinder, which thus forms a common cloacal chamber. The combined exhalent currents emerging from the open end act as a jet propulsive force. The tropical forms of *Pyrosoma* are brilliantly luminescent and can reach a length of 2 meters or more. Luminescence is evoked by tactile stimulation. Fertilized oozoids begin budding in the atrial cavity of each parent and are released at the four-bud stage to form a new free-living colony following further budding.

The remaining groups, Doliolida and Salpida, are solitary with inhalent and exhalent openings forming the 'top' and 'bottom' of their barrel-shaped body. Conspicuous muscle bands encircle the body; their contraction produces a current of water that is both propulsive and nutritive. The pharynx in *Salpa* is pierced by only two large slits. The egg in this genus develops within the parent blastozoid, and here there is an alternation of generations. The developed salpid egg is released as an oozoid, which is an asexual stage producing a trailing stolon which bears buds that are finally shed as sexually reproducing adults. The situation in doliolids is even further complicated. The sexually mature barrel-shaped adult blastozoid produces three eggs, each of which forms a tadpole larva that metamorphoses into an adult asexual oozoid which is indistinguishable save in minor details from the sexual form. The oozoid produces two complex chains of feeding buds which support the parent oozoid as it finally gives rise to a third chain of sexual blastozoid buds for eventual release.

The appendicularians are the most complex of tunicates. The adults exhibit neoteny, a condition where sexual maturity is attained by the larval form (hence Larvacea). This group appear as miniature transparent tadpoles with a head area retaining a rudimentary tunicate pharynx and a locomotory tail complete with notochord and nerve cord. Common genera are *Oikopleura* and *Fritillaria*.

The most fascinating aspect of these animals is that they build an external protective enclosure for themselves which has become known as their 'house.' This construction of

The common solitary sea squirt *Halocynthia papillosa* is one of the large species that can sometimes be adapted to the marine aquarium (G. Marcuse photo).

hardened mucus is designed to operate as a filter-feeding device. Water drawn in by muscular movement of the tail passes through first a coarse then a fine filter. Particles enmeshed on the fine filter pass by ciliary and tail currents to the pharynx for ingestion. As pressure within the house increases, the excess water escapes via a posterior valve, producing a force driving the animal forward. When the filtration membranes become clogged the house is evacuated and a new one constructed within the hour. It is worth noting that in a few groups such as *Fritillaria* the house is protective only, the animal filter-feeding unaided outside the house.

Cephalochordates, represented by the lancelets, are universally known 'archetypal' ancestral chordates which were formerly known under the generic name *Amphioxus* (subphylum Cephalochordata). They are now more correctly classified as *Branchiostoma*, with amphioxus being retained as the common collective name. They are of worldwide distribution, frequenting inshore sand and gravel deposits into which they burrow. Adults can reach a size of 150 mm in a species found off the Amoy coast of China, but are more commonly less than 50 mm in length.

They tend to be buried, head projecting from burrow, although if necessary they can swim by lateral undulations of the body. The degree and depth of burrowing are determined by the particle size of the substrate, coarse deposits favoring deeper burrows.

Externally the animals are laterally compressed and pointed at either end. The body muscles are clearly visible as a series of V-shaped units known as myotomes; these do not, however, represent metameric segments. The muscle blocks on either side act as antagonists for each other, with the notochord, which stretches the length of the animal, acting as a flexible yet non-compressible skeletal rod. Dorsal to the notochord lies the hollow nerve cord, which for much of its length gives off dorsal and ventral nerve roots, while at its anterior end may be found a terminal enlargement, the cerebral vesicle.

Anteriorly the mouth is protected by an oral hood of stiff buccal cirri which form a coarse sieve. Following on from the mouth the finely perforated pharynx forms a branchial basket occupying more than half the length of the animal. A vigorous current of water produced by lateral cilia on the pharyngeal bars draws particles in through the cirri to be further strained by velar tentacles surrounding the mouth. A complex system of ciliary tracts, the wheel organ which receives mucus from a pit located dorsally in the oral cavity, traps the suspended food particles. From here they are swept in through the mouth. The water current passes out through the pharyngeal gill slits into the atrium. This atrial cavity is a secondary body cavity surrounding the pharynx and formed by an infolding in the ventral body wall. It opens just posterior to the pharynx by the atriopore, the exit for the inhalent water current.

The small, new-hatched ciliated neurula larva eventually gives rise to a planktonic and free-swimming asymmetrical larva. The mouth lies on the left, while the gill slits, still only a few in number, arise on the right. As the animal slowly grows these openings gradually move into the mid line and the slits subdivide to produce the multiperforate pharynx. The subgenus *Asymmetron* retains much of the asymmetry of these late larval or 'amphioxides' forms of *Branchiostoma* and tends to remain planktonic as an adult.

Selected Reading

General

Barnes, R.D. 1973. *Invertebrate Zoology.* W.B. Saunders.

Bayer, F.M. and H.B. Owre. 1968. *The Free-living Lower Invertebrates.* McMillian.

Buchsbaum, R. and L.J. Milne. 1960. *The Lower Animals.* Doubleday.

Colin, P.I. 1978. *Caribbean Reef Invertebrates and Plants.* T.F.H.

Friese, U.E. 1973. *Marine Invertebrates.* T.F.H.

George, D. and J. George. 1979. *Marine Life.* L. Leventhal Ltd.

Gosner, K.L. 1971. *Guide to Identification of Marine and Estuarine Invertebrates (Cape Hatteras to the Bay of Fundy).* Wiley-Interscience.

Grese, A.C. and J.S. Pearse. 1974—. *Reproduction of Marine Invertebrates.* Academic Press. (Multivolume set)

Hickman, C.P. 1967. *Biology of the Invertebrates.* C.V. Mosby.

Hyman, L.H. 1940—. *The Invertebrates.* McGraw-Hill. (Multivolume set, basic reference)

Kozloff, E.N. 1973. *Seashore Life of Puget Sound.* Univ. Washington Press.

Milne, L. and M. Milne. 1976. *Invertebrates of North America.* Doubleday.

Miner, R.W. 1950. *Field Book of Seashore Life.* G.P. Putnam's Sons.

Smith, R.I. and J.T. Carlton. 1975. *Light's Manual. Intertidal Invertebrates of the Central California Coast.* Univ. California Press.

Voss, G.L. 1976. *Seashore Life of Florida and the Caribbean.* E.A. Seemann

Walls, J.G. 1974. *Starting with Marine Invertebrates.* T.F.H.

Zann, L.P. 1980. *Living Together in the Sea.* T.F.H. (Good study of symbiosis)

Zeiller, W. 1974. *Tropical Marine Invertebrates of Southern Florida and the Bahama Islands.* Wiley.

Cnidaria

Burgess, W.E. 1979. *Corals.* T.F.H.

Devaney, D.M. and L.G. Eldredge. 1977. *Reef and Shore Fauna of Hawaii. 1: Protozoa through Ctenophora.* Bishop Museum Press.

Friese, U.E. 1972. *Sea Anemones.* T.F.H.

Mollusca

Baba, K. 1949. *Opisthobranchia of Sagami Bay*. Iwanami Shoten, Japan.

Lane, F. 1957. *The Kingdom of the Octopus*. Jarrold.

MacFarland, F.M. 1966. *Studies of Opisthobranchiate Mollusks of the Pacific Coast of North America*. Calif. Acad. Sci.

Morton, J.E. 1967. *Molluscs*. Hutchinson.

Robson, G.C. 1957. "Cephalopoda." *Encyclopedia Brittanica*. Vol. 5: 148-156.

Solem, A. 1974. *The Shell Makers*. Wiley-Interscience.

Thompson, T.E. 1976. *Biology of Opisthobranch Molluscs*. Ray Society, London.

Thompson, T.E. 1976. *Nudibranchs*. T.F.H.

Thompson, T.E. and G.H. Brown. 1976. *British Opisthobranch Molluscs*. Academic Press.

Walls, J.G. 1981. *Shell Collecting*. T.F.H.

Wells, M.J. 1962. *Brain and Behaviour in Cephalopods*. Heinemann.

Yonge, C.M. and T.E. Thompson. 1976. *Living Marine Molluscs*. Collins.

Arthropoda

Giwojna, P. 1978. *Marine Hermit Crabs*. T.F.H.

Green, J. 1961. *A Biology of Crustacea*. Witherby Ltd.

Kaestner, A. 1970. *Invertebrate Zoology. Crustacea. Vol. III*. Wiley-Interscience.

Schmitt, W.L. 1965. *Crustaceans*. Univ. Michigan Press.

Taylor, H. 1975. *The Lobster: Its Life Cycle*. Sterling.

Warner, G.F. 1977. *The Biology of Crabs*. Elek Science.

Echinodermata

Boolootian, R.A. 1966. *Physiology of Echinoderms*. Wiley-Interscience.

Clark, A.M. 1977. *Starfishes and Related Echinoderms*. T.F.H.

Millott, N. 1967. *Echinoderm Biology*. Academic Press.

Moore, R.C. 1966. *Treatise on Invertebrate Zoology. Part U. Echinodermata*. Geological Soc. America and Univ. Kansas Press. (Multivolume set, mostly fossils)

Nichols, D. 1969. *Echinoderms*. Hutchinson.

Invertebrate Chordates

Barrington, E.J.W. 1965. *The Biology of Hemichordata and Protochordata*. Oliver and Boyd.

Barrington, E.J.W. and R.P.S. Jefferies. 1975. *Protochordates*. Academic Press.

Romer, A.S. 1971. *The Vertebrate Body*. Holt-Saunders.

724

Index to Text Pages

Black and white photos and line drawings are indicated in **bold** type. References such as 123 ff. indicate that page 123 and the pages that follow contain information on the subject.

725

727

Index to Color Photos